IDENTITIES

FUNDAMENTAL IDENTITIES

1. $\sec\theta = \dfrac{1}{\cos\theta}$

2. $\csc\theta = \dfrac{1}{\sin\theta}$

3. $\cot\theta = \dfrac{1}{\tan\theta}$

4. $\tan\theta = \dfrac{\sin\theta}{\cos\theta}$

5. $\cot\theta = \dfrac{\cos\theta}{\sin\theta}$

6. $\cos^2\theta + \sin^2\theta = 1$

7. $1 + \tan^2\theta = \sec^2\theta$

8. $\cot^2\theta + 1 = \csc^2\theta$

COFUNCTION IDENTITIES

9. $\cos(\tfrac{\pi}{2} - \theta) = \sin\theta$

10. $\sin(\tfrac{\pi}{2} - \theta) = \cos\theta$

11. $\tan(\tfrac{\pi}{2} - \theta) = \cot\theta$

OPPOSITE-ANGLE IDENTITIES

12. $\cos(-\theta) = \cos\theta$

13. $\sin(-\theta) = -\sin\theta$

14. $\tan(-\theta) = -\tan\theta$

ADDITION LAWS

15. $\cos(\alpha + \beta) = \cos\alpha\cos\beta - \sin\alpha\sin\beta$

16. $\cos(\alpha - \beta) = \cos\alpha\cos\beta + \sin\alpha\sin\beta$

17. $\sin(\alpha + \beta) = \sin\alpha\cos\beta + \cos\alpha\sin\beta$

18. $\sin(\alpha - \beta) = \sin\alpha\cos\beta - \cos\alpha\sin\beta$

19. $\tan(\alpha + \beta) = \dfrac{\tan\alpha + \tan\beta}{1 - \tan\alpha\tan\beta}$

20. $\tan(\alpha - \beta) = \dfrac{\tan\alpha - \tan\beta}{1 + \tan\alpha\tan\beta}$

DOUBLE-ANGLE IDENTITIES

21. $\cos 2\theta = \cos^2\theta - \sin^2\theta$
$$= 2\cos^2\theta - 1$$
$$= 1 - 2\sin^2\theta$$

22. $\sin 2\theta = 2\sin\theta\cos\theta$

23. $\tan 2\theta = \dfrac{2\tan\theta}{1 - \tan^2\theta}$

HALF-ANGLE IDENTITIES

24. $\cos\tfrac{1}{2}\theta = \pm\sqrt{\dfrac{1 + \cos\theta}{2}}$

25. $\sin\tfrac{1}{2}\theta = \pm\sqrt{\dfrac{1 - \cos\theta}{2}}$

26. $\tan\tfrac{1}{2}\theta = \dfrac{1 - \cos\theta}{\sin\theta}$
$$= \dfrac{\sin\theta}{1 + \cos\theta}$$

PRODUCT IDENTITIES

27. $2\cos\alpha\cos\beta = \cos(\alpha - \beta) + \cos(\alpha + \beta)$

28. $2\sin\alpha\sin\beta = \cos(\alpha - \beta) - \cos(\alpha + \beta)$

29. $2\sin\alpha\cos\beta = \sin(\alpha + \beta) + \sin(\alpha - \beta)$

30. $2\cos\alpha\sin\beta = \sin(\alpha + \beta) - \sin(\alpha - \beta)$

SUM IDENTITIES

31. $\cos\alpha + \cos\beta = 2\cos\left(\dfrac{\alpha + \beta}{2}\right)\cos\left(\dfrac{\alpha - \beta}{2}\right)$

32. $\cos\alpha - \cos\beta = -2\sin\left(\dfrac{\alpha + \beta}{2}\right)\sin\left(\dfrac{\alpha - \beta}{2}\right)$

33. $\sin\alpha + \sin\beta = 2\sin\left(\dfrac{\alpha + \beta}{2}\right)\cos\left(\dfrac{\alpha - \beta}{2}\right)$

34. $\sin\alpha - \sin\beta = 2\sin\left(\dfrac{\alpha - \beta}{2}\right)\cos\left(\dfrac{\alpha + \beta}{2}\right)$

INVERSE IDENTITIES

35. $\operatorname{Arccot} x = \begin{cases} \operatorname{Arctan}\dfrac{1}{x} & \text{if } x > 0 \\[2mm] \dfrac{\pi}{2} & \text{if } x = 0 \\[2mm] \pi + \operatorname{Arctan}\dfrac{1}{x} & \text{if } x < 0 \end{cases}$

36. $\operatorname{Arcsec} x = \operatorname{Arccos}\dfrac{1}{x}$ if $x \geq 1$ or $x \leq -1$

37. $\operatorname{Arccsc} x = \operatorname{Arcsin}\dfrac{1}{x}$ if $x \geq 1$ or $x \leq -1$

REDUCTION IDENTITY

38. $a\sin\theta + b\cos\theta = \sqrt{a^2 + b^2}\,\sin(\theta + \alpha)$,
where α is chosen so that

$$\cos\alpha = \dfrac{a}{\sqrt{a^2 + b^2}} \quad \text{and}$$

$$\sin\alpha = \dfrac{b}{\sqrt{a^2 + b^2}}$$

FIFTH EDITION

TRIGONOMETRY FOR COLLEGE STUDENTS

SUPPLEMENTS AVAILABLE FOR THIS BOOK

ANSWER BOOK: provides the answers to all problems in the text.

EXPTEST QUESTION BANK: provides approximately 800 electronically accessible test items.

INSTRUCTOR'S MANUAL: provides suggested assignments, teaching suggestions, daily quizzes with answers, transparency masters.

SOLUTIONS AND STUDY GUIDE: provides detailed step-by-step solutions to selected odd-numbered problems, and section-by-section hints and suggestions.

TRIGPAK: provides 200 electronic function files keyed to the text for access to complete graphs, tutorials with examples, and hundreds of activities.

OTHER BROOKS/COLE TITLES BY THE SAME AUTHOR

Algebra Study Guide
Karl J. Smith and Patrick J. Boyle

Beginning Algebra for College Students, Fourth Edition
Karl J. Smith and Patrick J. Boyle

Primer for College Algebra
Karl J. Smith and Patrick J. Boyle

Intermediate Algebra for College Students, Fourth Edition
Karl J. Smith and Patrick J. Boyle

Essentials of Trigonometry, Second Edition
Karl J. Smith

College Algebra, Fourth Edition
Karl J. Smith and Patrick J. Boyle

Precalculus Mathematics: A Functional Approach, Fourth Edition
Karl J. Smith

Mathematics: Its Power and Utility, Third Edition
Karl J. Smith

The Nature of Mathematics, Sixth Edition
Karl J. Smith

Finite Mathematics, Second Edition
Karl J. Smith

Calculus with Applications
Karl J. Smith

College Mathematics and Calculus
Karl J. Smith

FIFTH EDITION

TRIGONOMETRY FOR COLLEGE STUDENTS

KARL J. SMITH

Santa Rosa Junior College

BROOKS/COLE PUBLISHING COMPANY
Pacific Grove, California

Brooks/Cole Publishing Company
A Division of Wadsworth, Inc.

Printed in the United States of America

10 9 8 7 6 5 4 3 2 1

LIBRARY OF CONGRESS CATALOGING IN PUBLICATION DATA

Smith, Karl J.
    Trigonometry for college students/Karl J. Smith.—5th ed.
    Includes index.
    ISBN 0-534-13728-8
    1. Trigonometry.   I. Title.
    QA531.S653   1991
    516.24—dc20
                                                    90-43580
                                                    CIP

Sponsoring Editor: Paula-Christy Heighton
Editorial Assistant: Sarah Wilson
Production Editor: Joan Marsh
Production Service: Susan L. Reiland
Interior Design: Janet Bollow
Cover Design: Vernon T. Boes
Cover Photo: Mario Grassi
Interior Illustration: Lori Heckelman and Reese Thornton
Photo Researcher: Mary Kay Hancharick
Typesetting: Polyglot Compositors Pte., Ltd.
Cover Printing: Lehigh Press
Printing and Binding: Arcata Graphics/Fairfield

About the cover: The length of the day at different times during the year will, when plotted, trace out a sine curve (see the Application for Further Study on page 143). The tides are also related to sine and cosine curves as seen by the sun and moon curves in Problems 28 and 29 on page 88. This cover photograph is taken from Eagle Summit in interior Alaska, and depicts 15 exposures of the sun at 15-minute intervals at about the 65th parallel on June 21, the day of the summer solstice. The exposure is looking north toward the Crazy Mountains. Beyond that is the Yukon River, Point Barrow, Prudhoe Bay, and the Arctic Ocean.

To my mother,
Rosamond,
with love and thanks.

PREFACE

There is perhaps nothing which so occupies,
as it were, the middle position of mathematics,
as trigonometry.

J. F. HERBART, 1890

WHAT IS TRIGONOMETRY?

Trigonometry is usually considered to be the transition course between high school algebra and calculus. It changes the focus from the development of manipulative skills to the understanding of mathematical concepts. Quite simply, six functions, called the *trigonometric functions*, are defined in the first chapter, and then the remainder of the course is devoted to investigating and understanding these functions, their graphs, their relationships to one another, and ways in which they can be used in a variety of applications.

ABOUT THIS BOOK

This book was written for a two- or three-hour college course in trigonometry. The prerequisite is a second-year high school algebra course or equivalent. Even though a course in geometry is also a desirable prerequisite, I have found that more and more students are taking trigonometry without having taken high school geometry, so, in addition to a review of geometry in Appendix C, I've included all of the necessary ideas from geometry as they are needed.

TO THE STUDENT

Each new course begins with a sense of anticipation, and this course is probably no exception. You are enrolling in a course and your instructor has adopted this textbook. What can you expect? You will learn about the *trigonometric functions* and their properties. As you read the description of trigonometry in the first paragraph you see that the course depends on one important definition. If you have had a right-triangle definition of these functions in a previous course, you will notice that the definition in Chapter 1 is much more general (we will get to the right-triangle definition in Chapter 5). It is important that you thoroughly understand this definition as it is discussed in Chapter 1.

You will find that the course will be much easier for you if you *read the textbook prior to discussing it in class*. Do not assume you are finished with the day's work as soon as you have finished the day's homework assignment. Read, study, and *understand* the important ideas. There will be some things you will need to *memorize*. I have tried to make it easy for you to know what you need to do. Definitions and formulas are enclosed in boxes; new terms are shown in boldface and are again listed at the end of each chapter. Each chapter begins with an overview and ends with a list of objectives, so you should know exactly what you need to be able to do. In addition to the answers to the odd-numbered problems at the back of the book, I have included the answers to all the problems in the sample and cumulative tests to help you prepare for examinations.

Use your book and make it your own. You probably were shocked by the price of this book, and Brooks/Cole has done everything possible to keep the cost down. Nevertheless, books are expensive, and I hope you see the money you pay for this book as an investment that will pay off not only for this course, but also as a valuable reference book in your future mathematics courses. If you intend to continue your study in mathematics, you will want to make this book part of your professional library so you will have access to the notation and formulas developed in this course.

TO THE INSTRUCTOR

Overview

Trigonometry for College Students, Fifth Edition, defines the circular functions on a unit circle in the third section of the book and then almost immediately generalizes this definition in terms of any point on the terminal side of the angle. You will also notice that radian measure is defined in the second section and then the book emphasizes working with radian measure of angles as real numbers. I have found that most students have had an introduction to degree measure and an early introduction of radians and real numbers builds a solid foundation for more advanced mathematics courses.

If you are familiar with the fourth edition, then you will notice I've made several changes in this edition. The fundamental identities are now all presented in the first chapter so that multiplication with the different forms of these identities is separated from the techniques of proving identities presented in Chapter 3. I have combined Chapters 3 and 4 into one chapter, and have added sections on inverse functions and conics in polar form. In addition, I've added many new problems, including sets of Miscellaneous Problems at the end of each chapter.

If you are not familiar with prior editions of this book, I would like you to look at the following features:

CHAPTER 1:   The emphasis is on understanding that we are indeed dealing with *functions* of an angle, which can be measured in degrees, radians, or real numbers.

CHAPTER 2:   Graphing is introduced early in my text and I give it special emphasis. You will find an innovative and unique method, called *framing*. You will find that with this technique you can quickly and efficiently communicate the concepts of amplitude, period, and phase displacement without the usual difficulties of plotting "ugly coordinates." Using this technique, I'm convinced your students will have a better understanding of graphing more complicated equations—with less class time. It also changes the focus of graphing from what it has been in the past (plotting points) to what it needs to be in the 1990s. Today's (and certainly tomorrow's) technology makes it easier to graph the trigonometric functions *if* we know how to choose the scale and scaling factors appropriately. My framing technique emphasizes the choice of the scaling numbers.

CHAPTER 3:   One of the core topics for any college trigonometry course is proving identities. You will find a multitude of graded identity problems in this chapter, as well as a summary of all identities listed inside the front cover of the book. Notice that two entire sections are devoted to using the eight fundamental identities to prove other identities, and well as one section devoted to recognizing when a given equation is not an identity.

CHAPTER 4:   One of the topics that often distinguishes one trigonometry book from another is the treatment of inverse trigonometric functions and solving identities. These topics are often cluttered and I have taken great care in separating the solving of equations with inverse relations from the limitations necessary for the inverse trigonometric functions for the students.

CHAPTER 5:   The other core topic for a trigonometric course is solving triangles. I have emphasized concepts rather than mechanics and thus have taken time to help the student formulate which method to choose. I think that the immediate introduction to the law of cosines eliminates the confusion that sometimes arises with the law of sines and its ambiguous case. Notice that I've devoted an entire section to the ambiguous case. If you would prefer to solve right triangles as an introductory topic, you might look at my other trigonometry textbook, *Essentials of Trigonometry, Second Edition*.

Chapters 6, 7, and 8 are optional.

CHAPTER 6: An entire chapter is devoted to the complex numbers and De Moivre's Formula. I have found that students often confuse complex numbers and polar forms, so in this book I've separated these ideas into different chapters.

CHAPTER 7: You will find one of the most complete introductions to polar-form curves in this chapter. I was able to spend more time with the topic by separating Chapters 6 and 7 into two chapters.

CHAPTER 8: In many schools, trigonometry is the last precalculus course, and in such cases it is desirable to introduce the conic sections prior to beginning calculus. I have included this chapter for those schools that find it worthwhile not only to review the conic sections, but also to apply trigonometry by doing rotations of the conics and conics in polar form.

Problems

I provide abundant examples and problems for both drill and theory. The problems are divided into A, B, and C categories by order of difficulty, and in matched pairs so one problem has an answer and the other does not. The student is generally required to work at the A and B levels, whereas C problems are designed for an honors class or to present a challenge to most students.

In making up assignments, you need to have some idea about the problems before you make the assignment, and you do not want any surprises with a difficult problem embedded within a list of simple problems. For this reason, I've taken care in developing a *standard mathematics assignment*. This means, I've written the problem sets so they are all about the same length, and by using a, b, and c parts for particular problems, I have been able to distribute the different types of problems and difficulty levels within each problem set in basically the same way. This means, for example, that you could assign the same problems day after day and have problem sets with consistent length and difficulty. For example, when I teach from this book I assign Problems 3–30, multiples of 3, so that I am sure the student will have a uniform assignment each day and will have practice with all of the important concepts. Because some instructors like to make spiral assignments, I've designed the problems so that they are also easy to assign; for example, Problems 5–30, multiples of 5 of the current section, and Problems 7, 14, 21, and 28 of the previous section.

TYPES OF PROBLEMS:

A and B Problems: matched pairs; both drill and applied.

C Problems: challenging problems.

Applications for Further Study: These problems are included to illustrate some of the useful applications at length.

Historical Questions: These problems are included to put trigonometry in its historical perspective.

Calculator and Computer Problems: Today's technology cannot be ignored, nor can it be forced. It is assumed that all students have inexpensive calculators, but the more advanced graphics calculators and computers are treated as optional. You will notice that I've included these problems in specially designed boxes.

Supplementary Materials

There is an *Instructor's Manual* that contains suggested assignments, teaching suggestions, sample daily quizzes, and transparency masters. In addition, a computer supplement called TRIGPAK using an IBM format is available. This supplement uses the framing technique I use in this book as well as an excellent tutorial on solving triangles. Finally, Brooks/Cole has an $\mathbb{EXP}$ testing system (also in IBM format) which makes it easy to use or modify questions available on disk to make your own examinations.

ACKNOWLEDGMENTS

I am grateful to the following persons who reviewed the manuscript for this fifth edition:

P. D. Briggs, East Central University

Dawson Carr, Sandhills Community College

Gerald Goff, Oklahoma State University

Libby Holt, Florida Community College at Jacksonville

Paula Kemp, Southwest Missouri State University

Pamela Littleton, Tarleton State University

Lee Minor, West Carolina University

Joseph Moser, San Diego State University

V. Merriline Smith, California State Polytechnic University

I would also like to thank the reviewers of the previous editions:

Donald Adlong, University of Central Arkansas

Susan Barker, Atlantic Community College

Roger Breen, Florida Junior College

Wayne Britt, Louisiana State University

Louis Chatterly, Brigham Young University

Grace De Velbiss, Sinclair Community College

Arthur Dull, Diablo Valley College

Marjorie Freeman, University of Houston Downtown College

Charles R. Friese, North Dakota State University

Carolyn Funk, Thornton Community College

Gerald Goff, Oklahoma State University

Paul W. Haggard, East Carolina University

Franz X. Heirgeist, West Virginia University

Louis F. Hoelzle, Bucks County Community College

Stephen Johnson, Linn-Benton Community College

Elton Lacey, Texas A & M University

Michael Morgan, Linn-Benton Community College

Mustapha Munem, Macomb County Community College

Carla B. Oviatt, Montgomery College

Charlene Pappin, Wright College

William L. Perry, Texas A & M University

Beverly Rich, Illinois State University

Dick Schulz, Charles County Community College

Edwin Schulz, Texas Tech University

Thomas Seremet, Charles County Community College

Richard Slinkman, Bemidji State University

John Spellmann, Southwest Texas State University

Monty Strauss, Texas Tech University

Dorothy Sulock, University of North Carolina

Donna Szott, Community College of Allegheny County

Robert J. Wisner, New Mexico State University

Karen Zobrist, Louisiana State University

CONTENTS

This book presents a unit-circle approach to trigonometry. For a right-triangle treatment of the subject, see the author's other book, Essentials of Trigonometry, Second Edition, *also published by Brooks/Cole Publishing Company. The* Essentials of Trigonometry *text also includes a chapter on logarithms in place of the chapter on conic sections contained in this book.*

* Optional sections.

CHAPTER 4

TRIGONOMETRIC EQUATIONS AND INVERSE FUNCTIONS

CHAPTER 5

SOLVING TRIANGLES

CHAPTER 6

COMPLEX NUMBERS

* Optional sections.

* Optional sections.

FIFTH EDITION

TRIGONOMETRY FOR COLLEGE STUDENTS

CHAPTER 1

CIRCULAR FUNCTIONS

CHAPTER OVERVIEW

This chapter introduces the basic ideas of trigonometry. The ideas of angles, triangles, and the trigonometric functions are defined. The origins of trigonometry are obscure, but we do know that it began more than 2,000 years ago with the Mesopotamian, Babylonian, and Egyptian civilizations. Much of the knowledge from these civilizations was passed on by the Greeks, who formally developed many of the ideas in this book. The word *trigonometry* comes from the words "triangle" (*trigon*) and "measurement" (*metry*), and the ancients used trigonometry in a very practical way to measure triangles as a means of surveying land. We will look at this in Chapter 5. However, today it is used in a much more theoretical way, not only in mathematics but also in electronics, engineering, and computer science. Today we can characterize trigonometry as the branch of mathematics that deals with the properties and applications of six functions: the cosine, sine, tangent, secant, cosecant, and cotangent. These functions are defined in this chapter.

1.1 ANGLES AND DEGREE MEASURE

Angles

Essential to the definition of the trigonometric functions is the idea of an angle. Recall from geometry that an **angle** is defined as two rays or line segments with a common endpoint. The rays or line segments are called the **sides** of the angle, and the common endpoint is called the **vertex**. In geometry you named angles by naming the vertex or the three points as shown in Figure 1.1. In trigonometry it is also common to use lowercase Greek letters to denote angles. A list of the common Greek letters is included as part of Figure 1.1.*

FIGURE 1.1 Ways to denote an angle

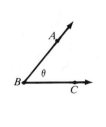

The symbol $\angle$ denotes angle. The given angle can be denoted in the following ways:

$\angle ABC$ $\angle B$
$\angle CBA$ θ

└ *The vertex must be listed between the other two points.*

Symbol	Name
α	alpha
β	beta
γ	gamma
δ	delta
θ	theta
λ	lambda
μ	mu
ϕ or φ	phi
ω	omega

* The Greek letter π (pi) is a lowercase letter that will not be used to represent an angle, because it denotes an irrational number approximately equal to 3.141592654.

1

In trigonometry and more advanced mathematics this geometric definition of an angle is generalized.

GENERALIZED
DEFINITION OF
AN ANGLE

> An **angle** is formed by rotating a ray about its endpoint (called the **vertex**) from some initial position (called the **initial side**) to some terminal position (called the **terminal side**). The measure of an angle is the amount of rotation. An angle is also formed if a line segment is rotated about one of its endpoints.

If the rotation of the ray is in a counterclockwise direction, the measure of the angle is called **positive**. If the rotation is in a clockwise direction, the measure of the angle is called **negative**. A small curved arrow is used to denote the direction and amount of rotation. In geometry, you did not need to show a rotation for an angle, but now with this generalized definition the curved arrow is essential.

Degree Measurement of Angles

Several units of measurement are used for measuring angles. Historically, the most common scheme uses *degrees*. Draw a circle with any nonzero radius r and center at O. An **arc** is part of a circle, so, if P and Q are any two distinct points on the circle, then $\overset{\frown}{PQ}$ (read as arc PQ) is an arc of the circle. Angle POQ is called a **central angle** that is **subtended** by $\overset{\frown}{PQ}$ as shown in Figure 1.2.

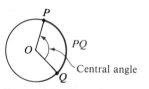

FIGURE 1.2　Central angle and an arc

The distance around a circle is called the **circumference** of the circle. A **degree** is an angle measurement that is defined as the size of a central angle subtended by an arc equal to $\frac{1}{360}$ of the length of the circumference. This definition of degree measure is independent of the size of the radius of the circle.

To measure angles you can use a protractor showing degree measure (see Figure 1.3).

FIGURE 1.3　Protractor for degree measure

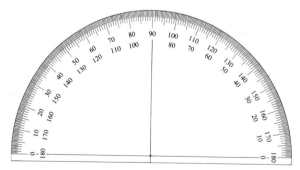

You will not need to use a protractor in this class, but you will need to know the approximate size of various angles measured in degrees. The degree symbol is ° and is used as shown:

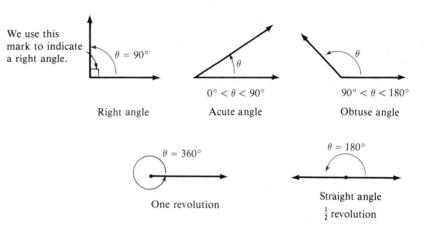

We use this mark to indicate a right angle.

$\theta = 90°$ Right angle

$0° < \theta < 90°$ Acute angle

$90° < \theta < 180°$ Obtuse angle

$\theta = 360°$ One revolution

$\theta = 180°$ Straight angle $\frac{1}{2}$ revolution

Other common angles are shown in Example 1. When we write $\angle B = 45°$ we mean the measure of angle B is $45°$, and when we write $\theta = 30°$ we mean the measure of angle θ is $30°$.

Example 1

Draw angles whose measures are given.

a. $\alpha = 30°$ **b.** $\angle G = 45°$ **c.** $\beta = -60°$ **d.** $\gamma = 270°$

Solution
WARNING! You should remember the approximate sizes of the angles $0°, 30°, 45°, 60°, 90°, 180°, 270°,$ and $360°$ because they are frequently used in mathematics.

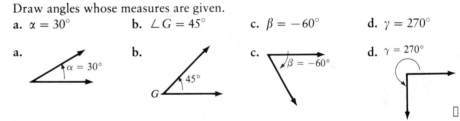

a. $\alpha = 30°$ **b.** $45°$, G **c.** $\beta = -60°$ **d.** $\gamma = 270°$

A degree can be divided into 60 equal parts, each called a **minute**, denoted by $(1° = 60')$. Furthermore, a minute can be divided into 60 equal parts, each called a **second**, denoted by $''$ $(1' = 60''; 1° = 60' = 3,600'')$. With the widespread use of calculators, it is becoming more and more common to represent minutes and seconds as decimal degrees. In this book you should use decimal degrees rather than minutes and seconds unless you are directed to do otherwise.

Example 2

Convert the degree measures to decimal degrees.

a. $25°30'$ **b.** $128°14'$ **c.** $42°13'40''$

Solution

a. Since $1° = 60'$, the number of minutes should be divided by 60 in order to convert it to decimal degrees.

$$30' = \frac{30°}{60} = \frac{1°}{2} = .5°$$

Thus, $25°30' = 25.5°$.

b. $14' = \dfrac{14°}{60} = .2\overline{3}° \approx .23°$

If an angle is measured in minutes, then round the converted decimal to the nearest hundredth of a degree.* Thus, $128°14' \approx 128.23°$.

CALCULATOR CONVERSION:

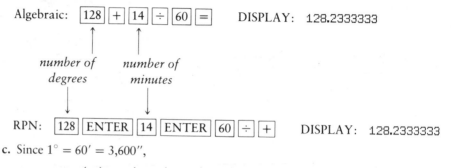

c. Since $1° = 60' = 3,600''$,

$$13'40'' = \left(\frac{13}{60}\right)° + \left(\frac{40}{3,600}\right)° = \left(\frac{820}{3,600}\right)°$$

By division, $\frac{820}{3,600} = .22\overline{7}$. If an angle is measured in seconds, then round the converted decimal to the nearest thousandth of a degree. Thus, $42°13'40'' \approx 42.228°$.

CALCULATOR CONVERSION:

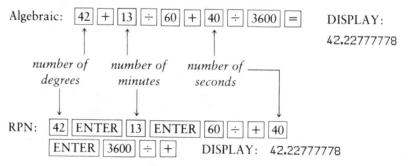

Standard-Position Angles

It is often useful to superimpose a Cartesian system on an angle so that the vertex is at the origin and the initial side is along the positive x-axis. The angle is then said to be in **standard position.** In this book we will assume that angles are in standard position

---

* Accuracy of measurements will be important in trigonometry. With calculators it is easy to assume much more accuracy than we have. Pay attention to agreements about proper rounding as they are mentioned. Also, note the difference between the exact representation of an irrational number such as $\sqrt{2}$, π, and $\frac{\pi}{2}$, and their approximations to the nearest hundredth: 1.41, 3.14, and 1.57.

unless otherwise noted. Angles in standard position that have the same terminal sides are **coterminal angles**. Given any angle α, there is an unlimited number of coterminal angles (some positive and some negative). In Figure 1.4, β is coterminal with α. Can you find other angles coterminal with α?

FIGURE 1.4 Standard-position angles α and β. α is a positive angle, and β is a negative angle. α and β are coterminal angles.

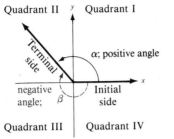

If the terminal side of θ is on one of the coordinate axes, then θ is called a **quadrantal angle**. If θ is not a quadrantal angle, then it is said to be in a certain **quadrant** if its terminal side lies in that quadrant. Thus, in Figure 1.4, both α and β would be called Quadrant II angles.

Example 3

Find a positive angle less than one revolution that is coterminal with the given angle. Also, classify the angle by quadrant.

a. $-30°$ **b.** $500°$ **c.** $-2,000°$

Solution

a. $-30°$; Quadrant IV   *$-30°$ and $330°$ are coterminal.*

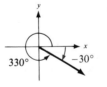

b. $500°$; Quadrant II   *$500°$ and $140°$ are coterminal.*

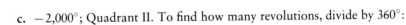

c. $-2,000°$; Quadrant II. To find how many revolutions, divide by $360°$:

$$2,000 \div 360 = 5.\overline{5}$$

Thus, it is between 5 and 6 revolutions. To find how much more than 5 revolutions, subtract 5 revolutions $(5 \cdot 360°)$ from $2,000°$:

$$2,000° - 5(360°) = 2,000° - 1,800° = 200°$$

Now draw the negative angle (5 revolutions clockwise, plus 200° clockwise):

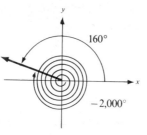

$-2,000°$ and $160°$ are coterminal.

Problem Set 1.1*

A

1. Name the given Greek letter.
    a. θ　　　　b. α　　　　c. ϕ　　　　d. ω　　　　e. μ

2. Name the given Greek letter.
    a. γ　　　　b. β　　　　c. δ　　　　d. λ　　　　e. φ

3. Write the symbol for the given Greek letter.
    a. delta　　　b. phi　　　c. theta　　　d. omega　　　e. mu

4. Write the symbol for the given Greek letter.
    a. alpha　　　b. beta　　　c. gamma　　　d. lambda　　　e. pi

Without using a protractor, draw an angle that approximates each angle with the measure indicated in Problems 5–10.

5. a. $60°$　　　b. $180°$　　　c. $45°$　　　d. $360°$　　　e. $300°$

6. a. $90°$　　　b. $30°$　　　c. $0°$　　　d. $270°$　　　e. $150°$

7. a. $120°$　　　b. $210°$　　　c. $135°$　　　d. $330°$　　　e. $240°$

8. a. $-30°$　　　b. $-135°$　　　c. $-120°$　　　d. $-270°$　　　e. $-200°$

9. a. $-60°$　　　b. $315°$　　　c. $-330°$　　　d. $-45°$　　　e. $-90°$

10. a. $-300°$　　　b. $-210°$　　　c. $-240°$　　　d. $-360°$　　　e. $-150°$

Find a positive angle less than one revolution that is coterminal with each of the angles with the measures indicated in Problems 11–16.

11. a. $400°$　　　b. $540°$　　　c. $800°$　　　d. $750°$　　　e. $3,000°$

12. a. $370°$　　　b. $900°$　　　c. $550°$　　　d. $1,050°$　　　e. $2,500°$

13.	a. 600°	b. 730°	c. 580°	d. 1,000°	e. 4,000°
14.	a. −30°	b. −200°	c. −100°	d. −300°	e. −140°
15.	a. −55°	b. −320°	c. −270°	d. −400°	e. −530°
16.	a. −75°	b. −450°	c. −360°	d. −500°	e. −750°

+330°

B   *Change the given measures of the angles in Problems 17–26 to decimal degrees to the nearest hundredth.*

17.	65°40′	18.	146°50′	19.	85°20′	20.	127°10′	21.	240°30′
22.	315°25′	23.	16°42′	24.	29°17′	25.	143°23′	26.	165°58′

Change the given measures of the angles in Problems 27–36 to decimal degrees to the nearest thousandth.

27.	128°10′40″	28.	13°30′50″	29.	48°28′10″	30.	281°31′36″	31.	210°40′15″
32.	94°21′31″	33.	38′42″	34.	12′24″	35.	6′7″	36.	2′6″

C   37.   *Historical Question*   The division of one revolution (a *circle*) into 360 equal parts (called *degrees*) is no doubt due to the sexagesimal (base 60) numeration system used by the Babylonians. Several explanations have been put forward to account for the choice of this number. (For example, see Howard Eves' *In Mathematical Circles.*) One possible explanation is based on the fact that the radius of a circle can be applied exactly six times to its circumference as a chord. Illustrate this geometric fact. Notice that sexagesimal division (that is, division into 60 parts) makes this geometric construction easy. It then seems natural that each of these chords should be further divided into 60 equal parts, resulting in the division of the circle into 360 equal parts. Write a short research paper on the history of degree measure for an angle.

1.2   RADIAN MEASURE OF ANGLES

Radian Measurement

In the last section the degree measure of an angle was discussed. There is a second commonly used measure for an angle that, instead of being based on 60 ($60'' = 1'$, $60' = 1°$, $360° = 1$ revolution), is based on a real number measurement.

Draw a circle with any positive radius r. Next, measure out an arc whose length is r. Figure 1.5a (page 8) shows the case in which $r = 1$, and Figure 1.5b shows $r = 2$. Regardless of your choice of r, the size of the angle determined by this arc of length equal to the radius is the same (it is labeled θ in Figure 1.5). This fact can be proved from plane geometry, but you can easily convince yourself by drawing a few circles of *different* radii and verifying that every time the angle is the same size. Because it is the same size for every radius, we use this angle as a standard of comparison. Its measure is called one **radian**.

FIGURE 1.5 Angles of measure 1. The size of the angle is independent of the size of the circle.

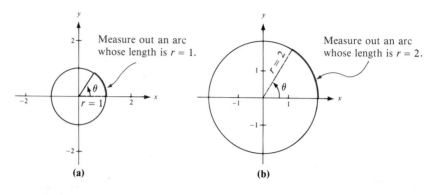

(a)   (b)

RADIAN

> One **radian** is the measure of a central angle intercepting an arc equal to the radius of the circle.

Notice that the circumference generates an angle of one revolution. Since $C = 2\pi r$, and since the radian is measured in terms of r, **an angle of one revolution has a radian measure of 2π**.

Angle	Radian measure
One revolution	2π
Straight angle $\left(\dfrac{1}{2}\,\text{rev}\right)$	π
Right angle $\left(\dfrac{1}{4}\,\text{rev}\right)$	$\dfrac{\pi}{2}$

Notice that, when measuring angles in radians, we are using *real numbers*. Because radian measure is used so frequently in more advanced work, we agree that radian measure is understood when *no units* of measure for an angle are indicated.

WARNING! Note the use of degree and radian measures in this book.

> In this book, when measuring in degrees we will always say *degrees* or use the degree symbol, as with $\theta = 5°$. When measuring in radians we will **omit** the word *radians*. That is, *when no unit is given, radian measure is understood*. For example,
>
> $$\theta = 5$$
>
> means the measure of θ is 5 radians.

Figure 1.6 shows a protractor for measuring angles using radian measure. Notice that the maximum *direct* measurement you can make with this protractor is a straight angle (π radians). Therefore, if the angle you wish to measure is more than π radians, you must subtract the measure of a straight angle (see Example 1c).

FIGURE 1.6 Protractor for radian measure

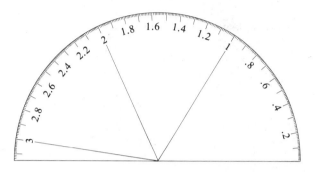

Just as with degree measure, you should have the size of 1 radian memorized so that you can draw angles of various sizes from memory with a reasonable amount of accuracy. You will have to practice thinking in terms of radian measure. You should memorize the approximate size of an angle of measure 1 in much the same way you have memorized the approximate size of an angle of 45°.

Example 1

Use Figure 1.6 to draw the angles whose measures are given. Do not attempt to change the measures in these problems to degrees. Work them directly in radians; do not make comparisons back to degree measure.

    a. $\theta = 2$ **b.** $\theta = -3$ **c.** $\theta = 4$ **d.** $\theta = -\frac{\pi}{4}$

Solution

    a. $\theta = 2$

    b. $\theta = -3$

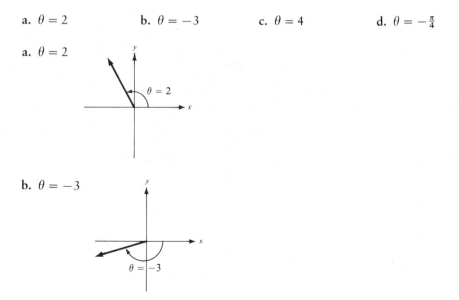

c. $\theta = 4$

You cannot measure values greater than π (about 3.14) *directly* because a protractor measures angles up to a straight angle. Thus, **subtract the measure of a straight angle** (or a multiple) to give a residual less than π; use the protractor to measure the residual:

$$4 - \pi \approx .86$$

Thus,

$$4 \approx \pi + .86$$

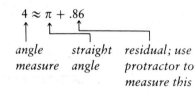

angle   straight   residual; use
measure   angle   protractor to
                     measure this

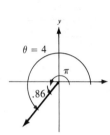

d. $\theta = -\frac{\pi}{4}$

$-\frac{\pi}{4}$ is called an *exact measurement* of the angle.

To use a protractor, write an approximation for the measure of the angle:

$$-\frac{\pi}{4} \approx -.79$$

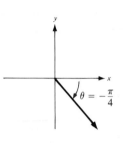

Degree–Radian Conversions

We are led to the following relationships:

1 revolution is measured by 360°.

1 revolution is measured by 2π.

Thus,

$$\text{number of revolutions} = \frac{\text{angle in degrees}}{360}$$

$$\text{number of revolutions} = \frac{\text{angle in radians}}{2\pi}$$

Therefore, we can state the following equivalency:

RELATIONSHIP BETWEEN
DEGREE AND RADIAN
MEASURE

$$\frac{\text{angle in degrees}}{360} = \frac{\text{angle in radians}}{2\pi}$$

To change from degree measure to radian measure, let θ be the angle measure in radians. Multiply both sides by 2π to obtain

$$\theta = \frac{2\pi}{360} \text{ (angle in degrees)}$$

$$= \frac{\pi}{180} \text{ (angle in degrees)}$$

DEGREE TO RADIAN

> **Multiply the angle in degrees by the constant $\frac{\pi}{180}$ to obtain the radian measure.**

Example 2

Change the following degree measures to radian measures.

a. $45°$ b. $125°$ c. $1°$

Solution

a. $\theta = \frac{\pi}{180}(45)$

$\quad = \frac{\pi}{4}$ *Exact value*

An alternate method for converting this angle measure is to remember that π radians is $180°$. That is, because $45°$ is $\frac{1}{4}$ of $180°$, you know that the radian measure is $\frac{\pi}{4}$.

b. $\theta = \frac{\pi}{180}(125)$

$\quad = \frac{25\pi}{36}$ *Exact*

If you want a calculator approximation of θ, press:

$\boxed{25}\,\boxed{\times}\,\boxed{\pi}\,\boxed{\div}\,\boxed{36}\,\boxed{=}$

DISPLAY: 2.1816616

$125° \approx 2.1817$ *Approximate*

c. $\theta = \frac{\pi}{180}(1)$

$\quad = \frac{\pi}{180}$ *Exact*

If you want a calculator approximation of θ, press:

$\boxed{\pi}\,\boxed{\div}\,\boxed{180}\,\boxed{=}$

DISPLAY: .01745329

$1° \approx .0175$ *Approximate*   ☐

To change from radian measure to degree measure, let θ be the angle measured in degrees. Multiply both sides of the equivalency formula by 360 to obtain

$$\theta = \frac{360}{2\pi} \text{ (angle in radians)}$$

$$= \frac{180}{\pi} \text{ (angle in radians)}$$

RADIAN TO DEGREE

> **Multiply the angle in radians by the constant $\frac{180}{\pi}$ to obtain the degree measure.**

Example 3

Change the following radian measures to degree measures.

a. $\frac{\pi}{9}$ b. 1   c. 2.30

Solution

a. $\theta = \dfrac{180}{\pi}\left(\dfrac{\pi}{9}\right)$

$\quad = 20$

Thus, $\frac{\pi}{9}$ is an angle with measure of $20°$.

c. $\theta = \dfrac{180}{\pi}(2.30)$

$\quad \approx (57.296)(2.3)$

$\quad = 131.7803$ To the nearest hundredth, the angle is $131.78°$.

b. $\theta = \dfrac{180}{\pi}(1)$

$\quad = \dfrac{180}{\pi}$

On a calculator,
$\theta \approx 57.29577951°$.

If you have a calculator, you can obtain a much more accurate answer by using a better approximation for $\frac{180}{\pi}$. For example, if 2.3 is exact, then by calculator:

Algebraic logic: $\boxed{180}\ \boxed{\div}\ \boxed{\pi}\ \boxed{\times}\ \boxed{2.3}\ \boxed{=}$

RPN logic: $\boxed{180}\ \boxed{\text{ENTER}}\ \boxed{\pi}\ \boxed{\div}\ \boxed{2.3}\ \boxed{\times}$ DISPLAY: 131.7802929

For the more common measures of angles, it is a good idea to memorize the equivalent degree and radian measures shown in Table 1.1. If you keep in mind that $180°$ equals π in radian measure, the rest of the table will be very easy to remember.

TABLE 1.1 Relationship between Degree and Radian Measure of Angles

Degrees	Radians
$0°$	0
$30°$	$\dfrac{\pi}{6}$
$45°$	$\dfrac{\pi}{4}$
$60°$	$\dfrac{\pi}{3}$
$90°$	$\dfrac{\pi}{2}$
$180°$	π ←
$270°$	$\dfrac{3\pi}{2}$
$360°$	2π

Quadrantal angles

Common first quadrant angles

If you remember

$\quad 180° = \pi$

you will be able to easily remember the other entries in this table.

Arc Length

In the preceding discussion we used an arc to find the measure of an angle, but we can also use the measure of an angle to find the length of an arc of a circle, called **arc length**. To do this, use the formula for the circumference of a circle with radius r.

ARC LENGTH

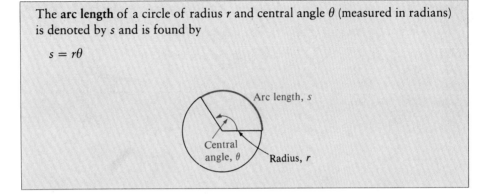

The **arc length** of a circle of radius r and central angle θ (measured in radians) is denoted by s and is found by

$$s = r\theta$$

Arc length, s

Central angle, θ

Radius, r

To see where this formula comes from, recall the definition of radian measure to notice that the angle is 2π times that portion of the circle subtended by the angle θ. For example,

1 revolution, $\dfrac{s}{C} = 1$, so $\theta = 1(2\pi) = 2\pi$

$\dfrac{1}{2}$ revolution, $\dfrac{s}{C} = \dfrac{1}{2}$, so $\theta = \dfrac{1}{2}(2\pi) = \pi$

$\dfrac{1}{4}$ revolution, $\dfrac{s}{C} = \dfrac{1}{4}$, so $\theta = \dfrac{1}{4}(2\pi) = \dfrac{\pi}{2}$

In general, $\theta = \dfrac{s}{C}(2\pi)$. Now, algebraically solve for s:

$$s = \frac{\theta C}{2\pi}$$

But $C = 2\pi r$, so

$$s = \frac{\theta(2\pi r)}{2\pi}$$

$$s = r\theta$$

Example 4

What is the length of the arc of a circle of radius 10 cm if the central angle is 2?

Solution

$$s = r\theta$$
$$= (10 \text{ cm})(2)$$
$$= 20 \text{ cm}$$

Example 5

Find the arc length if the central angle is $36°$ and the radius of the circle is 20 centimeters (cm).

Solution

First, change the degree measure to radian measure:

$$\theta = \frac{\pi}{180}(36) \quad \text{where } \theta \text{ is the radian measure of the angle}$$

$$= \frac{\pi}{5}$$

Thus,

$$s = r\theta$$

$$= 20\left(\frac{\pi}{5}\right)$$

$$= 4\pi$$

The length of the arc is 4π cm or about 12.6 cm.

Locations on the earth are given by *latitude* and *longitude*. Latitude is measured north or south of the equator by angles between $0°$ and $90°$. Longitude is measured by angles between $0°$ and $180°$ either west or east of the Greenwich meridian.

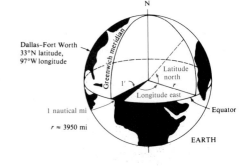

Example 6

The Dallas–Fort Worth airport is located at approximately $33°$N latitude and $97°$W longitude. How far north (to the nearest hundred miles) is it from the equator?

Solution

The location is $33°$N latitude, so first convert $33°$ to radians:

$$\theta \approx 33° = 33\left(\frac{\pi}{180}\right) \approx .576$$

Next, use the arc length formula:

$$s = r\theta$$
$$\approx 3,950\theta$$
$$\approx 2,300$$

The Dallas–Fort Worth airport is about 2,300 miles north of the equator.   ⬜

Problem Set 1.2

A

1. From memory, give the radian measure for each of the angles whose degree measure is stated.
   a. 30°   **b.** 90°   **c.** 270°   **d.** 45°   **e.** 360°   **f.** 60°   **g.** 180°

2. From memory, give the degree measure for each of the angles whose radian measure is stated.
   a. π **b.** $\frac{\pi}{4}$ **c.** $\frac{\pi}{3}$ **d.** 2π **e.** $\frac{\pi}{2}$ **f.** $\frac{3\pi}{2}$ **g.** $\frac{\pi}{6}$

Use the radian protractor in Figure 1.6 to help you sketch each of the angles with the measures indicated in Problems 3–6. Do not change them to degrees, but try to think in radians.

3. **a.** $\frac{\pi}{2}$ **b.** $\frac{\pi}{6}$ **c.** 2.5   **d.** 6
4. **a.** $\frac{2\pi}{3}$ **b.** $\frac{3\pi}{4}$ **c.** -1 **d.** 5
5. **a.** $-\frac{\pi}{4}$ **b.** $\frac{7\pi}{6}$ **c.** -2.76 **d.** $\sqrt{17}$
6. **a.** $-\frac{3\pi}{2}$ **b.** $\frac{13\pi}{3}$ **c.** -1.2365 **d.** $\sqrt{23}$

Find a positive angle less than one revolution that is coterminal with each of the angles with the measures indicated in Problems 7–10.

7. **a.** 3π **b.** $\frac{13\pi}{6}$ **c.** $-\pi$ **d.** 7
8. **a.** $-\frac{\pi}{4}$ **b.** $\frac{17\pi}{4}$ **c.** $\frac{11\pi}{3}$ **d.** -2
9. **a.** 9   **b.** -5 **c.** $\sqrt{50}$ **d.** -6
10. **a.** 6.2832   **b.** -3.1416 **c.** 30   **d.** $3\sqrt{5}$

Change each of the angles with the indicated measures in Problems 11–13 to radians using exact values.

11. **a.** 40°   **b.** 20°   12. **a.** $-64°$ **b.** $-220°$ 13. **a.** 254°   **b.** 85°

Change each of the angles with the indicated measures in Problems 14–16 to radians correct to the nearest hundredth.

14. **a.** 112°   **b.** 314°   15. **a.** $-62.8°$ **b.** 350°

16. **a.** $-205.2°$ **b.** 535°

Change each of the angles with the indicated measures in Problems 17–19 to degrees correct to the nearest hundredth.

17. **a.** $\frac{2\pi}{9}$ **b.** $\frac{\pi}{10}$ 18. **a.** $-\frac{11\pi}{12}$ **b.** 2   19. **a.** -3 **b.** $-.25$

B

In Problems 20–25 find the length of each intercepted arc to the nearest hundredth, if the measure of the central angle and radius are as given.

20. Angle 1, radius 1 m   21. Angle 2.34, radius 6 cm   22. Angle $\frac{\pi}{3}$, radius 4 m

23. Angle $\frac{3\pi}{2}$, radius 15 cm   24. Angle 40°, radius 7 km   25. Angle 72°, radius 10 dm

26. How far (to the nearest hundredth) does the tip of an hour hand measuring 2.00 cm move in three hours?

27. A 50-cm pendulum on a clock swings through an angle of 100°. How far (to the nearest centimeter) does the tip travel in one arc?

28. *Geography.* If New York City is at 40°N latitude and 74°W longitude, what is its approximate distance (to the nearest hundred miles) from the equator?

29. *Geography.* If Chicago is located at 42°N latitude and 88°W longitude, what is its approximate distance (to the nearest hundred miles) from the equator?

30. *Geography.* If Columbia, South Carolina, is located at 34°N latitude and 81°W longitude, what is its approximate distance (to the nearest hundred miles) from Disneyland, which is located at 34°N latitude and 118°W longitude? [*Hint:* The radius at 34° is approximately 3,250 mi.]

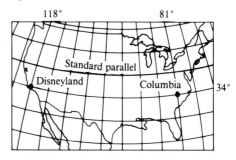

*31. *Physics.* Through how many radians does a pulley with a 20.0-cm diameter turn when 1.00 m of cable is pulled through it without slippage? (1 m = 100 cm)

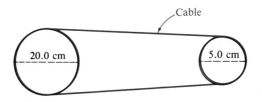

*32. *Physics.* Answer the question in Problem 31 for a pulley with a 5.0-cm diameter.

C

33. *Historical Question* In about 230 B.C. a mathematician named Eratosthenes estimated the radius of the earth using the following information. Syene and Alexandria in Egypt are on the same line of longitude. They are also 800 km apart. At noon on the longest day of the year, when the sun was

---

* See the Application for Further Study at the end of this chapter for further applications involving linear velocity and angular velocity.

directly overhead in Syene, Eratosthenes measured the sun to be 7.2° from the vertical in Alexandria. Because of the distance of the sun from the earth, he assumed that the sun's rays were parallel. Thus he concluded that the central angle subtending rays from the center of the earth to Syene and Alexandria was also 7.2°. Using this information, find the approximate radius of the earth.

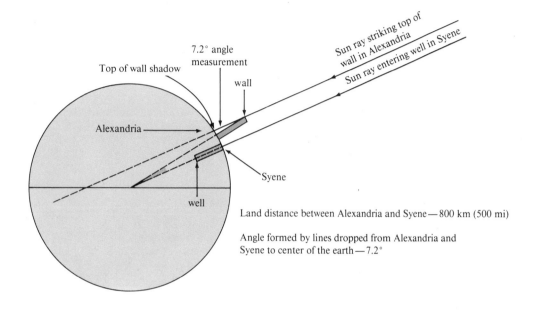

Land distance between Alexandria and Syene—800 km (500 mi)

Angle formed by lines dropped from Alexandria and Syene to center of the earth—7.2°

34. *Surveying.* Omaha, Nebraska, is located at approximately 97°W longitude, 41°N latitude; Wichita, Kansas, is located at approximately 97°W longitude, 37°N latitude. Notice that these two cities have about the same longitude. If we know that the radius of the earth is about 6,370 kilometers (km), what is the distance between these cities to the nearest 10 km?

35. *Surveying.* Entebbe, Uganda, is located at approximately 33°E longitude, and Stanley Falls in Zaire is located at 25°E longitude. Also, both these cities lie approximately on the equator. If we know that the radius of the earth is about 6,370 km, what is the distance between the cities to the nearest 10 km?

36. *Astronomy.* Suppose it is known that the moon subtends an angle of 45.75' at the center of the earth. It is also known that the center of the moon is 384,417 km from the surface of the earth. If you assume that the radius of the earth is 6,370 km, what is the diameter of the moon to the nearest 10 km?

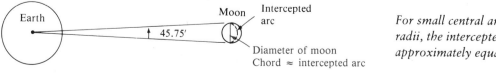

For small central angles with large radii, the intercepted arc is approximately equal to its chord.

1.3   TRIGONOMETRIC FUNCTIONS ON A UNIT CIRCLE

Unit Circle

To introduce you to the trigonometric functions, we will consider a relationship between angles and circles. The **unit circle** is the circle centered at the origin with radius one. The equation of the unit circle is

$$x^2 + y^2 = 1$$

Draw a unit circle with an angle θ in standard position, as shown in Figure 1.7.

FIGURE 1.7   The unit circle

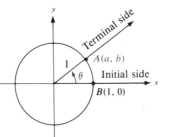

Cosine and Sine Functions

The initial side of θ intersects the unit circle at $(1, 0)$, and the terminal side intersects the unit circle at (a, b). Define functions of θ as follows:

$$c(\theta) = a \quad \text{and} \quad s(\theta) = b$$

Example 1

Find:

a. $s(90°)$ **b.** $c(90°)$ **c.** $s(-270°)$ **d.** $s(-\frac{\pi}{2})$ **e.** $c(-3.1416)$

Solution

a. $s(90°)$; this is the second component of the ordered pair (a, b), where (a, b) is the point of intersection of the terminal side of a $90°$ standard-position angle and the unit circle. By inspection, it is 1. Thus, $s(90°) = 1$.

b. $c(90°) = 0$

c. $s(-270°) = 1$; this is the same as part a because the angles $90°$ and $-270°$ are coterminal.

d. $s(-\frac{\pi}{2}) = -1$

e. $c(-3.1416) \approx c(-\pi) = -1$ ⬜

Trigonometric Functions

The function $c(\theta)$ is called the **cosine function**, and the function $s(\theta)$ is called the **sine function**. These functions, along with four additional functions, make up what are called the **trigonometric functions** or **trigonometric ratios**.

UNIT CIRCLE DEFINITION
OF THE TRIGONOMETRIC
FUNCTIONS

Let θ be an angle in standard position with the point (a, b) the intersection of the terminal side of θ and the unit circle. Then the six trigonometric functions are defined as follows:

cosine:	$\cos \theta = a$	**secant:**	$\sec \theta = \dfrac{1}{a} \quad a \neq 0$
sine:	$\sin \theta = b$	**cosecant:**	$\csc \theta = \dfrac{1}{b} \quad b \neq 0$
tangent:	$\tan \theta = \dfrac{b}{a} \quad a \neq 0$	**cotangent:**	$\cot \theta = \dfrac{a}{b} \quad b \neq 0$

WARNING! cos, sin, tan, ... are not variables. They cannot stand alone, and must be associated with an angle.

The angle θ is an important part of the definitions above. The words *cos, sin, tan,* and so on as defined in the preceding box are meaningless without the θ. You can speak of the *cosine function* or of *cos* θ, but you cannot simply say *cos*.

Notice the condition on the tangent, secant, cosecant, and cotangent functions. These exclude division by 0; for example, $a \neq 0$ means that θ cannot be 90°, 270°, or any angle coterminal to these angles. We summarize this condition by saying that the *tangent and secant are not defined for 90° or 270°*. If $b \neq 0$, then $\theta \neq 0°$, 180°, or any angle coterminal to these angles; thus, *cosecant and cotangent are not defined for 0° or 180°*. In this book we will assume all values that cause division by 0 are excluded from the domain.

Fundamental Identities

There are certain relationships among the trigonometric functions that you will need in order to evaluate these functions. These relationships are commonly referred to as the **fundamental identities**. Recall from algebra that an **identity** is an open equation (has at least one variable) which is true for all values in the domain.

The first three identities are called the **reciprocal identities**. Remember from arithmetic that 5 and $\frac{1}{5}$ are reciprocals, as are $\frac{2}{3}$ and $\frac{3}{2}$; in general, two numbers are reciprocals if their product is 1.

RECIPROCAL IDENTITIES

1. $\sec \theta = \dfrac{1}{\cos \theta}$ The cosine and secant are reciprocal functions.

2. $\csc \theta = \dfrac{1}{\sin \theta}$ The sine and cosecant are reciprocal functions.

3. $\cot \theta = \dfrac{1}{\tan \theta}$ The tangent and cotangent are reciprocal functions.

It is easy to prove that these are identities by looking at the definition of secant, cosecant, and cotangent. Also, notice that these can be rewritten in other forms:

$$\cos\theta\sec\theta = 1 \quad \text{or} \quad \cos\theta = \frac{1}{\sec\theta}$$

Example 2

Write the second identity two different ways.

Solution

$$\sin\theta\csc\theta = 1 \quad \text{and} \quad \sin\theta = \frac{1}{\csc\theta}$$

Example 3

Write each given function in terms of its reciprocal function.

a. $\dfrac{1}{\sin 81°}$ **b.** $\sec 2$ **c.** $\tan\frac{\pi}{4}$ **d.** $\cot 35°$ **e.** $\csc 3$

Solution

a. $\dfrac{1}{\sin 81°} = \csc 81°$ **b.** $\sec 2 = \dfrac{1}{\cos 2}$ **c.** $\tan\frac{\pi}{4} = \dfrac{1}{\cot\frac{\pi}{4}}$

d. $\cot 35° = \dfrac{1}{\tan 35°}$ **e.** $\csc 3 = \dfrac{1}{\sin 3}$

The next two identities are called the **ratio identities** and they express the tangent and cotangent functions using sine and cosine functions.

RATIO IDENTITIES

$$4. \ \tan\theta = \frac{\sin\theta}{\cos\theta}$$

$$5. \ \cot\theta = \frac{\cos\theta}{\sin\theta}$$

To prove these identities we also use the definition.

$$\tan\theta = \frac{b}{a} \qquad \textit{Definition of tangent}$$

$$= \frac{\sin\theta}{\cos\theta} \quad \begin{array}{l}\textit{Substitution of values from the definition:} \\ \sin\theta = b \textit{ and } \cos\theta = a\end{array}$$

The ratio identity for the cotangent is proved in exactly the same way.

Example 4

If $\sin\theta = -\frac{3}{5}$ and $\cos\theta = \frac{4}{5}$, find the values of $\tan\theta$ and $\cot\theta$.

Solution

$$\tan\theta = \frac{\sin\theta}{\cos\theta} = \frac{-\frac{3}{5}}{\frac{4}{5}} = -\frac{3}{5}\cdot\frac{5}{4} = -\frac{3}{4} \qquad \cot\theta = \frac{1}{\tan\theta} = -\frac{4}{3}$$

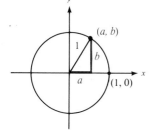

The last three identities are called the **Pythagorean identities** because they follow directly from the Pythagorean theorem. Since (a, b) is on a unit circle, the Pythagorean theorem tells us

$$a^2 + b^2 = 1$$

From the definition of sine and cosine (and by substitution), we obtain

$$(\cos \theta)^2 + (\sin \theta)^2 = 1$$

In trigonometry, we simplify the notation involving the square of a trigonometric function by writing the previous equation as follows:

$$\cos^2\theta + \sin^2\theta = 1$$

PYTHAGOREAN
IDENTITIES

> 6. $\cos^2\theta + \sin^2\theta = 1$
>
> 7. $1 + \tan^2\theta = \sec^2\theta$
>
> 8. $\cot^2\theta + 1 = \csc^2\theta$

In order to prove identities 7 and 8 we will need only to divide both sides of

$$a^2 + b^2 = 1$$

by a^2 (for identity 7) or b^2 (for identity 8):

$$\frac{a^2}{a^2} + \frac{b^2}{a^2} = \frac{1}{a^2} \qquad \text{Dividing both sides by } a^2 \ (a \neq 0)$$

$$1 + \left(\frac{b}{a}\right)^2 = \left(\frac{1}{a}\right)^2 \qquad \text{Properties of exponents}$$

$$1 + (\tan \theta)^2 = (\sec \theta)^2 \qquad \text{Definition of } \tan \theta \text{ and } \sec \theta, \text{ and substitution}$$

$$1 + \tan^2\theta = \sec^2\theta$$

The proof of identity 8 is left as an exercise.

These Pythagorean identities are frequently written in a different form. For example, we can solve identity 7 for $\tan \theta$:

$$1 + \tan^2\theta = \sec^2\theta \qquad \text{Identity 7}$$

$$\tan^2\theta = \sec^2\theta - 1 \qquad \text{Subtract 1 from both sides.}$$

$$\tan \theta = \pm\sqrt{\sec^2\theta - 1} \qquad \text{Do not forget the } \pm \text{ using the square root property from algebra.}$$

Example 5

Write all the trigonometric functions in terms of $\sin \theta$ by using the eight fundamental identities.

Solution

a. $\sin \theta = \sin \theta$

b. $\cos \theta$:   Use identity 6:

$$\cos^2\theta + \sin^2\theta = 1$$
$$\cos^2\theta = 1 - \sin^2\theta$$
$$\cos \theta = \pm\sqrt{1 - \sin^2\theta}$$

c. $\tan \theta = \dfrac{\sin \theta}{\cos \theta}$ *From identity 4*

$$= \dfrac{\sin \theta}{\pm\sqrt{1 - \sin^2\theta}}$$ *From part b**

d. $\cot \theta = \dfrac{1}{\tan \theta}$ *From identity 3*

$$= \dfrac{1}{\dfrac{\sin \theta}{\pm\sqrt{1 - \sin^2\theta}}}$$ *From part c*

$$= \dfrac{\pm\sqrt{1 - \sin^2\theta}}{\sin \theta}$$ *Invert and multiply.*

e. $\csc \theta = \dfrac{1}{\sin \theta}$ *Identity 2*

f. $\sec \theta = \dfrac{1}{\cos \theta}$ *Identity 1*

$$= \dfrac{1}{\pm\sqrt{1 - \sin^2\theta}}$$ *From part b*   ☐

Evaluation of Trigonometric Functions

In many applications, you will know the angle measure and will want to find one or more of its trigonometric functions. To do this you will carry out a process called **evaluation of the trigonometric functions**. In order to help you see the relationship between the angle and the function, consider Figure 1.8 and Example 6.

---

* In algebra you needed to rationalize the denominator; however, since we work with many reciprocals in trigonometry, we generally relax the requirement that expressions with a radical in a denominator need to be rationalized.

Evaluation Using the Definition

Example 6

Evaluate the trigonometric functions of $\theta = 110°$ by drawing the unit circle and approximating the point (a, b).

Solution

Draw a circle with $r = 1$ as shown in Figure 1.8.

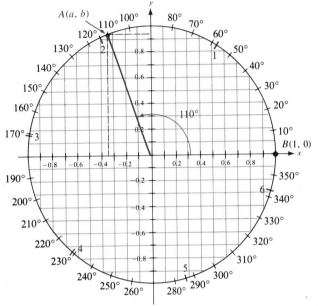

FIGURE 1.8 Approximate values of circular functions

Next, draw the terminal side of the angle 110°. Estimate (by counting squares) $a \approx -.35$ and $b \approx .95$. Thus,

$$\cos 110° \approx -.35 \qquad\qquad \sec 110° \approx \frac{1}{-.35} \approx -2.9$$

$$\sin 110° \approx .95 \qquad\qquad \csc 110° \approx \frac{1}{.95} \approx 1.1$$

$$\tan 110° \approx \frac{.95}{-.35} \approx -2.7 \qquad \cot 110° \approx \frac{-.35}{.95} \approx -.37 \qquad\qquad □$$

Using Figure 1.8, it is easy to derive a very important property of the trigonometric functions. Notice if θ is in Quadrant I, then all of the trigonometric functions are positive since a and b are both positive.

In Quadrant II, $a < 0$ and $b > 0$, so

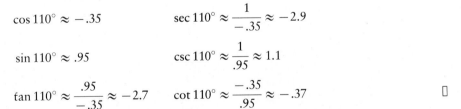

$\cos \theta = a < 0$

$\sin \theta = b > 0$

$\tan \theta < 0$ *Do you see why tangent is negative in Quadrant II? If not, use the definition:* $\tan \theta = \frac{b}{a}$, *which is a positive divided by a negative.*

The signs of $\sec\theta$, $\csc\theta$, and $\cot\theta$ are the same as their respective reciprocals. In Quadrant III, $a < 0$ and $b < 0$, so

$$\begin{cases} \cos\theta < 0 \\ \sin\theta < 0 \\ \tan\theta > 0 \text{ } \textit{Remember, a negative divided by a negative is positive.} \end{cases}$$

$$\begin{cases} \sec\theta < 0 \\ \csc\theta < 0 \\ \cot\theta > 0 \end{cases}$$

Finally, in Quadrant IV, $a > 0$ and $b < 0$, so

$$\begin{cases} \cos\theta > 0 \\ \sin\theta < 0 \\ \tan\theta < 0 \end{cases}$$

$$\begin{cases} \sec\theta > 0 \\ \csc\theta < 0 \\ \cot\theta < 0 \end{cases}$$

It will be important in your work in trigonometry to remember which of the trigonometric functions are positive in each of the quadrants. However, this information is difficult to remember using the above notation, so it is summarized below. Remember that the values of $\sec\theta$, $\csc\theta$, and $\cot\theta$ have the same signs as their reciprocals.

SIGNS OF THE TRIGONOMETRIC FUNCTIONS　　EASY-TO-REMEMBER FORM:

SINE POSITIVE	ALL POSITIVE	SMART	A
TANGENT POSITIVE	COSINE POSITIVE	TRIG	CLASS

Example 7

State the quadrant or quadrants in which θ may lie to make each of the expressions true.

a. $\tan\theta = -.9$　　　　**b.** $\sec\theta = -\frac{5}{3}$　　　　**c.** $\cot\theta = .5$
d. $\sin\theta = .3$ and $\tan\theta < 0$　　　　**e.** $\tan\theta = -2$ and $\sin\theta > 0$
f. $\sin\theta = \sqrt{1 - \cos^2\theta}$　　　　**g.** $\csc\theta = -\sqrt{\cot^2\theta + 1}$

Solution

a. Quadrant II or IV (tangent is negative)
b. The reciprocal of secant is cosine and cosine is negative in Quadrants II and III, so the secant is negative in Quadrants II and III.
c. Quadrant I or III　　(Think about where the tangent is positive.)
d. The function $\sin\theta$ is positive in Quadrants I and II and the tangent is negative in Quadrants II and IV, so there is one quadrant that makes both statements true; θ is in Quadrant II.
e. $\tan\theta$ is negative in Quadrants II and IV and $\sin\theta$ is positive in Quadrants I and II; answer is Quadrant II.

f. Radical symbols are positive so sine is positive in Quadrants I and II.

g. Since radicals are positive, opposites of radicals must be negative, so θ is in Quadrant III or IV (think about the reciprocal of cosecant, namely sine).   ☐

In practice you will not use the definition and Figure 1.8 to evaluate trigonometric functions, but this exercise is important because it will help you to understand what the values you obtain from a table or a calculator mean.

Evaluation by Calculator

Evaluating the trigonometric ratios by calculator is very straightforward, but you must note several details.

CALCULATOR
EVALUATION OF THE
TRIGONOMETRIC
FUNCTION

1. *Unit of measure used.*   Most calculators measure angles in both degrees and radians. You must make certain that your calculator is in the *proper mode.* Some calculators simply have a switch (often labeled R–D) that changes the angle-measure mode. However, there are other ways, so you should consult the owner's manual to find out how to make sure your calculator is in either degree or radian mode as needed. From now on, we will assume that you are working in the appropriate angle-measure mode. Remember, if later in the course you suddenly start obtaining "strange" answers and have no idea what you are doing wrong, double-check to make sure your calculator is in the proper mode.
2. With most calculators you *enter the angle first* and then press the button corresponding to the trigonometric function.
3. Your calculator has keys for $\boxed{\cos}$, $\boxed{\sin}$, and $\boxed{\tan}$, but not for secant, cosecant, or cotangent. Keys for the latter three functions are unnecessary as they are readily obtained using the reciprocal identities and the $\boxed{1/x}$ key on your calculator. If $x \neq 0$, you can enter x, press the reciprocal button, and see the reciprocal of x displayed in decimal form.

Example 8

Find the values of the trigonometric functions for $\theta = 37.8°$. Round to four decimal places.

Solution

Press	Display	Value
$\boxed{37.8}$ $\boxed{\cos}$	.7901550124*	$\cos 37.8° \approx .7902$
$\boxed{37.8}$ $\boxed{\sin}$	.6129070537	$\sin 37.8° \approx .6129$
$\boxed{37.8}$ $\boxed{\tan}$	.7756795108	$\tan 37.8° \approx .7757$
$\boxed{37.8}$ $\boxed{\cos}$ $\boxed{1/x}$	1.265574456	$\sec 37.8° \approx 1.2656$
$\boxed{37.8}$ $\boxed{\sin}$ $\boxed{1/x}$	1.631568758	$\csc 37.8° \approx 1.6316$
$\boxed{37.8}$ $\boxed{\tan}$ $\boxed{1/x}$	1.289192232	$\cot 37.8° \approx 1.2892$

☐

---

* Your calculator may have fewer digits in the display than are shown here. Also, if you obtained .9949151, your calculator is set to radian mode.

Evaluation by Table

In Appendix D you will find a table entitled "Trigonometric Functions." This table gives values of the trigonometric ratios for acute angles. A portion of this table is reproduced here.

Rad	Deg	cos	sin	tan		
.65	37.5	.7934	.6088	.7673	52.5	.92
.66	37.6	.7923	.6101	.7701	52.4	.91
.66	37.7	.7912	.6115	.7729	52.3	.91
.66	37.8	.7902	.6129	.7757	52.2	.91
.66	37.9	.7891	.6143	.7785	52.1	.91
.66	38.0	.7880	.6157	.7813	52.0	.91
.66	38.1	.7869	.6170	.7841	51.9	.91
.67	38.2	.7859	.6184	.7869	51.8	.90
.67	38.3	.7848	.6198	.7898	51.7	.90
.67	38.4	.7837	.6211	.7926	51.6	.90
		sin	cos	cot	Deg	Rad

Example 9

Find $\sin 37.8°$ and $\cos 52.0°$.

Solution

First, look down the column headed "Deg" until you find 37.8; then look across until you come to the column headed "sin" to find:

$$\sin 37.8° \approx .6129$$

To find values of trigonometric functions that are larger than 45°, use the table headings at the bottom of the table. Thus,

$$\cos 52.0° \approx .6157$$ ◻

You must be aware of certain relationships in order to understand completely the use of Appendix D Table. Notice that the column at the left headed "Deg" gives angles from 0° to 45.0° and that the column at the right headed "Deg" gives angles from 45.0° to 90.0°. Also notice that the degree measures on the left and on the right in any one line add up to 90°.

Two angles are **complementary** if the sum of their measures is 90°, or $\frac{\pi}{2}$. Now notice the relationship between the trigonometric functions named at the top and at the bottom of any single column. These relationships are called the **complementary relationships**.

COMPLEMENTARY
RELATIONSHIPS

For any complementary angles α and β (that is, $\alpha + \beta = 90°$ or $\alpha + \beta = \frac{\pi}{2}$),

$$\cos \alpha = \sin \beta \qquad \sec \alpha = \csc \beta$$
$$\sin \alpha = \cos \beta \qquad \csc \alpha = \sec \beta$$
$$\tan \alpha = \cot \beta \qquad \cot \alpha = \tan \beta$$

To help you remember these relationships, we use the word *cofunction*:

Sine and cosine are called cofunctions.

Tangent and cotangent are called cofunctions.

Secant and cosecant are called cofunctions.

Now, we can summarize these complementary relationships by saying that **cofunctions of complementary angles are equal.**

Example 10

Write each given function using both a reciprocal and a complementary relationship.

a. $\sin 18°$　　　　**b.** $\cos \frac{\pi}{3}$　　　　**c.** $\tan 1.2$

Solution

GIVEN FUNCTION	RECIPROCAL RELATIONSHIP	COMPLEMENTARY RELATIONSHIP	COMMENTS
a. $\sin 18°$	$\dfrac{1}{\csc 18°}$	$\cos 72°$	$90° - 18° = 72°$
b. $\cos \frac{\pi}{3}$	$\dfrac{1}{\sec \frac{\pi}{3}}$	$\sin \frac{\pi}{6}$	$\frac{\pi}{2} - \frac{\pi}{3} = \frac{3\pi}{6} - \frac{2\pi}{6} = \frac{\pi}{6}$
c. $\tan 1.2$	$\dfrac{1}{\cot 1.2}$	$\cot .37$	$\frac{\pi}{2} \approx 1.57$ so use $1.57 - 1.2 \approx .37$

Problem Set 1.3

A   1.   State the unit circle definition of the trigonometric functions, from memory.

        2.   Name the reciprocal of each given trigonometric function.
            a. cosine   **b.** cotangent   **c.** sine   **d.** secant   **e.** cosecant

        3.   Solve $\cos^2\theta + \sin^2\theta = 1$ for $\cos\theta$.

        4.   Solve $\cos^2\theta + \sin^2\theta = 1$ for $\sin\theta$.

        5.   Solve $\cot^2\theta + 1 = \csc^2\theta$ for $\cot\theta$.

Use Figure 1.9 and the definition of the trigonometric functions to evaluate the functions given in Problems 6–9 to one decimal place.

FIGURE 1.9   Using a unit circle
definition of the trigonometric
functions

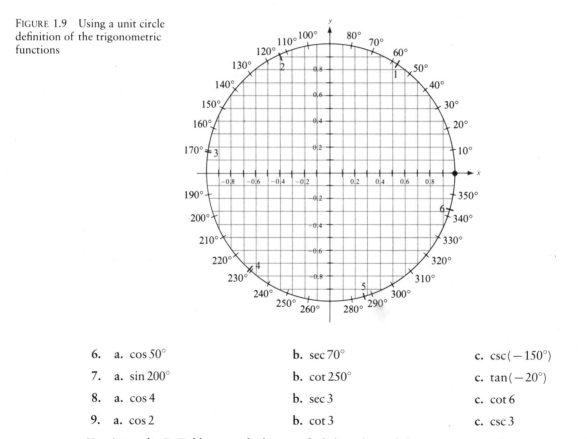

    6.   **a.** $\cos 50°$ **b.** $\sec 70°$ **c.** $\csc(-150°)$

    7.   **a.** $\sin 200°$ **b.** $\cot 250°$ **c.** $\tan(-20°)$

    8.   **a.** $\cos 4$ **b.** $\sec 3$ **c.** $\cot 6$

    9.   **a.** $\cos 2$ **b.** $\cot 3$ **c.** $\csc 3$

Use Appendix D Table or a calculator to find the values of the trigonometric functions given in Problems 10–15. Round to four decimal places.

   10.   **a.** $\sin 34°$ **b.** $\sec 56°$ **c.** $\tan 15°$

   11.   **a.** $\sin 61.5°$ **b.** $\sec 18.6°$ **c.** $\cot 75.4°$

12. **a.** $\cos 6.8°$ **b.** $\csc 39.8°$ **c.** $\tan 41.8°$

13. **a.** $\sin 1$ **b.** $\sec 1.5$ **c.** $\cot .9$

14. **a.** $\cos .5$ **b.** $\csc .25$ **c.** $\cot .43$

15. **a.** $\sin .1$ **b.** $\sec .3$ **c.** $\cot .65$

In Problems 16–18 use the definition of the trigonometric functions to write each given function in terms of its reciprocal function.

16. **a.** $\dfrac{1}{\csc 25°}$ **b.** $\dfrac{1}{\cos 125°}$ **c.** $\dfrac{1}{\cot 320°}$ **d.** $\dfrac{1}{\sin 210°}$ **e.** $\dfrac{1}{\tan 18°}$ **f.** $\dfrac{1}{\sec 35°}$

17. **a.** $\dfrac{1}{\cos 3}$ **b.** $\dfrac{1}{\sec 1}$ **c.** $\dfrac{1}{\cot 2}$ **d.** $\sin 19°$ **e.** $\sec 181°$ **f.** $\cot 210°$

18. **a.** $\sin 3$ **b.** $\cos 1$ **c.** $\cot 6$ **d.** $\tan 2$ **e.** $\sec 3$ **f.** $\csc 4$

In each of Problems 19–21 state the quadrant or quadrants in which θ may lie to make the expression true.

19. **a.** $\cos \theta = .2345$ **b.** $\sin \theta = -.43$
      c. $\sin \theta = \sqrt{1 - \cos^2\theta}$ **d.** $\cos \theta = -\sqrt{1 - \sin^2\theta}$

20. **a.** $\tan \theta = 2$ **b.** $\sec \theta = 4.1258$
      c. $\cot \theta = \sqrt{\csc^2\theta - 1}$ **d.** $\sin \theta = -\sqrt{1 - \cos^2\theta}$ and $\tan \theta > 0$

21. **a.** $\csc \theta = \frac{3}{2}$ and $\tan \theta < 0$ **b.** $\sec \theta = -\sqrt{1 + \tan^2\theta}$ and $\sin \theta > 0$
      c. $\cot \theta = \sqrt{5}$ and $\sin \theta < 0$ **d.** $\csc \theta = \sqrt{1 + \cot^2\theta}$ and $\tan \theta < 0$

Write each of the expressions in Problems 22–25 as a single trigonometric function of some angle by using one of the eight fundamental identities.

22. **a.** $\dfrac{\sin 50°}{\cos 50°}$ **b.** $1 - \cos^2 18°$

23. **a.** $\dfrac{\cos(A + B)}{\sin(A + B)}$ **b.** $\dfrac{1}{\sec 75°}$

24. **a.** $\dfrac{1}{\cot \frac{\pi}{15}}$ **b.** $-\sqrt{1 - \sin^2 127°}$

25. **a.** $-\sqrt{1 - \cos^2 210°}$ **b.** $\sqrt{1 + \tan^2 210°}$

Evaluate each of the expressions in Problems 26–31 by first using one of the eight fundamental identities.

26. $\cos 128° \sec 128°$ 27. $\sin^2 112° + \cos^2 112°$ 28. $\sec^2 \frac{\pi}{12} - \tan^2 \frac{\pi}{12}$

29. $\cot^2 37° - \csc^2 37°$ 30. $\tan^2 135° - \sec^2 135°$ 31. $\csc 85° \sin 85°$

B

32. Prove that $\csc \theta = \dfrac{1}{\sin \theta}$.   33. Prove that $\cot \theta = \dfrac{\cos \theta}{\sin \theta}$.

34. Prove that $1 + \tan^2\theta = \sec^2\theta$.

35. Prove that $\cot^2\theta + 1 = \csc^2\theta$.

36. *Historical Question* There is evidence (see Eves' *In Mathematical Circles*) that the Babylonians had constructed tables of the trigonometric functions as early as 1900 to 1600 B.C. A tablet known as the Plimpton 322 can be used as a table of secant values from 45° to 31°. There were probably companion tablets for angles ranging from 30° to 16° and from 15° to 1°. The Babylonians used primitive Pythagorean triplets and a right triangle to build their table. Today, we would use a computer. Write a computer program that will output a table of trigonometric functions for the sine, cosine, and tangent for every degree from 0° to 90°.

37. *Historical Question* Do some research on primitive Pythagorean triplets (see Problem 36), and write a paper on this topic.

1.4   TRIGONOMETRIC FUNCTIONS OF ANY ANGLE

It is not always convenient to work with the unit circle definition of the trigonometric functions developed in the last section. In this section we will make use of a result from geometry concerning similar triangles that will allow us to generalize the unit circle definition.

Similar Triangles

Suppose we choose *any* point $P(x, y)$ on the terminal side of a standard-position angle θ so that its distance r from the origin is not zero, as shown in Figure 1.10.

FIGURE 1.10   Similar triangles

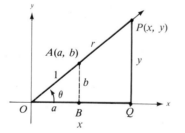

As before, let $A(a, b)$ be the point of intersection of the terminal side and a unit circle. If we drop perpendicular lines from A and P to the x-axis and label the points B and Q as shown in Figure 1.10, two right triangles AOB and POQ are formed. Because $\triangle AOB$ and $\triangle POQ$ are **similar triangles**, it follows that the corresponding sides of the triangles are proportional:

$$\frac{a}{1} = \frac{x}{r} \qquad \frac{b}{1} = \frac{y}{r} \qquad \frac{b}{a} = \frac{y}{x}$$

$$\frac{1}{a} = \frac{r}{x} \qquad \frac{1}{b} = \frac{r}{y} \qquad \frac{a}{b} = \frac{x}{y}$$

Generalized Definition of the Trigonometric Functions

Using these ratios, we can now state the generalized definition of the trigonometric functions that does not depend on a unit circle.

GENERALIZED
DEFINITION OF THE
TRIGONOMETRIC
FUNCTIONS

> Let θ be any angle in standard position with a point $P(x, y)$ on the terminal side a distance of r from the origin $(r \neq 0)$. Then the six trigonometric functions are defined as follows:
>
> $$\cos \theta = \frac{x}{r} \qquad\qquad \sec \theta = \frac{r}{x} \quad (x \neq 0)$$
>
> $$\sin \theta = \frac{y}{r} \qquad\qquad \csc \theta = \frac{r}{y} \quad (y \neq 0)$$
>
> $$\tan \theta = \frac{y}{x} \quad (x \neq 0) \qquad \cot \theta = \frac{x}{y} \quad (y \neq 0)$$

This definition of the trigonometric functions should be memorized because a great deal of your success in this course will depend on a quick recall of this definition. Note from the Pythagorean theorem that r is determined by x and y:*

$$r^2 = x^2 + y^2 \quad \text{or} \quad r = \sqrt{x^2 + y^2}$$

Historical Note

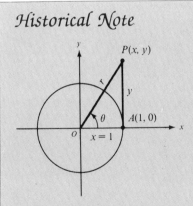

$\tan \theta$ and $\sec \theta$ are the same as the lengths of the tangent and secant lines drawn on a unit circle:

$$\tan \theta = \frac{y}{x} = \frac{y}{1} = |PA| \quad \overleftrightarrow{PA} \text{ is tangent}$$

$$\sec \theta = \frac{r}{x} = r = |PO| \quad \overleftrightarrow{PO} \text{ is a secant.}$$

The words *tangent* and *secant* come from the relationship of the ratios to the lengths of the tangent and secant lines drawn on a circle. However, the word *sine* has a more interesting origin. According to Howard Eves in his book *In Mathematical Circles*, Aryabhata called it *jya-adga* (chord half) and abbreviated it *jya*, which the Arabs first wrote as *jiba* but later shortened to *jb*. Later writers saw *jb*, and since *jiba* was meaningless to·them, they substituted *jaib* (cove or bay). The Latin equivalent for *jaib* is *sinus*, from which our present word *sine* is derived. Finally, Edmund Gunter (1581–1626) first used the prefix *co* to invert the words *cosine* and *cotangent*. In 1620, Gunter published a seven-place table of the common logarithms of the sines and tangents of angles of intervals of a minute of an arc. Gunter originally entered the ministry but later decided on astronomy as a career. He was such a poor preacher that Eves states that Gunter left the ministry in 1619 to the "benefit of both occupations."

---

* From algebra the square root property says $r = \pm\sqrt{x^2 + y^2}$ but we include only the positive value because r represents a distance.

Example 1

Find the values of the trigonometric functions for an angle θ in standard position whose terminal side passes through $(-5, 5)$.

Solution

$x = -5$ and $y = 5$ (given); $r = \sqrt{x^2 + y^2} = \sqrt{(-5)^2 + (5)^2} = 5\sqrt{2}$.

$$\cos\theta = \frac{-5}{5\sqrt{2}} \qquad \sin\theta = \frac{5}{5\sqrt{2}} \qquad \tan\theta = \frac{5}{-5}$$

$$= \frac{-\sqrt{2}}{2} \qquad\qquad = \frac{\sqrt{2}}{2} \qquad\qquad = -1$$

$$\sec\theta = \frac{5\sqrt{2}}{-5} \qquad \csc\theta = \frac{5\sqrt{2}}{5} \qquad \cot\theta = \frac{-5}{5}$$

$$= -\sqrt{2} \qquad\qquad = \sqrt{2} \qquad\qquad = -1 \qquad\qquad □$$

It is also important to understand that, even though the trigonometric functions are defined in terms of x, y, and r, they are indeed functions of θ. The relationship between x, y, r, and θ is shown in Figure 1.11.

FIGURE 1.11   Relationship between θ and x, y, and r

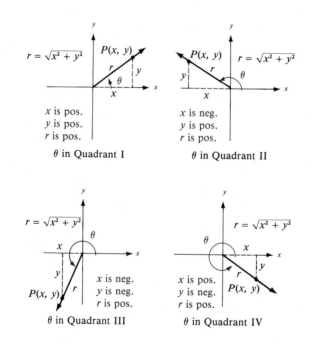

Reference Angles and Reduction Principle

Related to every angle is another angle that is of utmost importance in trigonometry. It is called the reference angle and, given some angle θ, you will need to be able to find its **reference angle**, called θ', routinely and effortlessly.* For this reason, you should consider the following definition with care.

REFERENCE ANGLE

> Given any angle θ, the **reference angle** θ' is defined as the acute angle the terminal side of θ makes with the x-axis. This means that θ' is positive and is between $0°$ and $90°$ or between 0 and $\frac{\pi}{2}$. (Note that $\frac{\pi}{2} \approx 1.57$.)

The procedure for finding the reference angle varies, depending on the quadrant of θ, as shown in Figure 1.12. Notice that the reference angle θ', shown in color, is always drawn to the x-axis and *never* to the y-axis.

FIGURE 1.12   Reference angles

Quadrant I   Quadrant II   Quadrant III   Quadrant IV

Example 2

Find the reference angle and draw both the given angle and the reference angle.

a. $210°$ **b.** $150°$

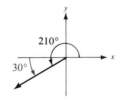   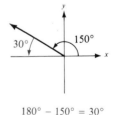

$210° - 180° = 30°$ $180° - 150° = 30°$

Reference angle is $30°$.   Reference angle is $30°$.

---

* We will use θ' to denote the reference angle of θ, α' to denote the reference angle of α, and so on throughout this book. That is, if you see the notation β', you should know that it is the reference angle for β.

c. $-\frac{5\pi}{3}$

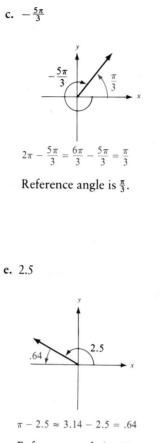

$$2\pi - \frac{5\pi}{3} = \frac{6\pi}{3} - \frac{5\pi}{3} = \frac{\pi}{3}$$

Reference angle is $\frac{\pi}{3}$.

e. 2.5

$\pi - 2.5 \approx 3.14 - 2.5 = .64$

Reference angle is .64.

d. 812°

If the angle is more than one revolution, first find a nonnegative coterminal angle less than one revolution.
812° is coterminal with 92°.

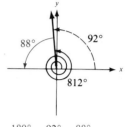

$180° - 92° = 88°$

Reference angle is 88°.

f. 30

30 is coterminal with 4.867.
$(30 - 8\pi \approx 4.867)$

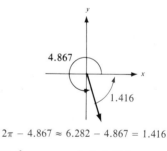

$2\pi - 4.867 \approx 6.282 - 4.867 = 1.416$

Reference angle is 1.416.   □

Reference angles are used in a procedure that reduces the value of any trigonometric
function to a function of its reference angle. Since the reference angle is between 0° and
90°, you can find its value in Appendix D Table. To carry out this procedure, called
the **reduction principle,** you will also need to remember in which quadrants the trigonometric functions are positive and in which quadrants they are negative.

REDUCTION PRINCIPLE

> A trigonometric function of an angle θ equals that function of the reference
> angle θ' with the sign (positive or negative) appropriate to the quadrant of θ.

Example 3   Evaluate the given functions using Appendix D Table.

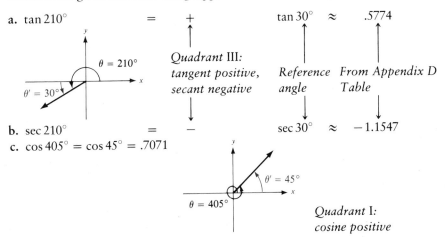

a. $\tan 210°$ =   + $\tan 30°$ ≈   .5774

Quadrant III: *Reference* *From Appendix D*
tangent positive,   angle   Table
secant negative

b. $\sec 210°$ =   − $\sec 30°$ ≈ -1.1547
c. $\cos 405° = \cos 45° = .7071$

Quadrant I:
cosine positive

Problem Set 1.4

A   *Find the reference angle in each of Problems 1–6.*

1. **a.** $30°$ **b.** $60°$ **c.** $120°$ **d.** $240°$ **e.** $50°$

2. **a.** $300°$ **b.** $330°$ **c.** $390°$ **d.** $400°$ **e.** $450°$

3. **a.** $140°$ **b.** $70°$ **c.** $200°$ **d.** $310°$ **e.** $510°$

4. **a.** $-60°$ **b.** $-120°$ **c.** $-150°$ **d.** $-180°$ **e.** $-75°$

5. **a.** $-210°$ **b.** $-240°$ **c.** $-270°$ **d.** $-300°$ **e.** $-400°$

6. **a.** $-315°$ **b.** $-268°$ **c.** $-418°$ **d.** $-910°$ **e.** $-520°$

Use a calculator to find the values of the trigonometric functions in Problems 7–12 correct to six decimal places.

7. **a.** $\cos 50°$ **b.** $\sin 20°$ **c.** $\tan 440°$ **d.** $\sin 148°$ **e.** $\cos 85°$

8. **a.** $\sin 70°$ **b.** $\tan 28°$ **c.** $\cos 409°$ **d.** $\sin 185°$ **e.** $\cos 215°$

9. **a.** $\sec 35°$ **b.** $\csc 48°$ **c.** $\csc 13.3°$ **d.** $\sec 794.6°$ **e.** $\cot 18°$

10. **a.** $\cot 25°$ **b.** $\sec 75°$ **c.** $\csc 52.8°$ **d.** $\cot 399.8°$ **e.** $\csc 183.2°$

11. **a.** $\tan(-20°)$ **b.** $\sin(-40°)$ **c.** $\cos(-50°)$ **d.** $\sec(-18°)$ **e.** $\cot(-38°)$

12. **a.** $\sin(-128°)$ **b.** $\csc(-213°)$ **c.** $\cot(-170°)$ **d.** $\sec(-335°)$ **e.** $\cot(-415°)$

Use Appendix D Table to evaluate each of the functions given in Problems 13–18 correct to four decimal places.

13. **a.** $\cos 50°$ **b.** $\sin 20°$ **c.** $\tan 440°$ **d.** $\sin 148°$ **e.** $\cos 75°$

14. **a.** $\sin 70°$ **b.** $\tan 28°$ **c.** $\cos 409°$ **d.** $\sin 185°$ **e.** $\cos 365°$

15. **a.** $\cos 60°$ **b.** $\tan 52°$ **c.** $\sin 76°$ **d.** $\cos 498°$ **e.** $\tan 89°$

16. **a.** $\sec 35°$ **b.** $\csc 48°$ **c.** $\csc 13.3°$ **d.** $\sec 794.6°$ **e.** $\cot 125.3°$

17. **a.** $\cot 25°$ **b.** $\sec 15°$ **c.** $\csc 388°$ **d.** $\cot 165°$ **e.** $\sec 193°$

18. **a.** $\tan(-20°)$ **b.** $\sin(-40°)$ **c.** $\cos(-50°)$ **d.** $\sec(-18°)$ **e.** $\cot(-80°)$

B   *Find the values of the six trigonometric functions for an angle θ in standard position whose terminal side passes through the points in Problems 19–24. Draw a picture showing θ and the reference angle θ'.*

19. $(3, 4)$ 20. $(3, -4)$ 21. $(-5, -12)$

22. $(-5, 12)$ 23. $(-6, 1)$ 24. $(-2, -3)$

Use your calculator to evaluate the expressions given in Problems 25–30, and round to four decimal places.

25. **a.** $\cos(2 \cdot 30°)$ **b.** $2\cos 30°$ **c.** $\dfrac{\cos(2 \cdot 60°)}{2}$ **d.** $\dfrac{2\cos 60°}{2}$

26. **a.** $(\cos 30°)^2$ **b.** $(\sin 30°)^2$ **c.** $\csc(\frac{1}{2} \cdot 60°)$ **d.** $\dfrac{\csc 60°}{2}$

27. **a.** $\sin 17° \csc 17°$ **b.** $\cos 38° \sec 38°$
     c. $\tan 25° \cot 25°$

28. **a.** $\cos(80° - 50°)$ **b.** $\cos 80° - \cos 50°$
     c. $\cos 80° \cos 50° + \sin 80° \sin 50°$

29. **a.** $\sin(70° - 40°)$ **b.** $\sin 70° - \sin 40°$
     c. $\sin 70° \cos 40° - \cos 70° \sin 40°$

30. **a.** $\sin 18.6° \csc 18.6°$ **b.** $\tan 134.2° \cot 134.2°$
     c. $\cos 214.3° \sec 214.3°$

C   31. *Historical Question* In this book we have measured angles in degrees and in radians. Another measurement, called a *grad*, was proposed in France and is used in Europe. A grad, also called a *new degree*, is found on many calculators and is defined as a unit of measurement that divides one revolution into 400 equal parts. These new degrees are denoted by 1^g. Notice that 100^g is a right angle and 200^g is a straight angle. Write a formula relating grads and degrees.

32. Use Problem 31 to evaluate (correct to four decimal places) the following functions.
     a. $\cos 50^g$ **b.** $\sin 150^g$ **c.** $\tan 25^g$

33. Use Problem 31 to evaluate (correct to four decimal places) the following functions.
     a. $\sin 14^g$ **b.** $\cos 83^g$ **c.** $\tan 261^g$

1.5 Trigonometric Functions of Real Numbers

Circular Functions

In Section 1.3, the trigonometric functions were defined using a unit circle, and this definition was extended in Section 1.4 to include any angle. In this section we will extend the definition of the trigonometric functions to include real-number domains. We will use the term **circular function** to include both angle and real-number domains.

The extension of the definition is easy because radian measure is already in terms of real numbers. That is, for

$$\cos \frac{\pi}{2} \quad \text{or} \quad \cos 2$$

it does not matter whether $\frac{\pi}{2}$ and 2 are considered as radian measures of angles or simply as real numbers—the functional values are the same.

TRIGONOMETRIC
FUNCTIONS OF REAL
NUMBERS

> For any real number t, let $t = \theta$, where θ is a standard-position angle measured in radians. Then
>
> $$\cos t = \cos \theta \qquad \sin t = \sin \theta \qquad \tan t = \tan \theta$$
> $$\sec t = \sec \theta \qquad \csc t = \csc \theta \qquad \cot t = \cot \theta$$

Example 1

Find the functional value of each function using a calculator.

a. $\cos(-2)$ **b.** $\sin \frac{\pi}{12}$ **c.** $\sec 2.68$ **d.** $\cot \sqrt{3}$

Solution

Switch to radian mode (see your owner's manual if you are not sure how to do this).

a. Algebraic and RPN: $\boxed{2}\ \boxed{+/-}\ \boxed{\cos}$ DISPLAY: $-.4161468365$
Note the negative value because the angle -2 is in Quadrant III (cosine negative).

b. Algebraic: $\boxed{\pi}\ \boxed{\div}\ \boxed{12}\ \boxed{=}\ \boxed{\sin}$ DISPLAY: $.2588190451$
RPN: $\boxed{\pi}\ \boxed{\text{ENTER}}\ \boxed{12}\ \boxed{\div}\ \boxed{\sin}$ DISPLAY: $.258819045$
Note the positive value because $\frac{\pi}{12}$ is in Quadrant I (sine positive).

c. Algebraic and RPN: $\boxed{2.68}\ \boxed{\cos}\ \boxed{1/x}$ DISPLAY: -1.116888769
Note the negative value because 2.68 is in Quadrant II (secant negative).

d. Algebraic and RPN: $\boxed{3}\ \boxed{\sqrt{x}}\ \boxed{\tan}\ \boxed{1/x}$ DISPLAY: $-.1626668737$
Note the negative value because $\sqrt{3} \approx 1.71$ is in Quadrant II (cotangent negative). ▫

Example 2

Find the functional value of each function using Appendix D Table.

a. $\tan 1.50$ **b.** $\sin 5.20$ **c.** $\cos(-4.30)$

Solution

a. From Appendix D Table, $\tan 1.50 \approx 14.101$.

b. Notice that Appendix D Table gives values of θ between 0 and 1.57. Thus, you need to use the reduction principle.

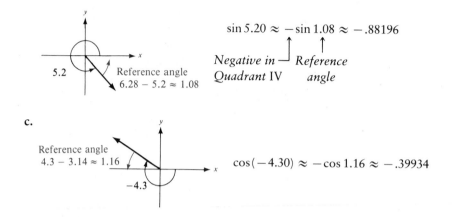

$$\sin 5.20 \approx -\sin 1.08 \approx -.88196$$

Negative in — *Reference*
Quadrant IV   *angle*

c.

Reference angle
$4.3 - 3.14 \approx 1.16$

$$\cos(-4.30) \approx -\cos 1.16 \approx -.39934$$

-4.3

Exact Values

Calculators and tables give most values for the trigonometric functions as approximations. However, many applications in this course, as well as in calculus and more advanced courses, require **exact values**.

Example 3

Find the exact values for the trigonometric functions of $\frac{\pi}{4}$.

Solution

$$\cos\frac{\pi}{4} = \frac{x}{r} \qquad \text{From the definition (this is true for any angle)}$$

$$= \frac{x}{\sqrt{x^2 + y^2}} \quad \text{By substitution } (r = \sqrt{x^2 + y^2}).$$

If $\theta = \frac{\pi}{4}$, then $x = y$ since $\frac{\pi}{4}$ bisects Quadrant I. By substitution,

$$\cos\frac{\pi}{4} = \frac{x}{\sqrt{x^2 + x^2}}$$

$$= \frac{x}{\sqrt{2x^2}}$$

$$= \frac{x}{x\sqrt{2}} \qquad \sqrt{x^2} = |x| = x \quad \text{since } x \text{ is positive in Quadrant I}$$

$$= \frac{1}{\sqrt{2}}$$

$$= \frac{1}{2}\sqrt{2} \qquad \frac{1}{\sqrt{2}} = \frac{1}{\sqrt{2}} \cdot \frac{\sqrt{2}}{\sqrt{2}} = \frac{\sqrt{2}}{2} = \frac{1}{2}\sqrt{2}$$

Similarly,

$$\sin\frac{\pi}{4} = \frac{\sqrt{2}}{2}, \tan\frac{\pi}{4} = 1, \sec\frac{\pi}{4} = \sqrt{2}, \csc\frac{\pi}{4} = \sqrt{2}, \text{ and } \cot\frac{\pi}{4} = 1 \qquad \Box$$

Example 4

Find the exact values for the trigonometric functions of $30°$.

Solution

Consider not only the standard-position angle $30°$, but also the standard-position angle $-30°$. Choose $P_1(x, y)$ and $P_2(x, -y)$, respectively, on the terminal sides.

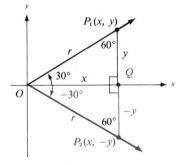

Angles OQP_1 and OQP_2 are right angles, so $\angle OP_1Q = 60°$ and $\angle OP_2Q = 60°$. Thus, $\triangle OP_1P_2$ is an equiangular triangle (all angles measure $60°$). From geometry, an equiangular triangle has sides that are the same length. Thus, $2y = r$. Notice the following relationship between x and y:

$$r^2 = x^2 + y^2 \quad \textit{Pythagorean theorem}$$
$$(2y)^2 = x^2 + y^2 \quad \textit{Since } r = 2y$$
$$3y^2 = x^2 \quad \textit{Subtract } y^2 \textit{ from both sides.}$$
$$\sqrt{3}\,|y| = |x|$$

For $30°$, x and y are both positive, so $x = \sqrt{3}\,y$.

THE RECIPROCAL FUNCTIONS ARE EASIER:

$$\cos 30° = \frac{x}{r} = \frac{\sqrt{3}\,y}{2y} = \frac{\sqrt{3}}{2} \qquad\qquad \sec 30° = \frac{2}{\sqrt{3}} = \frac{2}{3}\sqrt{3}$$

$$\sin 30° = \frac{y}{r} = \frac{y}{2y} = \frac{1}{2} \qquad\qquad \csc 30° = 2$$

$$\tan 30° = \frac{y}{x} = \frac{y}{\sqrt{3}\,y} = \frac{1}{\sqrt{3}} = \frac{\sqrt{3}}{3} \qquad \cot 30° = \sqrt{3} \qquad \Box$$

Example 5

Find the exact values for the trigonometric functions of $\frac{\pi}{3}$.

Solution

Notice $30° = \frac{\pi}{6}$ and $\frac{\pi}{6} + \frac{\pi}{3} = \frac{\pi}{6} + \frac{2\pi}{6} = \frac{3\pi}{6} = \frac{\pi}{2}$. Thus, $\frac{\pi}{3}$ is complementary to $30°$. In Section 1.3 we developed complementary relationships that we can use here in order to capitalize on the work shown in Example 4.

Cofunctions of complementary angles	*Relate to Example 4*	*Result from Example 4*	RECIPROCAL FUNCTIONS:
$\cos\dfrac{\pi}{3} = \sin\dfrac{\pi}{6}$	$=$	$\sin 30° = \dfrac{1}{2}$	$\sec\dfrac{\pi}{3} = 2$
$\sin\dfrac{\pi}{3} = \cos\dfrac{\pi}{6}$	$=$	$\cos 30° = \dfrac{\sqrt{3}}{2}$	$\csc\dfrac{\pi}{3} = \dfrac{2}{\sqrt{3}} = \dfrac{2\sqrt{3}}{3}$
$\tan\dfrac{\pi}{3} = \cot\dfrac{\pi}{6}$	$=$	$\cot 30° = \sqrt{3}$	$\cot\dfrac{\pi}{3} = \dfrac{1}{\sqrt{3}} = \dfrac{\sqrt{3}}{3}$

The results of Examples 3–5 are summarized in Table 1.2. You should now be able to verify all the entries in this table.

TABLE 1.2 Exact Trigonometric Values

Function \ Angle θ	$0 = 0°$	$\dfrac{\pi}{6} = 30°$	$\dfrac{\pi}{4} = 45°$	$\dfrac{\pi}{3} = 60°$	$\dfrac{\pi}{2} = 90°$	$\pi = 180°$	$\dfrac{3\pi}{2} = 270°$
$\cos\theta$	1	$\dfrac{\sqrt{3}}{2}$	$\dfrac{\sqrt{2}}{2}$	$\dfrac{1}{2}$	0	-1	0
$\sin\theta$	0	$\dfrac{1}{2}$	$\dfrac{\sqrt{2}}{2}$	$\dfrac{\sqrt{3}}{2}$	1	0	-1
$\tan\theta$	0	$\dfrac{\sqrt{3}}{3}$	1	$\sqrt{3}$	undef.	0	undef.
$\sec\theta$	$1.$	$\dfrac{2}{\sqrt{3}} = \dfrac{2}{3}\sqrt{3}$	$\dfrac{2}{\sqrt{2}} = \sqrt{2}$	$\dfrac{2}{1} = 2$	undef.	$\dfrac{1}{-1} = -1$	undef.
$\csc\theta$	undef.	$\dfrac{2}{1} = 2$	$\dfrac{2}{\sqrt{2}} = \sqrt{2}$	$\dfrac{2}{\sqrt{3}} = \dfrac{2}{3}\sqrt{3}$	1	undef.	-1
$\cot\theta$	undef.	$\dfrac{3}{\sqrt{3}} = \sqrt{3}$	1	$\dfrac{1}{\sqrt{3}} = \dfrac{\sqrt{3}}{3}$	0	undef.	0

$\sec\theta$, $\csc\theta$, and $\cot\theta$ are the reciprocals (which is why the exact values are given in reciprocal form as well as in simplified form).

This table of values is used extensively, and you should memorize it as you did multiplication tables in elementary school. It is used along with the reduction principle as shown in Example 6.

Example 6

Find the exact values.

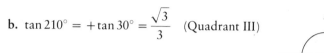

a. $\cos 135° = -\cos 45° = -\dfrac{\sqrt{2}}{2}$ reference angle

Quadrant II, so cosine is negative

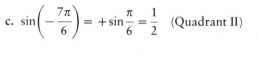

b. $\tan 210° = +\tan 30° = \dfrac{\sqrt{3}}{3}$ (Quadrant III)

c. $\sin\left(-\dfrac{7\pi}{6}\right) = +\sin\dfrac{\pi}{6} = \dfrac{1}{2}$ (Quadrant II)

d. $\tan\dfrac{5\pi}{3} = -\tan\dfrac{\pi}{3} = -\sqrt{3}$ (Quadrant IV)

e. $\csc\left(-\dfrac{5\pi}{3}\right) = +\csc\dfrac{\pi}{3} = \dfrac{2}{\sqrt{3}} = \dfrac{2}{3}\sqrt{3}$ (Quadrant I)

f. $\cot\left(-\dfrac{11\pi}{6}\right) = +\cot\dfrac{\pi}{6} = \dfrac{3}{\sqrt{3}} = \sqrt{3}$ (Quadrant I)

Problem Set 1.5

A   *Find the functional values correct to four decimal places in Problem 1–8 by calculator or by using Appendix D Table.*

1. **a.** $\sin 1$ **b.** $\cos .2$ **c.** $\tan .3$ **d.** $\sec .5$ **e.** $\csc .8$

2. **a.** $\cos 1.5$ **b.** $\sin .75$ **c.** $\tan .6$ **d.** $\cot 1.2$ **e.** $\sec .95$

3. **a.** $\cos .45$ **b.** $\tan 1.35$ **c.** $\sin 1.08$ **d.** $\sec .66$ **e.** $\cot .52$

4. a. $\sin 2.5$ b. $\cos 4$ c. $\cos 5$ d. $\sec 4$ e. $\csc 6$

5. a. $\tan(-2)$ b. $\sin(-1.5)$ c. $\cos(-.85)$ d. $\cot(-.55)$ e. $\csc(-.25)$

6. a. $\tan(-2.4)$ b. $\cos(-7)$ c. $\sin(-8)$ d. $\cot(-15)$ e. $\sec\sqrt{2}$

7. a. $\csc\frac{\pi}{15}$ b. $\cos 5.8$ c. $\sin 10$ d. $\tan 23$ e. $\sec(-\frac{7\pi}{5})$

8. a. $\sec(-\frac{3\pi}{10})$ b. $\sin\frac{5\pi}{3}$ c. $\cos\frac{2\pi}{7}$ d. $\tan\frac{\pi}{15}$ e. $\cot(-\frac{3\pi}{11})$

In Problems 9–15 give the exact functional value of each from memory.

9. a. $\tan\frac{\pi}{4}$ b. $\cos 0$ c. $\sin 60°$ d. $\csc\frac{\pi}{6}$ e. $\csc 0$

10. a. $\cos 270°$ b. $\cos 30°$ c. $\sec\frac{\pi}{4}$ d. $\tan 180°$ e. $\sin 45°$

11. a. $\sin\frac{\pi}{2}$ b. $\sec 0°$ c. $\tan 0$ d. $\sec\frac{\pi}{6}$ e. $\cos\frac{\pi}{4}$

12. a. $\cot\pi$ b. $\sec\frac{\pi}{4}$ c. $\tan 90°$ d. $\tan 60°$ e. $\cos\frac{\pi}{3}$

13. a. $\cot 45°$ b. $\cos\pi$ c. $\sin\frac{3\pi}{2}$ d. $\sin 0°$ e. $\sec\frac{\pi}{2}$

14. a. $\csc\frac{\pi}{2}$ b. $\cos 90°$ c. $\sec\pi$ d. $\tan 270°$ e. $\sin\frac{\pi}{6}$

15. a. $\tan\frac{\pi}{6}$ b. $\sin\pi$ c. $\cot 90°$ d. $\sec\frac{\pi}{3}$ e. $\sec 270°$

B   *Use Table 1.2 and the reduction principle to find the exact functional value in Problems 16–21.*

16. a. $\cos 300°$ b. $\sin 120°$ c. $\tan 120°$ d. $\cot 240°$ e. $\sec 120°$

17. a. $\tan 135°$ b. $\sin(-135°)$ c. $\cos 225°$ d. $\cot 225°$ e. $\tan 315°$

18. a. $\sin 240°$ b. $\cos(-300°)$ c. $\tan(-120°)$ d. $\sin 300°$ e. $\cos 120°$

19. a. $\csc\frac{3\pi}{2}$ b. $\sin 390°$ c. $\sin\frac{17\pi}{4}$ d. $\cos(-6\pi)$ e. $\sin(-675°)$

20. a. $\cos(-\frac{5\pi}{6})$ b. $\sin\frac{5\pi}{6}$ c. $\cot\frac{5\pi}{6}$ d. $\tan\frac{7\pi}{6}$ e. $\sec(-\frac{7\pi}{6})$

21. a. $\sin\frac{7\pi}{3}$ b. $\cos(-\frac{11\pi}{3})$ c. $\tan(-\frac{7\pi}{3})$ d. $\sec\frac{5\pi}{3}$ e. $\sec(-\frac{5\pi}{3})$

Simplify each of the expressions in Problems 22–31 by substituting the exact values for the trigonometric functions and then algebraically simplifying.

22. a. $\sin\frac{\pi}{6}\csc\frac{\pi}{6}$ b. $\csc\frac{\pi}{2}\sin\frac{\pi}{2}$ 23. a. $\sin^2\frac{\pi}{3}+\cos^2\frac{\pi}{3}$ b. $\sin^2\frac{\pi}{6}+\cos^2\frac{\pi}{3}$

24. a. $\sin(\frac{\pi}{4}-\frac{\pi}{2})$ b. $\sin\frac{\pi}{4}-\sin\frac{\pi}{2}$ 25. a. $\tan(2\cdot 30°)$ b. $2\tan 30°$

26. a. $\csc(\frac{1}{2}\cdot 60°)$ b. $\dfrac{\csc 60°}{2}$

27. a. $\cos(\frac{\pi}{2}-\frac{\pi}{6})$ b. $\cos\frac{\pi}{2}\cos\frac{\pi}{6}+\sin\frac{\pi}{2}\sin\frac{\pi}{6}$

28. a. $\tan(2\cdot 60°)$ b. $\dfrac{2\tan 60°}{1-\tan^2 60°}$

29. a. $\cos(\frac{1}{2}\cdot 60°)$ b. $\sqrt{\dfrac{1+\cos 60°}{2}}$

30. a. $\sin(\frac{1}{2}\cdot 60°)$ b. $\sqrt{\dfrac{1-\cos 60°}{2}}$

31. **a.** $\sin\left(\frac{1}{2} \cdot 120°\right)$ **b.** $\sqrt{\dfrac{1 - \cos 120°}{2}}$

C   **32.** You will learn in calculus that (for x near 0)

$$\sin x = x - \frac{x^3}{3!} + \frac{x^5}{5!} - \frac{x^7}{7!} + \cdots$$

where $n! = n(n-1)(n-2) \cdot \cdots \cdot 3 \cdot 2 \cdot 1$. Find $\sin 1$ correct to four decimal places by using this equation.

33. You will learn in calculus that

$$\cos x = 1 - \frac{x^2}{2!} + \frac{x^4}{4!} - \frac{x^6}{6!} + \cdots$$

where $n! = n(n-1)(n-2) \cdot \cdots \cdot 3 \cdot 2 \cdot 1$. Find $\cos 1$ correct to four decimal places by using this equation.

1.6 SUMMARY AND REVIEW

	OBJECTIVES	PAGES/EXAMPLES
1.1 ANGLES AND DEGREE MEASURE	1. Know the Greek letters used to name angles in trigonometry.	p. 1; Figure 1.1.
	2. Draw angles whose measures are given in degrees; know the approximate size of angles 0°, 30°, 45°, 60°, 90°, and multiples of these angles.	p. 3; Example 1; Problems 5–10.
	3. Convert degree–minutes–seconds to decimal degrees.	p. 3; Example 2; Problems 17–36.
	4. Find a positive angle less than one revolution that is coterminal with a given angle.	p. 5; Example 3; Problems 11–16. Also, Section 1.2, Problems 7–10.
1.2 RADIAN MEASURE OF ANGLES	5. Draw angles whose measures are given in radians.	p. 9; Example 1; Problems 3–6.
	6. Convert degree measure to radian measure.	p. 11; Example 2; Problems 1, 11–16.
	7. Convert radian measure to degree measure.	p. 12; Example 3; Problems 2, 17–19.
	8. Find arc length.	p. 13; Examples 4–6; Problems 20–32.

1.3 TRIGONOMETRIC
FUNCTIONS ON A UNIT
CIRCLE

9. Know the unit circle definition of the trigonometric functions.

p. 19; Problem 1.

10. State the eight fundamental identities.

pp. 19–22; Example 5; Problem 2.

11. Prove the eight fundamental identities.

pp. 20–22; Problems 32–35.

12. Use the fundamental identities to simplify trigonometric expressions and to evaluate trigonometric functions.

p. 19; Examples 2–4, 10; Problems 3–5, 16–18, 22–31.

13. Find the approximate values of the trigonometric functions using a unit circle and the
definition.

p. 23; Example 6; Problems 6–9.

14. Know the signs of the trigonometric functions by
quadrant.

p. 24; Example 7; Problems 19–21.

15. Evaluate the trigonometric functions using a
calculator or a table.

pp. 25–26; Examples 8 and 9; Problems 10–15, 26–31. Also, Section 1.4, Problems 7–12, 25–30.

1.4 TRIGONOMETRIC
FUNCTIONS OF ANY
ANGLE

16. Know the generalized definition of the trigonometric functions.

p. 31.

17. Evaluate the trigonometric functions given a
point on the terminal side.

p. 32; Example 1; Problems 19–24.

18. Find the reference angle for any given angle.

p. 33; Example 2; Problems 1–6.

19. Use the reduction principle to evaluate trigonometric functions.

pp. 34–35; Example 3; Problems 13–18.

1.5 TRIGONOMETRIC
FUNCTIONS OF REAL
NUMBERS

20. Evaluate trigonometric functions of real numbers by calculator or by table.

p. 37; Examples 1–2; Problems 1–8.

21. Know the exact values for the trigonometric
functions of $0, \frac{\pi}{6}, \frac{\pi}{4}, \frac{\pi}{3}, \frac{\pi}{2}, \pi$, and $\frac{3\pi}{2}$.

p. 40; see Table 1.2 and Examples 3–6; Problems 9–30.

Terms

Acute angle [1.1]
Alpha (α) [1.1]
Angle [1.1]
Arc [1.1]
Arc length [1.2]
Beta (β) [1.1]

Central angle [1.1]
Circular functions [1.5]
Circumference [1.1]
Cofunction [1.3]
Complementary
relationships [1.3]

Cosecant [1.3]
Cosine [1.3]
Cotangent [1.3]
Coterminal angles [1.1]
Degree [1.1]
Delta (δ) [1.1]

Evaluation [1.3]
Exact values [1.5]
Fundamental identity
    [1.3]
Gamma (γ) [1.1]
Identity [1.3]
Initial side [1.1]
Lambda (λ) [1.1]
Minute [1.1]
Mu (μ) [1.1]
Negative angle [1.1]
Obtuse angle [1.1]
Omega (ω) [1.1]
Phi (ϕ) [1.1]

Positive angle [1.1]
Pythagorean identities
    [1.3]
Quadrant [1.1]
Quadrantal angle [1.1]
Radian [1.2]
Ratio identities [1.3]
Reciprocal identities [1.3]
Reduction principle [1.4]
Reference angle [1.4]
Right angle [1.1]
Secant [1.3]
Second [1.1]
Side of an angle [1.1]

Similar triangles [1.4]
Sine [1.3]
Standard-position angle
    [1.1]
Straight angle [1.1]
Subtended angle [1.1]
Tangent [1.3]
Terminal side [1.1]
Theta (θ) [1.1]
Trigonometric functions
    [1.3]
Unit circle [1.3]
Vertex [1.1]

Study Hints

In the following review of objectives, there is one question for each objective in this
chapter. If you have forgotten how to do a particular problem or you miss a review
problem, you can refer to the pages given in this section to find discussion, examples,
and similar problems. (*All* answers for the chapter review of objectives are provided in
the answer section.)

Chapter 1 Review of Objectives

1. Name the following Greek letters.
    a. α b. θ c. γ d. λ e. β
2. Draw the angle from memory whose measure is given.
    a. $30°$ b. $45°$ c. $60°$ d. $180°$ e. $300°$
3. Change $50°36'$ to decimal degrees correct to the nearest hundredth degree.
4. Find the positive angle less than one revolution that is coterminal with the given
    angle.
    a. $400°$ b. $-150°$ c. 7 d. $\frac{-5\pi}{6}$ e. $-\pi$
5. Draw the angle from memory whose measure is given.
    a. 1 b. 3 c. $\frac{3\pi}{4}$ d. $\frac{2\pi}{3}$ e. $\frac{5\pi}{6}$
6. Change $500°$ to radians correct to the nearest hundredth.
7. Change -3.8 to degrees correct to the nearest degree.
8. A curve on a highway is laid out as the arc of a circle of radius 500 m. If the curve
    subtends a central angle of $18.0°$, what is the distance around this section of
    road? Give the exact answer and an answer rounded to the nearest meter.
9. If α is an angle in standard position, what are the coordinates of P_α, which is the
    point of intersection of the terminal side of α and the unit circle?
10. State from memory the eight fundamental identities.
11. Prove $1 + \tan^2\theta = \sec^2\theta$ without using any of the fundamental identities.

12. Find the exact value of $\tan^2 114° - \sec^2 114°$ without using tables or a calculator.
13. Use the unit circle definition to approximate the values of the trigonometric functions of $340°$. Use Figure 1.9 on page 28.
14. Classify each of the six trigonometric functions as positive or negative for each of the quadrants.

	Quadrant I	Quadrant II	Quadrant III	Quadrant IV
Positive	a.	c.	e.	g.
Negative	b.	d.	f.	h.

15. Find the values of the trigonometric functions for an angle with measure $42.5°$. Round to four decimal places.
16. State the generalized definition of the trigonometric functions.
17. Using the definition of the trigonometric functions, find their exact values for an angle δ whose terminal side passes through $(-1, \sqrt{8})$.
18. Find the reference angle for each of the given angles.
    a. $500°$ b. $-60°$ c. $\frac{2\pi}{3}$ d. $-\frac{5\pi}{6}$ e. $82°$
19. Use a calculator or Appendix D Table to find the values of the trigonometric functions correct to four decimal places.
    a. $\cos 200°$ b. $\sin(-200°)$ c. $\tan(-192°)$
    d. $\cot 520°$ e. $\csc 300°$
20. Evaluate.
    a. $\cos 2$ b. $\sin 4$ c. $\tan .85$
    d. $\csc(-1.5)$ e. $\sec(-8)$
21. From memory, fill in the blanks indicated by lowercase letters. (This problem also involves objective 19.)

Angle in degrees	$60°$			$300°$	$-210°$			$225°$
Angle in radians		$\dfrac{\pi}{2}$	$\dfrac{5\pi}{6}$			$-\dfrac{11\pi}{3}$	3π	
cosine	a.	b.	c.	d.	e.	f.	g.	h.
sine	i.	j.	k.	l.	m.	n.	o.	p.
tangent	q.	r.	s.	t.	u.	v.	w.	x.

Miscellaneous Problems

1. If α is an angle in standard position, what are the coordinates of the point of intersection of the terminal side of α and a circle with radius 2?
2. If β is an angle in standard position, what are the coordinates of the point of intersection of the terminal side of β and a circle with radius 5?
3. If $23°$ is the measure of an angle in standard position, what are the coordinates of the point of intersection of the terminal side of this angle and a unit circle?

4. If 30° is the measure of an angle in standard position, what are the coordinates of the point of intersection of the terminal side of this angle and a unit circle?

Change the degree measures to radian measures (correct to the nearest hundredth).

5. 150°   6. −30°   7. −100°   8. 24.5°

Change the radian measures to degree measures (correct to the nearest degree).

9. $\frac{2\pi}{3}$ 10. $-\frac{5\pi}{4}$ 11. −2.5   12. .4589

Find the arc length (approximate answers correct to the nearest unit).

13. $r = 5$ in.; $\theta = 6$ 14. $r = \sqrt{50}$ ft; $\theta = \frac{\pi}{6}$
15. $r = 6.4$ ft; $\theta = 65°$ 16. $r = 34$ in.; $\theta = 125°$

Determine the quadrant of each angle and find a positive angle less than one revolution that is coterminal with the given angle.

17. 375°   18. −15°   19. 815°   20. −160°

Use a calculator or table to find the value of the trigonometric functions correct to four decimal places.

21. $\sin 85.2°$ 22. $\tan 106.9°$ 23. $\cos 2$
24. $\sec 3$ 25. $\cos(−3.5)$ 26. $-\sin(−18.5°)$
27. $\cot 6$ 28. $\csc 158°$ 29. $-\tan 34.5°$
30. $\sin .5°$ 31. $\sin .5$ 32. $\sec(−55.6°)$

Use the definition of the trigonometric functions to find the values for an angle θ whose terminal side passes through the given point.

33. $(4, 5)$ 34. $(−2, 3)$ 35. $(5, \sqrt{11})$ 36. $(\sqrt{55}, 3)$

Find the exact value of the following trigonometric functions.

37. $\sin 60°$ 38. $\cos(−60°)$ 39. $\tan \frac{\pi}{3}$ 40. $\cot \frac{5\pi}{3}$
41. $\cos 135°$ 42. $\sin(−135°)$ 43. $-\cot 45°$ 44. $-\tan 225°$
45. $\sec 0°$ 46. $\csc 2\pi$ 47. $\cot \frac{3\pi}{2}$ 48. $\tan \pi$
49. $\csc 240°$ 50. $\sec 210°$ 51. $\tan(−\frac{11\pi}{3})$ 52. $\sin(−330°)$

APPLICATION FOR FURTHER STUDY

ANGULAR AND LINEAR VELOCITY—FROM WINDMILLS TO PISTONS

Courtesy of Sara Hunsaker

In Section 1.2 we defined radian measure for angles and derived an equation for finding the length of an arc:

$s = r\theta$

where s is the arc length, r is the radius of the circle, and θ is the angle measured in radians. We can use this notion to define two concepts used extensively in engineering and physics, one giving a linear equation and the other giving a quadratic equation.

Suppose a bike wheel with a 24-in. diameter is turning completely around five times every second. We say that the wheel is rotating at the rate of 5 revolutions per second (rps), or 300 revolutions per minute (rpm). A spoke of this wheel will turn 2π radians for each rotation, so in each second the spoke will turn

$5 \cdot 2\pi$ radians per second = 10π radians per second or 10π radians/sec

We call this the **angular velocity**, denoted by ω.

ANGULAR VELOCITY

$$\omega = \text{angle (in radians) per unit of time} \quad \text{or} \quad \omega = \frac{\theta}{t}$$

Since the radius of the bike wheel is 12 in., a point on the circumference will move 12 in. for each radian. But since the wheel turns 10π radians per second, a point on the circumference moves

$10\pi \cdot 12$ in. per sec = 120π in. per sec

48

or

$$10\pi \cdot 1 \text{ ft per sec} = 10\pi \text{ ft per sec}$$

We call this the **linear velocity**, denoted by V.

Example 1

Calculate the linear velocity for this bike in miles per hour.

Solution

The angular velocity in radians per hour is

$$5\frac{\text{rev}}{\text{sec}} \cdot \underbrace{60\frac{\text{sec}}{\text{min}} \cdot 60\frac{\text{min}}{\text{hr}}}_{\substack{\textit{These convert}\\\textit{seconds to}\\\textit{hours.}}} \cdot \underbrace{2\pi\frac{\text{radians}}{\text{rev}}}_{\substack{\textit{This converts}\\\textit{revolutions to}\\\textit{radians.}}} = 36{,}000\pi\frac{\text{radians}}{\text{hr}}$$

Next, the radius (in miles) is

$$12 \text{ in.} = 1 \text{ ft} = \frac{1}{5{,}280} \text{ mi}$$

linear velocity = (angular velocity per radian)(radius)

$$= (36{,}000\pi)\left(\frac{1}{5{,}280}\right)$$

$$\approx 21.42$$

The bike is traveling at about 21.4 mph.　　　　　　　　　　　⬜

In general, if

$$s = r\theta \qquad \theta \textit{ is measured in radians.}$$

then

$$\frac{s}{t} = r \cdot \frac{\theta}{t} \qquad t \textit{ is the unit of time.}$$

Since

$$V = \frac{s}{t} \quad \text{and} \quad \omega = \frac{\theta}{t}$$

we have, by substitution,

$$V = r\omega$$

which is the number of radians per unit of time.

LINEAR VELOCITY

$$V = r\omega$$

Example 2

A belt runs a pulley of radius 40 cm at 50 rpm. Find the angular velocity of the pulley and the linear velocity of the belt.

Solution

For the angular velocity of the pulley:

IN RADIANS PER MINUTE

$$\omega = \frac{\theta}{t}$$

$$= \frac{50 \cdot 2\pi \text{ radians}}{1 \text{ min}}$$

$$= 100\pi \text{ radians/min}$$

IN RADIANS PER HOUR

$$\omega = \frac{100\pi \text{ radians}}{1 \text{ min}} \cdot \frac{60 \text{ min}}{1 \text{ hr}}$$

$$= \frac{6,000\pi \text{ radians}}{\text{hr}} \text{ or } 6,000\pi \text{ rad/hr}$$

For the linear velocity of the belt:

IN CENTIMETERS PER MINUTE

$$V = r\omega$$

$$= 40 \text{ cm} \cdot \frac{100\pi}{\text{min}}$$

$$= \frac{4,000\pi \text{ cm}}{\text{min}} \text{ or } 4,000\pi \text{ cm/min}$$

IN KILOMETERS PER HOUR

$$V = 40\text{cm} \cdot \frac{6,000\pi}{\text{hr}}$$

$$= \frac{240,000\pi \text{ cm}}{\text{hr}} \cdot \frac{1 \text{ km}}{100,000 \text{ cm}}$$

$$= \frac{2.4 \text{ km}}{\text{hr}} \approx 7.54 \text{ kph} \qquad \square$$

A second application is found by looking at the action of a piston, connecting rod, and crankshaft in an internal-combustion engine.

FIGURE 1.13

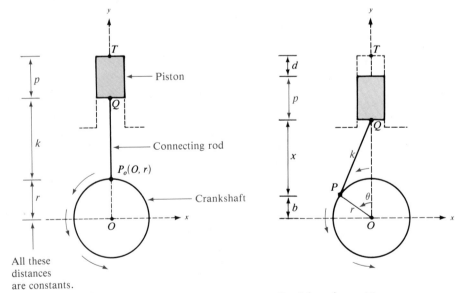

All these distances are constants.

Fully extended position

Partial-stroke position
(Arbitrary position)

We wish to determine the position of the piston indicated by the distance of its top from the fully extended position at any given time of the rotation of the crankshaft (see Figure 1.13).

Fully extended position: $|OT| = p + k + r$

Partial-stroke position: $|OT| = b + x + p + d$

Solve for d: $d = |OT| - b - x - p$

Substitute the relationship for the fully extended position into the equation that was solved for d. We obtain

$$d = (p + k + r) - b - x - p$$

Now, let θ be a rotation on the crankshaft as shown in Figure 1.14, and let P be a point on the terminal side of θ with coordinates (a, b). Then

$$\cos \theta = \frac{b}{r} \quad \text{or} \quad b = r \cos \theta$$

$$\sin \theta = \frac{a}{r} \quad \text{or} \quad a = r \sin \theta$$

Also, by the Pythagorean theorem,

$$a^2 + x^2 = k^2$$
$$x^2 = k^2 - a^2$$
$$x = \sqrt{k^2 - a^2} \quad \textit{Positive value,}$$
$$\textit{since } x \textit{ is a distance}$$
$$= \sqrt{k^2 - r^2 \sin^2\theta} \quad \textit{Since } a = r \sin \theta$$

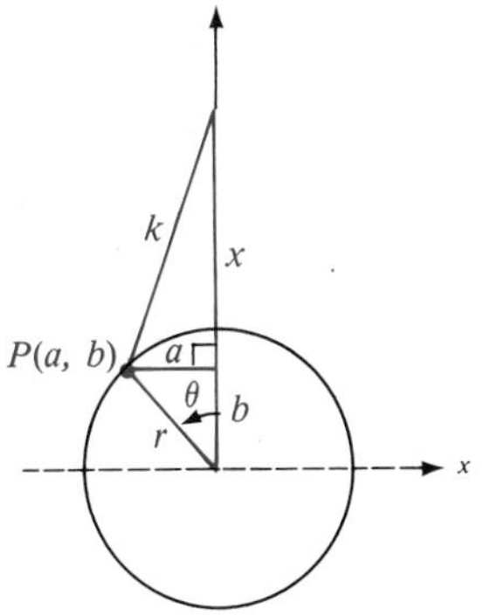

FIGURE 1.14

We now have an expression for d in terms of k, r, and θ:

$$d = (p + k + r) - b - x - p$$
$$= k + r - b - x$$
$$= k + r - r \cos \theta - \sqrt{k^2 - \sin^2\theta} \quad \textit{Substitute } b = r \cos \theta$$
$$\textit{and } x = \sqrt{k^2 - r^2 \sin^2\theta}$$

The distance d is the value you are looking for, but you will usually know the velocity of the crankshaft rather than θ, so perform one last substitution in order to make this a useful formula. If s is the arc length, then $s = r\theta$ and

arc length = (circumference)(number of revolutions)

arc length = (circumference)(revolutions per minute)(time in minutes)

$$s = (2\pi r)(v)(t) \quad \textit{Where } v = rpm,$$
$$t = \textit{time in minutes}$$

$$r\theta = 2\pi rvt \quad \textit{Since } s = r\theta$$
$$\theta = 2\pi vt \quad \textit{Divide both sides by } r.$$

51

$$d = k + r - r\cos(2\pi v t) - \sqrt{k^2 - r^2 \sin^2(2\pi v t)}$$

where d = depth of stroke of a piston

k = length of connecting rod

r = radius of the crankshaft

v = velocity of the crankshaft in rpm

t = time in minutes

Example 3

Find the position of a piston after .06 second if the connecting rod is 15 cm and the radius of the crankshaft is 5 cm. Also suppose that the crankshaft is turning at 600 rpm.

Solution

Given $k = 15$, $r = 5$, $v = 600$, and $t = .06/60 = .001$. Thus,

$$d = 15 + 5 - 5\cos\theta - \sqrt{15^2 - 5^2 \sin^2\theta}$$

where $\theta = 2\pi v t = 1{,}200\pi(.001)$. This calculator sequence is not a particularly easy one, so we will take it slowly, one step at a time.

PRESS   EXPLANATION

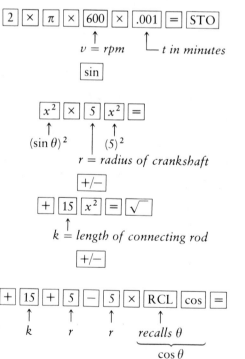

This sequence calculates and stores $\theta = 2\pi v t$.

sin θ; make sure calculator is in radian mode

Calculates $5^2 \sin^2\theta$

Changes subtraction to addition

Calculates $\sqrt{k^2 - r^2 \sin^2\theta}$

Changes subtraction of radical to an addition

Completes calculation

Final display: 9.335811964

The piston is about 9.3 cm from the top of the stroke after .06 sec.

Problems for Further Study: Angular and Linear Velocity—From Windmills to Pistons

1. *Physics.* A flywheel with radius 15 cm is spinning so that a point on the rim has a linear velocity of 120 cm/sec. Find the angular velocity.

2. *Physics.* The wheel of an automobile is 60 cm in diameter. What is the angular velocity of the wheel in radians per second for this car if it is traveling at 30 kph?

3. *Physics.* An airplane propeller measures 4 m from tip to tip and rotates at 1,800 rpm. Find the angular velocity and the linear velocity of a point on the tip of one of the blades. (Assume that the airplane itself is not moving.)

In Problems 4–6, find the position of the piston in Example 3 after the given times.

4. .01 second   5. 2 seconds   6. 3.852 seconds

7. *Space science.* An earth satellite travels in a circular orbit at 32,000 kph. If the radius of the orbit is 6,770 km, what is the angular velocity (in radians per hour)?

8. *Physics.* The biggest windmill in the world was built in Ulfborg, Denmark, to generate electricity. Its concrete tower soars 54 meters and is topped by three giant fiberglass propeller blades, each 27 meters long. If the propeller rotates at 30 rpm, find the angular velocity and the linear velocity of a point on the tip of one of the blades.

9. *Physics.* On January 28, 1978, the first federally funded commercial wind generator in the United States was dedicated in Clayton, New Mexico. The rotors span 125 feet each ($r = 125$) and generate enough electricity for about 60 homes. If the rotor rotates at 40 rpm, find the angular velocity and the linear velocity of a point on the tip of one of the rotors.

10. *Engineering.* Find the position of a piston after $\frac{1}{2}$ second if the crankshaft is turning at 60 revolutions per minute and its radius is 4 inches. Suppose that the connecting rod is 8 inches long.

11. *Engineering.* Find the position of the piston in Problem 10 if the crankshaft is turning at 200 rpm.

12. *Engineering.* The connecting rod in Problem 10 is twice the length of the radius of the crankshaft. Rewrite the general equation for d in terms of any radius r after t seconds, and then simplify for $\theta = 1$ to the nearest hundredth.

CHAPTER 2
GRAPHS OF TRIGONOMETRIC FUNCTIONS

CHAPTER OVERVIEW

This chapter introduces the idea of graphing the trigonometric functions. You can understand these functions more easily if you form a mental image or "picture" of them. A quick, simple method called *framing* is introduced to allow you to graph general cosine, sine, and tangent functions. This framing technique does not require background practice working on amplitude, period, and phase displacement, but instead graphs the functions very easily and efficiently in one step. Also, using available computer and calculator technology, you can more easily "see" many relationships and variations of those functions. The optional Application for Further Study at the end of this chapter gives you some insight into how you might use a computer as you progress through the material of this section.

You would probably expect to encounter the graph of a trigonometric curve in a science or physics course, but such curves are used to describe all sorts of periodic or cyclic phenomena. For example, the accompanying news clipping is from *The Wall Street Journal*.

Nikolai Kondratieff's Wave Is Running a Bit Behind Schedule (We Hope)

In the 1920s, Soviet economist Nikolai Kondratieff identified what he believed to be a long-term cycle governing economic activity. Based on studies of such factors as wages and commodity prices in 19th century Britain, France and America, he espoused the theory that capitalist economies expand and contract in cycles of 50 to 60 years.

The idealized wave below is often used as a simplified portrayal of Kondratieff's theory. The crests and troughs represent relative highs and lows in economic activity. Although the long wave uncannily pretraced the Great Depression and the post-war boom, it is dismissed by many modern economists as a statistical absurdity. So, they aren't worried that the wave now suggests the West is heading toward another economic crisis.

As for Kondratieff, he was branded a counterrevolutionary because his waves were inconsistent with the Marxist belief that capitalism would self-destruct. The economist was imprisoned in a Siberian labor camp where, according to Aleksandr Solzhenitsyn, he "became mentally ill and died."

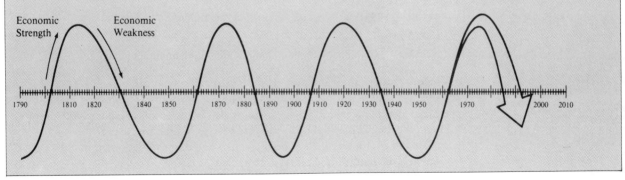

2.1   GRAPHS OF COSINE, SINE, AND TANGENT FUNCTIONS

In the last chapter we introduced the core of a trigonometry course—the general definition of the trigonometric functions. Our task is now to amplify, explain, discuss, and apply that definition in a variety of situations.

The nature of a function can often be more easily understood by looking at a picture or graph of it. In this chapter we will consider the graphs of these circular functions. With most functions we begin by plotting points to determine the general shape. We then generalize so you can graph the curve without a lot of calculations of individual points. In order to plot points we need to relate the trigonometric functions to the Cartesian coordinate system. For example, we may wish to write $y = \tan x$, but in so doing we change the meaning of x and y as they were used in the last chapter. Recall from the *definition* of the trigonometric functions that

$$\tan \theta = \frac{y}{x}$$

where (x, y) represented a point on the terminal side of θ. *Now*, when we write

$$y = \tan x$$

x represents the *angle*, and y is the value of the tangent function of x.

Graph of Standard Sine Function

We will begin by considering the graph of $y = \sin x$. To graph $y = \sin x$, begin by plotting familiar values for the sine:

$x = $ REAL NUMBER	0	$\frac{\pi}{2}$	π	$\frac{3\pi}{2}$	2π
$y = \sin x$	0	1	0	-1	1

These are exact values and they are plotted as colored points in Figure 2.1. You should plot additional points using more exact values, Appendix D table, or a calculator (see table in margin on page 57; these are plotted as black points in Figure 2.1). The plotted points are then connected by a smooth curve called the **sine curve**.

The domain for x is all real numbers, so what about values other than $0 \leq x \leq 2\pi$? Using the reduction principle, we know that

$$\sin x = \sin(x + 2\pi) = \sin(x - 2\pi) = \sin(x + 4\pi) = \sin(x + 6\pi) = \cdots$$

More generally,

$$\sin(\theta + 2n\pi) = \sin \theta$$

for any integer n. In other words, the values of the sine function repeat themselves

FIGURE 2.1   Graph of $y = \sin x$
by plotting points

x	y
-1	$-.84$
1	.84
2	.91
3	.14
4	$-.76$
5	$-.96$
6	$-.28$
7	.66

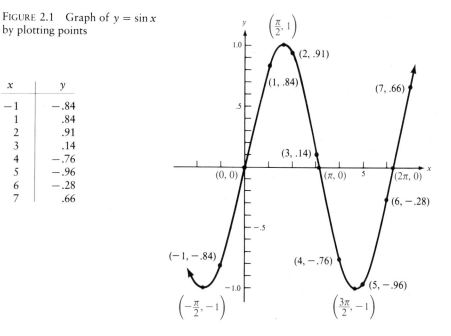

after 2π. We describe this by saying that sine is **periodic** with period 2π. The sine curve is shown in Figure 2.2. Notice that, even though the domain of the sine function is all real numbers, the range is restricted to values between -1 and 1 (inclusive).

FIGURE 2.2   Graph of $y = \sin x$
Domain: all real x
Range: $-1 \le y \le 1$

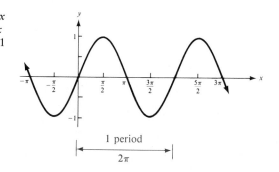

Framing a Sine Curve

Notice that, for the period labeled in Figure 2.2, the sine curve starts at $(0, 0)$, goes *up* to $(\frac{\pi}{2}, 1)$ and then *down* to $(\frac{3\pi}{2}, -1)$ passing through $(\pi, 0)$, and then goes back up to $(2\pi, 0)$, which completes one period. This graph shows that the **range** of the sine function is $-1 \le y \le 1$. We can summarize a technique for sketching the sine curve (see Figure 2.3) called **framing the curve**.

FRAMING A SINE CURVE

The standard sine function

$$y = \sin x$$

has domain $-\infty < x < \infty$ and range $-1 \le y \le 1$ and is periodic with period 2π. One period of this curve can be sketched by framing, as follows:

1. *Start* at the origin $(0, 0)$.

2. *Height* of this frame is two units: one unit up and one unit down from the starting point $(-1 \le y \le 1)$.

3. *Length* of this frame is 2π units (about 6.28) from the starting point (the period is 2π).

4. The curve is now framed. Plot five critical points within the frame:
   a. Endpoints (along axis)
   b. Midpoint (along axis)
   c. Quarterpoints (up first, then down)

5. Draw the curve through the critical points, remembering the shape of the sine curve.

FIGURE 2.3   Framing the sine curve

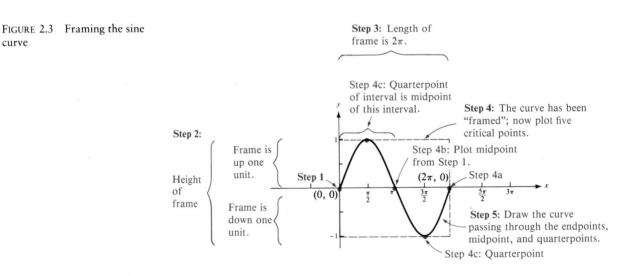

Step 3: Length of frame is 2π.

Step 4c: Quarterpoint of interval is midpoint of this interval.

Step 4: The curve has been "framed"; now plot five critical points.

Step 2:

Frame is up one unit.

Step 1

Step 4b: Plot midpoint from Step 1.

Height of frame

$(2\pi, 0)$ Step 4a

$(0, 0)$

Frame is down one unit.

Step 5: Draw the curve passing through the endpoints, midpoint, and quarterpoints.

Step 4c: Quarterpoint

You can draw a standard sine curve using any point as a starting point; the procedure is always the same.

Example 1

Draw one period of a standard sine curve using (3, 1) as the starting point for building a frame.

Solution

Plot the point (3, 1) as the starting point. Draw the standard frame as shown in the following figure.

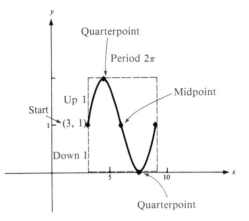

Framing a Cosine Curve

We can graph the cosine curve by plotting points, as we did with the sine curve. The details of plotting these points are left as an exercise. The cosine curve $y = \cos x$ is "framed" as described in Figure 2.4.

FIGURE 2.4 Framing the cosine curve

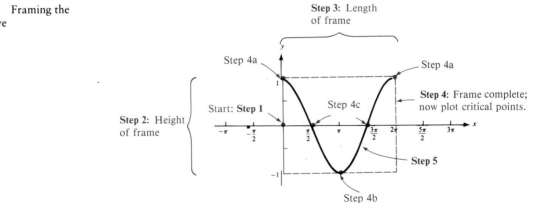

Notice that the only difference in the steps for graphing the cosine and the sine curves is in Step 4; the procedures for building the frame are identical.

FRAMING A COSINE CURVE

The standard cosine function

$$y = \cos x$$

has domain $-\infty < x < \infty$ and range $-1 \leq y \leq 1$ and is periodic with period 2π. One period of this curve can be sketched by framing, as follows:

1. *Start* at the origin $(0, 0)$.
2. *Height* of the frame is two units: one unit up and one unit down from the starting point $(-1 \leq y \leq 1)$.
3. *Length* of the frame is 2π units (about 6.28) from the starting point (the period is 2π).
4. The curve is now framed. Plot five critical points within the frame:
   a. Endpoints (at the top corners of the frame)
   b. Midpoint (at the bottom of the frame)
   c. Quarterpoints (along the axis)
5. Draw the curve through the critical points, remembering the shape of the cosine curve.

Since values for x greater than 2π or less than 0 are coterminal with those already considered, we see that **the period of the cosine function is 2π**. The cosine curve is shown in Figure 2.5. Notice that the domain and range of the cosine function are the same as they were for the sine function; D: all reals; R: $-1 \leq \cos x \leq 1$.

FIGURE 2.5 Graph of $y = \cos x$
Domain: all real x
Range: $-1 \leq y \leq 1$

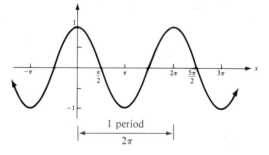

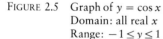

Example 2

Draw one period of a standard cosine curve using $(3, -\frac{1}{2})$ as the starting point for building a frame.

Solution

Plot the point $(3, -\frac{1}{2})$ as the starting point. Notice that the graph in this example is drawn using integers (compare with Example 1) so that you must approximate the distance 2π.

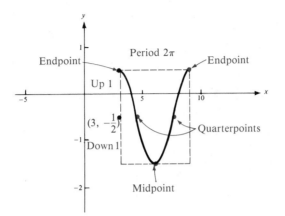

Framing a Tangent Curve

By setting up a table of values and plotting points (the details are left as an exercise), we notice that $y = \tan x$ does not exist at $\frac{\pi}{2}$, $\frac{3\pi}{2}$, or $\frac{\pi}{2} + n\pi$ for any integer n. The lines $x = \frac{\pi}{2}$, $x = \frac{3\pi}{2}, \ldots,$ $x = \frac{\pi}{2} + n\pi$ for which the tangent is not defined are called **asymptotes**. The procedure is summarized in Figure 2.6.

FIGURE 2.6   Framing the tangent curve

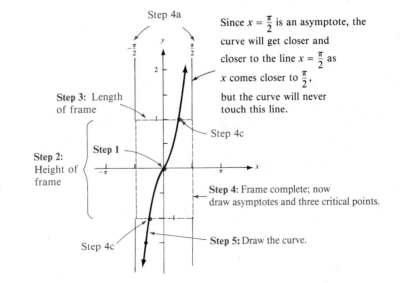

FIGURE 2.6   Framing the tangent curve

**FRAMING A
TANGENT CURVE**

The standard tangent function

$$y = \tan x$$

has as domain and range all real numbers except $x \neq \frac{\pi}{2} + n\pi$ and is periodic with period π. One period of this curve can be sketched by framing, as follows:

1. *Start* at the origin $(0, 0)$; for the tangent curve this is the *center* of the frame.
2. *Height* of this frame is two units: one unit up and one unit down from the starting point.
3. *Length* of this frame is π units (about 3.14) and is drawn so that it is $\frac{\pi}{2}$ (about 1.57) units on each side of the starting point (the period is π).
4. The curve is now framed. Draw the asymptotes and plot three critical points within the frame:
   a. Extend the vertical sides of the frame; these are the asymptotes.
   b. Midpoint (this was the starting point)
   c. Quarterpoints (down first; then up on the frame)
5. Draw the curve through the critical points, using the asymptotes as guides and remembering the shape of the tangent curve.

The tangent curve is indicated in Figure 2.7. Even though the curve repeats for values of x greater than 2π or less than 0, it also repeats after it has passed through an interval with length π. For this reason, $\tan(\theta + n\pi) = \tan\theta$ for any integer n, and we see that **the tangent has a period of π.** The domain of the tangent function is restricted so that multiples of π added to $\frac{\pi}{2}$ are excluded; this is because the tangent is not defined for these values. The range, on the other hand, is unrestricted; it is the set of all real numbers.

FIGURE 2.7   Graph of $y = \tan x$
Domain: all real x, $x \neq \frac{n\pi}{2}$
(n any integer)
Range: all real y.

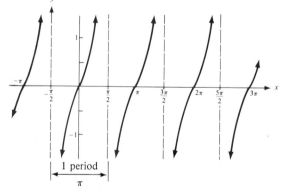

Example 3   Draw one period of a standard tangent curve using $(\pi, -2)$ as a starting point for building a frame.

Solution   Plot the point $(\pi, -2)$ and build the frame as shown.

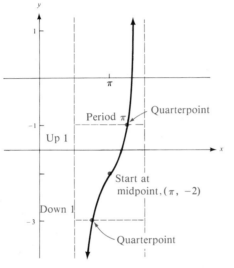

WARNING! Remember, the starting point is in the *center* of the frame for the tangent curve.

Problem Set 2.1

A   *In Problems 1–5 give the functional value of each from memory.*

1. **a.** $\sec 270°$ **b.** $\cot \frac{3\pi}{2}$ **c.** $\sin 0$ **d.** $\cos \frac{\pi}{6}$ **e.** $\csc 0$ **f.** $\tan 0°$

2. **a.** $\csc 30°$ **b.** $\tan \pi$ **c.** $\sec \frac{\pi}{2}$ **d.** $\sin \frac{\pi}{6}$ **e.** $\cot 180°$ **f.** $\sin 45°$

3. **a.** $\tan 90°$ **b.** $\sec \pi$ **c.** $\cot 90°$ **d.** $\cos 0$ **e.** $\csc 90°$ **f.** $\cos \frac{\pi}{3}$

4. **a.** $\cot \frac{\pi}{6}$ **b.** $\sin 60°$ **c.** $\cos 45°$ **d.** $\sec 0°$ **e.** $\tan \frac{\pi}{6}$ **f.** $\csc \pi$

5. **a.** $\sin \frac{\pi}{2}$ **b.** $\csc 45°$ **c.** $\tan 60°$ **d.** $\cot \frac{\pi}{4}$ **e.** $\cos \frac{\pi}{2}$ **f.** $\sin 270°$

6. Draw a quick sketch of $y = \sin x$ from memory by framing the curve.

7. Draw a quick sketch of $y = \cos x$ from memory by framing the curve.

8. Draw a quick sketch of $y = \tan x$ from memory by framing the curve.

In Problems 9–20 plot the given point; then draw a frame using the given point as a starting point to draw one period of the requested curve.

9. $(\pi, 1)$; sine curve   10. $(\pi, 1)$; cosine curve   11. $(\pi, 1)$; tangent curve

12. $(-\frac{\pi}{2}, 2)$; cosine curve   13. $(-\frac{\pi}{2}, 2)$; sine curve   14. $(-\frac{\pi}{2}, 2)$; tangent curve

15. $(-\frac{\pi}{4}, -2)$; cosine curve   16. $(-\frac{\pi}{4}, -2)$; sine curve   17. $(-\frac{\pi}{4}, -2)$; tangent curve

18. $(-1, 2)$; sine curve   19. $(-1, -2)$; cosine curve   20. $(-1, -2)$; tangent curve

B

21. Complete the table at the right by filling in values for $y = \cos x$:

x	y
-1	
0	
1	
2	
3	
4	
5	
6	
7	

22. Use the table in Problem 21, along with other values if necessary, to plot $y = \cos x$.

23. Complete a table of values like the one in Problem 21 for $y = \tan x$.

24. Use the table in Problem 23, along with other values if necessary, to plot $y = \tan x$.

25. Complete a table of values like the one in Problem 21 for $y = \sec x$.

26. Use the table in Problem 25, along with other values if necessary, to plot $y = \sec x$.

27. Complete a table of values like the one in Problem 21 for $y = \csc x$.

28. Use the table in Problem 27, along with other values if necessary, to plot $y = \csc x$.

29. Complete a table of values like the one in Problem 21 for $y = \cot x$.

30. Use the table in Problem 29, along with other values if necessary, to plot $y = \cot x$.

C

31. *Calculator problem.* Explain (in detail) how you would graph the trigonometric curves on your calculator.*

32. Plot points (with or without a graphics calculator) to graph $y = 3\cos(2x + \pi) - 3$.

33. Plot points (with or without a graphics calculator) to graph $y = 2\sin(3x + \frac{\pi}{4}) + 1$.

34. Plot points (with or without a graphics calculator) to graph $y = \dfrac{\sin x}{x}$.

35. Take a piece of paper and wrap it around a candle, as shown in the illustration. Make cuts at A and B, and unroll the paper.
    a. What do you think the edges A and B will look like?
    b. Perform the experiment to see if your guess was right.

For Problems 36–41 label the origin of a coordinate system O. Draw a unit circle with center at $(-1, 0)$. Let θ be an angle drawn with the vertex at the center of this circle, initial side the positive x-axis, and the point $P(x', y')$ the intersection of the terminal side of this angle and the unit circle. Let $|PQ|$ be the perpendicular drawn to the x'-axis and $|PR|$ the perpendicular drawn to the y'-axis. Finally, let S be the intersection of the line determined by the terminal side and the y-axis.

36. Show $\sin \theta = |PQ|$.   37. Show $\cos \theta = |PR|$.   38. Show $\tan \theta = |SO|$.

---

* This problem requires a calculator with graphing capabilities.

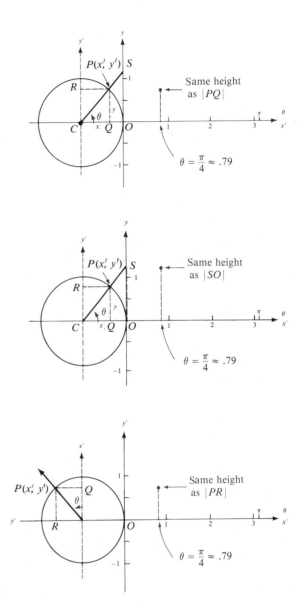

39. By Problem 36, $\sin\theta = |PQ|$, and you can use this fact to help you sketch $y = \sin\theta$. For example, when $\theta = \frac{\pi}{4}$, draw this angle, and then measure $|PQ|$ and *plot this height* at $\theta = \frac{\pi}{4}$ on the θ-axis as shown at the right. As P makes one revolution, it is easy to quickly plot the points on the curve $y = \sin\theta$.

40. By Problem 38, $\tan\theta = |SO|$, and you can use this fact to help you sketch $y = \tan\theta$. For example, when $\theta = \frac{\pi}{4}$, draw this angle, and then measure $|SO|$ and *plot this height* at $\theta = \frac{\pi}{4}$ on the θ-axis as shown at the right. As P makes one revolution, notice that $|SO|$ does not exist at $\theta = \frac{\pi}{2}$ and $\frac{3\pi}{2}$.

41. By Problem 37, $\cos\theta = |PR|$, and you can use this fact to help you sketch $y = \cos\theta$. To do this, rotate the unit circle described in the directions by $90°$. For example, when $\theta = \frac{\pi}{4}$, draw this angle, and then measure $|PR|$ and plot this height at $\theta = \frac{\pi}{4}$ on the θ-axis as shown at the right. As P makes one revolution, it is easy to quickly plot the points on the curve $y = \cos\theta$.

42. *Computer problem.* There are many computer graphics programs available for every type of computer. For example, TRIGPAK, which is available on IBM format from Brooks/Cole, graphs trigonometric curves using the framing technique. Obtain a graphics program, explore graphing trigonometric functions, and then write a short paper describing the program you are using.*

---

* See also the Application for Further Study at the end of this chapter.

2.2   GENERAL COSINE, SINE, AND TANGENT CURVES

Translations and Phase Shift

Once you are given a particular function, it is possible to shift the graph of that function to other locations. For example, let $y = f(x)$ be the function shown in Figure 2.8. It is possible to shift the entire curve up, down, right, or left, as shown in Figure 2.9.

FIGURE 2.8   Graph of $y = f(x)$

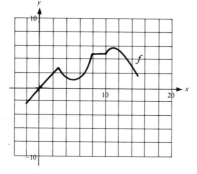

FIGURE 2.9   Shifting the graph of $y = f(x)$

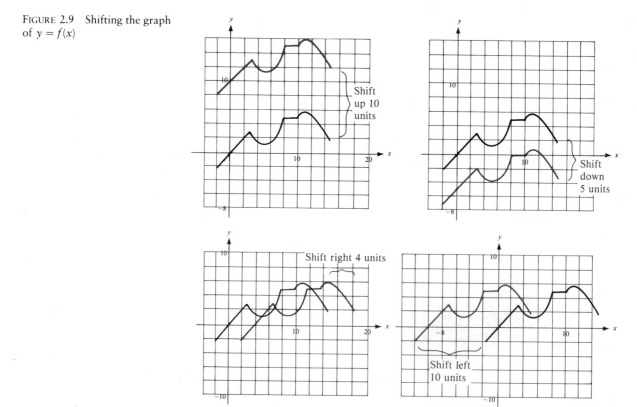

Instead of considering the curve shifting relative to fixed axes, consider the effect of shifting the axes. If the coordinate axes are shifted up k units, the origin of this new coordinate system would correspond to the point $(0, k)$ on the old coordinate system. If the axes are shifted h units to the right, the origin would correspond to the point $(h, 0)$ of the old system. A horizontal shift of h units followed by a vertical shift of k units would shift the new coordinate axes so that the origin corresponds to a point (h, k) on the old axes. Suppose a *new* coordinate system with origin at (h, k) is drawn, and the new axes are labeled x' and y', as shown in Figure 2.10.

FIGURE 2.10 Shifting the axes to (h, k)

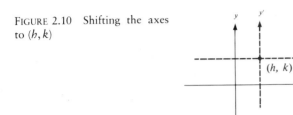

Every point on a given curve can now be denoted in two ways, as shown in Figure 2.11:

1. as (x, y) measured from the old origin, and
2. as (x', y') measured from the new origin.

FIGURE 2.11 Comparison of coordinate axes

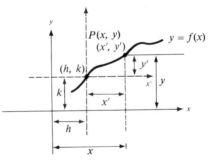

To find the relationship between (x, y) and (x', y'), consider the graph shown in Figure 2.11.

$$x = x' + h \qquad x' = x - h$$
$$\text{or}$$
$$y = y' + k \qquad y' = y - k$$

This relationship says that, if

$$y - k = f(x - h)$$

the graph of this function is the same as the graph of

$$y' = f(x')$$

where point (x', y') is measured from the new origin located at (h, k). This fact can greatly simplify the work of graphing, since $y' = f(x')$ is usually easier to graph than $y - k = f(x - h)$. This shifting is sometimes called a **phase shift**, but in this book we also call it a **translation**.

Example 1

Find (h, k) for each of the following equations.

a. $y - 5 = f(x - 7)$; $(h, k) = (7, 5)$
b. $y + 6 = f(x - \pi)$; $(h, k) = (\pi, -6)$. Did you notice that $y + 6 = y - (-6)$, so that $k = -6$?
c. $y + 1 = f(x + \frac{\pi}{3})$; $(h, k) = (-\frac{\pi}{3}, -1)$
d. $y = f(x)$; $(h, k) = (0, 0)$. This indicates no shift.
e. $y - 15 = f(x + \frac{1}{3})$; $(h, k) = (-\frac{1}{3}, 15)$
f. $y - 6 = f(x) + 15$. Write the equation as $y - 21 = f(x)$; thus, $(h, k) = (0, 21)$. This indicates no horizontal shift.

⬚

In the case of the graphs of the circular functions, it will simply be necessary to determine (h, k) and then build a frame at (h, k) rather than at the origin.

Example 2

Graph one period of $y = \sin(x + \frac{\pi}{2})$.

Solution

Step 1:   Frame the curve as shown in Figure 2.12a.

a. Plot $(h, k) = (-\frac{\pi}{2}, 0)$.
b. The period of the sine curve is 2π, and it has a high point up one unit and a low point down one unit.

Step 2:   Plot the five critical points (two endpoints, the midpoint, and two quarterpoints). For the sine curve, plot the endpoint (h, k) and use the frame to plot the other endpoint and the midpoint. For the quarterpoints, remember that the sine curve is "up–down"; use the frame to plot the quarterpoints as shown in Figure 2.12b.

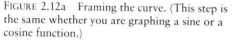
FIGURE 2.12a   Framing the curve. (This step is the same whether you are graphing a sine or a cosine function.)

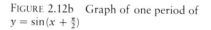
FIGURE 2.12b   Graph of one period of $y = \sin(x + \frac{\pi}{2})$

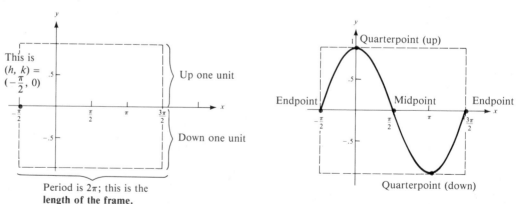

Step 3:   Remembering the shape of the sine curve, sketch one period of $y = \sin(x + \frac{\pi}{2})$ using the frame and the five critical points. If you wish to show more than one period, just repeat the same pattern.

⬚

Notice from Figure 2.12b that the graph of $y = \sin(x + \frac{\pi}{2})$ is the same as the graph of $y = \cos x$. Thus $\sin(x + \frac{\pi}{2}) = \cos x$.

Example 3

Solution

Graph one period of $y - 2 = \cos(x - \frac{\pi}{6})$.

Step 1: Frame the curve as shown in Figure 2.13. Notice that $(h, k) = (\frac{\pi}{6}, 2)$.

Step 2: Plot the five critical points. For the cosine curve the left and right endpoints are at the upper corners of the frame; the midpoint is at the bottom of the frame; the quarterpoints are on a line through the middle of the frame.

Step 3: Draw one period of the curve as shown in Figure 2.13.

FIGURE 2.13 Graph of one period of $y - 2 = \cos(x - \frac{\pi}{6})$

Example 4

Solution

Graph one period of $y + 3 = \tan(x + \frac{\pi}{3})$.

Step 1: Frame the curve as shown in Figure 2.14. Notice that $(h, k) = (-\frac{\pi}{3}, -3)$, and remember that the period of the tangent is π.

Step 2: For the tangent curve, (h, k) is the midpoint of the frame. The endpoints, each of which is a distance of $\frac{1}{2}$ the period from the midpoint, determine the location of the asymptotes. The top and the bottom of the frame are one unit from (h, k). Locate the quarterpoints at the top and bottom of the frame, as shown in Figure 2.14.

FIGURE 2.14 Graph of one period of $y + 3 = \tan(x + \frac{\pi}{3})$

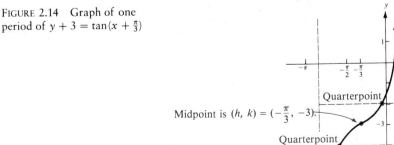

Step 3:   Sketch one period of the curve as shown in Figure 2.14. Remember that the tangent curve is not contained within the frame.   □

Changes in Amplitude and Period

We will discuss two additional changes for the function defined by $y = f(x)$. The first, $y = af(x)$, changes the scale on the y-axis; the second, $y = f(bx)$, changes the scale on the x-axis. For a function $y = af(x)$ it is clear that the y value is a times the corresponding value of $f(x)$, which means that $f(x)$ is stretched or shrunk in the y direction by the multiple of a. For example, if $y = f(x) = \cos x$, then $y = 3f(x) = 3\cos x$ is the graph of $\cos x$ that has been stretched so that the high point is at 3 units and the low point is at -3 units. The value $2|a|$ gives the height of the frame for f. To graph $y = 3\cos x$, frame the cosine using $a = 3$ rather than 1 (see Figure 2.15). For sine and cosine curves, $|a|$ is called the **amplitude** of the function. When $a = 1$, the amplitude is 1, so $y = \sin x$ and $y = \cos x$ are said to have amplitude 1.

FIGURE 2.15   Graph of one period of $y = 3\cos x$

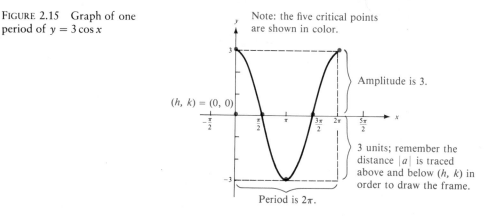

Note: the five critical points are shown in color.

Amplitude is 3.

$(h, k) = (0, 0)$

3 units; remember the distance $|a|$ is traced above and below (h, k) in order to draw the frame.

Period is 2π.

For a function $y = f(bx)$, $b > 0$, b affects the scale on the x-axis. Recall that $y = \sin x$ has a period of 2π ($f(x) = \sin x$, so $b = 1$). A function $y = \sin 2x$ ($f(x) = \sin x$ and $f(2x) = \sin 2x$) must complete one period as $2x$ varies from 0 to 2π. This statement means that one period is completed as x varies from 0 to π. (Remember that for each value of x the result is doubled *before* we find the sine of that number.) In general, **the period, p, of $y = \sin bx$ is $\frac{2\pi}{b}$, and the period of $y = \cos bx$ is $\frac{2\pi}{b}$**. However, since the period of $y = \tan x$ is π, $y = \tan bx$ has a **period of $\frac{\pi}{b}$**. Therefore, when framing the curve, use $\frac{2\pi}{b}$ for the sine and cosine, and $\frac{\pi}{b}$ for the tangent.

Example 5

Graph one period of $y = \sin 2x$.

Solution

The period is $p = \frac{2\pi}{2} = \pi$; thus the endpoints of the frame are $(0, 0)$ and $(\pi, 0)$, as shown in Figure 2.16.

FIGURE 2.16 Graph of one period of $y = \sin 2x$

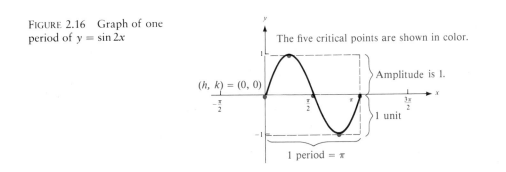

Graphs of General Trigonometric Curves

Summarizing all the preceding results, we have the *general* cosine, sine, and tangent curves:

GENERAL FORM OF COSINE, SINE, AND TANGENT CURVES

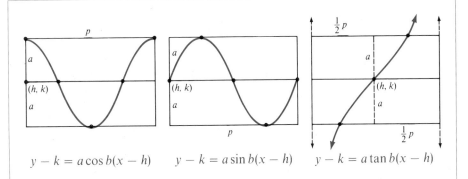

$$y - k = a \cos b(x - h) \qquad y - k = a \sin b(x - h) \qquad y - k = a \tan b(x - h)$$

1. Algebraically, put the equation into one of the general forms shown above.
2. Identify (by inspection) the following values: (h, k), a, and b. Then calculate p.
3. Draw the frame:
   a. Plot (h, k); this is the starting point.
   b. Draw a; the *height of the frame* is $2|a|$—up a units from (h, k) and down a units from (h, k).*
   c. Draw p; the *length of the frame* is p. Remember $p = \frac{2\pi}{b}$ for cosine and sine, and $p = \frac{\pi}{b}$ for tangent.
4. Locate the critical values using the frame as a guide, and then sketch the appropriate graphs. You do not need to know the coordinates of these critical values.

* In this section we have considered only the case where a is positive. The graphs of these functions where a is negative will be considered in Section 3.4.

Example 6

Graph $y + 1 = 2\sin\frac{2}{3}(x - \frac{\pi}{2})$.

Solution

Find (h, k), a, and p: $(h, k) = (\frac{\pi}{2}, -1)$, $a = 2$, and $b = \frac{2}{3}$; thus $p = 2\pi/\frac{2}{3} = 3\pi$. Now plot (h, k) and frame the curve. Then plot the five critical points (two endpoints, the midpoint, and two quarterpoints). Finally, after sketching one period, draw the other periods as shown in Figure 2.17.

FIGURE 2.17   Graph of $y + 1 = 2\sin\frac{2}{3}(x - \frac{\pi}{2})$

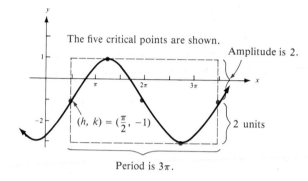

The five critical points are shown.

Amplitude is 2.

$(h, k) = (\frac{\pi}{2}, -1)$

2 units

Period is 3π.

Example 7

Graph $y = 3\cos(2x + \frac{\pi}{2}) - 2$.

Solution

Rewrite in standard form to obtain $y + 2 = 3\cos 2(x + \frac{\pi}{4})$. Notice that $(h, k) = (-\frac{\pi}{4}, -2)$; $a = 3$, $b = 2$, and $p = \frac{2\pi}{2} = \pi$. Plot (h, k) and frame the curve as shown in Figure 2.18.

FIGURE 2.18   Graph of $y + 2 = 3\cos 2(x + \frac{\pi}{4})$

Critical points are shown. Notice that (h, k) is not one of the critical points for the cosine curve.

$(h, k) = (-\frac{\pi}{4}, -2)$

Amplitude is 3.

3 units

Period is π.

One period inside the frame is drawn first; then the curve is extended outside the frame.

Example 8

Graph $y - 2 = 3\tan\frac{1}{2}(x - \frac{\pi}{3})$.

Solution

Notice that $(h, k) = (\frac{\pi}{3}, 2)$; $a = 3$, $b = \frac{1}{2}$, and $p = \pi/\frac{1}{2} = 2\pi$. Plot (h, k) and frame the curve as shown in Figure 2.19.

FIGURE 2.19   Graph of
$y - 2 = 3\tan\frac{1}{2}(x - \frac{\pi}{3})$

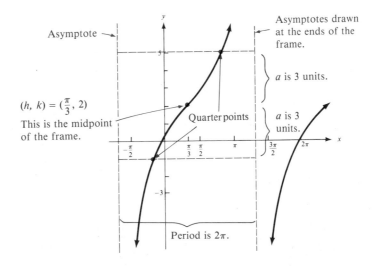

Asymptote →

Asymptotes drawn
at the ends of the
frame.

} *a* is 3 units.

$(h,\ k) = (\frac{\pi}{3},\ 2)$

This is the midpoint
of the frame.

Quarter points

} *a* is 3
units.

Period is 2π.

CALCULATOR GRAPHICS

Curve sketching by plotting points can efficiently be done on a calculator. Many people believe that it is not necessary to think if using a calculator, that the "machine" will do all of the work and you will no longer understand the principles involved if you begin using calculators for graphing. However, this is not true; with a calculator your attention must be focused on different matters. For example, in this section we are concerned with graphing $y = \cos x$. The first task is to decide on the domain and the range. If you know what the curve looks like, then it is easy to make an intelligent choice about the domain and range. However, if you do not know anything about the curve, you may have to make several attempts at setting the domain and the range before you obtain a "good-looking" graph. Without the work of this section prior to graphing a cosine or a sine curve on a calculator, the task would be very difficult—so difficult, in fact, that when you purchase a Casio fx 7000G calculator many graph functions are "preprogrammed" so that you will not become discouraged when you try graphing your own functions and probably obtain very disappointing results.

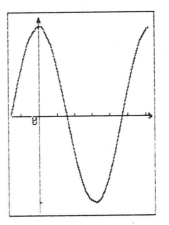

Graph of the function $y = \cos x$
on a Casio fx-7000G

For the graph shown above the following choices were made for the domain and range values:

Xmin: -1.57 Ymin: -1.1
  max: 6.28   max: 1.1
  scl: 1   scl: .1

Problem Set 2.2

A   *Graph one period of each of the functions given in Problems 1–21.*

1. $y = \sin(x + \pi)$
2. $y = \cos(x + \frac{\pi}{2})$
3. $y = \sin(x - \frac{\pi}{3})$

4. $y = \cos(x + \frac{3\pi}{2})$
5. $y = 2\cos x$
6. $y = 3\sin x$

7. $y = \cos 2x$
8. $y = \sin 3x$
9. $y = \tan(x + \frac{\pi}{6})$

10. $y = \tan(x - \frac{3\pi}{2})$
11. $y = \frac{1}{3}\tan x$
12. $y = \frac{1}{2}\sin x$

13. $y - 2 = \sin(x - \frac{\pi}{2})$
14. $y + 1 = \cos(x + \frac{\pi}{3})$
15. $y - 3 = \tan(x + \frac{\pi}{6})$

16. $y - \frac{1}{2} = \frac{1}{2}\cos x$
17. $y - 1 = 2\cos(x - \frac{\pi}{4})$
18. $y - 1 = \cos 2(x - \frac{\pi}{4})$

19. $y + 2 = 3\sin(x + \frac{\pi}{6})$
20. $y + 2 = \sin 3(x + \frac{\pi}{6})$
21. $y + 2 = \tan(x + \frac{\pi}{4})$

B   *Graph each of the curves given in Problems 22–36.*

22. $y = \sin(4x + \pi)$
23. $y = \sin(3x + \pi)$
24. $y = \tan(2x - \frac{\pi}{2})$

25. $y = \tan(\frac{x}{2} + \frac{\pi}{3})$
26. $y = \frac{1}{2}\cos(x + \frac{\pi}{6})$
27. $y = \cos(\frac{1}{2}x + \frac{\pi}{12})$

28. $y = 2\cos(3x + 2\pi) - 2$
29. $y = 4\sin(\frac{1}{2}x + 2)$
30. $y = \sqrt{2}\cos(x - \sqrt{2}) - 1$

31. $y = \sqrt{3}\sin(\frac{1}{3}x - \sqrt{\frac{1}{3}})$
32. $y = 2\sin 2\pi x$
33. $y = 3\cos 3\pi x$

34. $y = 4\tan\frac{\pi x}{5}$
35. $y + 2 = \frac{1}{2}\cos(\pi x + 2\pi)$
36. $y - 3 = 3\cos(2\pi x + 4)$

C   37.   *Electrical engineering.* The current I (in amperes) in a certain circuit is given by

$$I = 60\cos(120\pi t - \pi)$$

where t is time in seconds. Graph this equation for $0 \le t \le \frac{1}{30}$.

38.   *Engineering.* Suppose a point P on a waterwheel with a 30-ft radius is d units from the water as shown in the figure. If it turns at 6 revolutions per minute, then

$$d = 29 + 30\cos(\frac{\pi}{5}t - \pi)$$

Graph this equation for $0 \le t \le 20$.

39.   *Space science.* The distance that a certain satellite is north or south of the equator is given by

$$y = 3{,}000\cos(\frac{\pi}{60}t + \frac{\pi}{5})$$

where t is the number of minutes that have elapsed since lift-off.
  a.   Graph the equation for $0 \le t \le 120$.
  b.   What is the greatest distance that the satellite ever reaches north of the equator?
  c.   How long does it take to complete one period?

2.3 GRAPHS OF SECANT, COSECANT, AND COTANGENT FUNCTIONS

Graph of Cosecant Function

Procedures for graphing the reciprocal functions—namely, cosecant, secant, and cotangent—can be developed by plotting points as shown in Section 2.1. Instead, however, we will use the idea of a frame along with the idea of what it means to be a reciprocal. For example, the graph of $y = \csc x$ is related to the graph of $y = \sin x$ because

$$\csc x = \frac{1}{\sin x}$$

Remember the meaning of reciprocal:

Function $\sin x$	Reciprocal $\csc x$
1	1
$\frac{1}{2}$	2
$\frac{1}{3}$	3
$.1 = \frac{1}{10}$	10

Notice that as $\sin x$ approaches 0 (written as $\sin x \to 0$), its reciprocal $\csc x$ gets larger without limit (written as $\csc x \to \infty$). Thus, every t for which $\sin x = 0$ is an asymptote $x = t$ for the cosecant curve. The graph is shown in Figure 2.20.

FIGURE 2.20

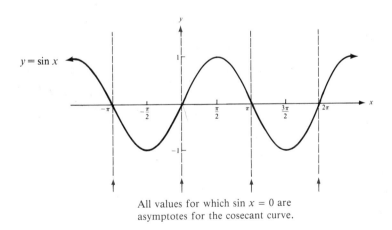

All values for which $\sin x = 0$ are
asymptotes for the cosecant curve.

Look at one period of $y = \csc x$. Begin by drawing a frame for $y = \sin x$ (that is, the reciprocal of the curve that you want to graph). See Figure 2.21 (page 76).

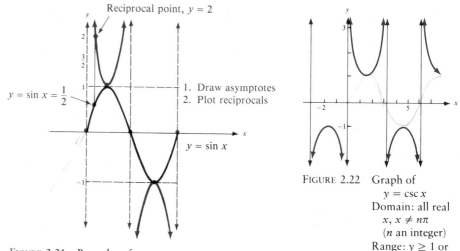

Reciprocal point, $y = 2$

$y = \sin x = \dfrac{1}{2}$

1. Draw asymptotes
2. Plot reciprocals

$y = \sin x$

FIGURE 2.21   Procedure for sketching one period of a cosecant curve

FIGURE 2.22   Graph of $y = \csc x$
Domain: all real $x, x \neq n\pi$ (n an integer)
Range: $y \geq 1$ or $y \leq -1$

Notice that the cosecant curve is not a continuous function. The domain for this curve is all real x, except those values corresponding to the asymptotes. These values are $x = n\pi$ for any integer n. The range is all values greater than (or equal to) 1 or less than (or equal to) -1. This curve is shown in Figure 2.22.

Graph of Secant Curve

The graph of the secant is also found by looking at the reciprocals, as shown in Figure 2.23.

FIGURE 2.23   Procedure for graphing one period of a secant curve

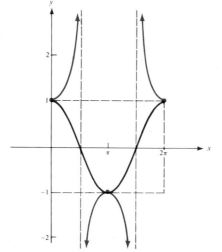

PROCEDURE FOR GRAPHING $y = \sec x$

1. Draw the reciprocal curve ($y = \cos x$). One period is shown here.
2. Draw asymptotes (values for which $y = \cos x$ is 0).
3. Plot reciprocals. (Notice that you do not need to plot a lot of points—not many more than the reciprocals of the critical points for the cosine curve.)
4. Draw the curve.

The entire curve is drawn by repeating the parts drawn for the frame shown in Figure 2.23. The result is Figure 2.24.

Figure 2.24   Graph of
$y = \sec x$
Domain: all real
$x, x \neq \frac{n\pi}{2}$
(*n* an integer)
Range: $y \leq -1$
or $y \geq 1$

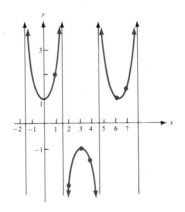

Graph of Cotangent Curve

The cotangent is, of course, the reciprocal of the tangent. But remember that $y = \tan x$ itself has asymptotes. That is,

$$\tan x \to \infty \quad \text{as } x \to \frac{\pi}{2}$$

Now, if the tangent values are getting large, then the values of the reciprocals must be approaching 0:

$\tan x$	$\cot x$
1	1
2	$\frac{1}{2}$
3	$\frac{1}{3}$
10	$\frac{1}{10}$
100	$\frac{1}{100}$
1,000	$\frac{1}{1,000}$

That is, $\cot x = 0$ for all values corresponding to the vertical asymptotes $x = t$ of the tangent curve. To draw one period of $y = \cot x$, look at the frame of the tangent curve.

FIGURE 2.25   One period of a cotangent curve

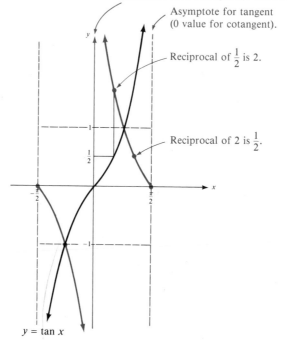

Asymptote for cotangent (0 value for tangent).

Asymptote for tangent (0 value for cotangent).

Reciprocal of $\frac{1}{2}$ is 2.

Reciprocal of 2 is $\frac{1}{2}$.

$y = \tan x$

It is difficult to "see" the cotangent curve based only on what is shown in Figure 2.25 because the cotangent curve is not continuous. However, if you repeat this period more than once and draw the entire curve as shown in Figure 2.26, it is much easier to visualize the cotangent curve.

FIGURE 2.26   Graph of
$y = \cot x$
Domain: all real
$x, x \neq n\pi$
(n any integer)
Range: all real y

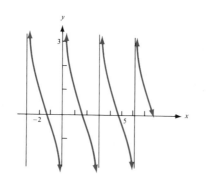

Graphs of General Trigonometric Curves

Since the graphs of the standard cosecant, secant, and tangent curves are related to the frames of the corresponding reciprocal functions, it is easy to extend the ideas of this section to the general curves as summarized in the following box.

GENERAL COSECANT,
SECANT, AND
COTANGENT FUNCTIONS

$y - k = a \csc b(x - h)$	Relate to the frame of the sine; starting point (h, k), height $2\|a\|$, length $\frac{2\pi}{b}$.
$y - k = a \sec b(x - h)$	Relate to the frame of the cosine; starting point (h, k), height $2\|a\|$, length $\frac{2\pi}{b}$.
$y - k = a \cot b(x - h)$	Relate to the frame of the tangent; starting point (h, k) is the center of the frame, height $2\|a\|$, length $\frac{\pi}{b}$.

Example 1

Graph $y - 1 = \frac{1}{2} \csc 2(x - \frac{\pi}{3})$.

Solution

$(h, k) = (\frac{\pi}{3}, 1)$, $a = \frac{1}{2}$, and $p = \frac{2\pi}{2} = \pi$. Draw the frame for sine.

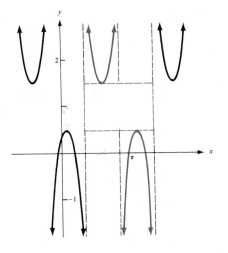

Example 2

Graph $y + 2 = \frac{1}{2} \sec \frac{1}{2}(x + \frac{\pi}{4})$.

Solution

$(h, k) = (-\frac{\pi}{4}, -2)$, $a = \frac{1}{2}$, $b = \frac{1}{2}$, and $p = 2\pi/(\frac{1}{2}) = 4\pi$. Relate to the frame of a cosine curve.

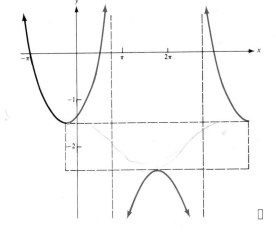

Example 3   Graph $y + 1 = \frac{2}{3}\cot 3(x - 2)$.

Solution $(h, k) = (2, -1)$, $a = \frac{2}{3}$, $b = 3$, and $p = \frac{\pi}{3}$. Relate to the frame of a tangent curve.

It is worthwhile to summarize the domain and range for each of the trigonometric functions. You should relate the information in Table 2.1 to the graph of each function. Remember that the domain is the set of replacements for the first component, x, and the range is the set of replacements for the second component, y. If you memorize the general shape of the trigonometric graphs, as well as the domain and range for each one, your future work will be simplified.

TABLE 2.1   Domain and Range for the Trigonometric Functions

Function	Domain LET n BE ANY INTEGER.	Range
$y = \cos x$	all reals	$-1 \leq y \leq 1$
$y = \sin x$	all reals	$-1 \leq y \leq 1$
$y = \tan x$	$x \neq \dfrac{\pi}{2} + n\pi$	all reals
$y = \sec x$	$x \neq \dfrac{\pi}{2} + n\pi$	$y \leq -1$ or $y \geq 1$
$y = \csc x$	$x \neq n\pi$	$y \leq -1$ or $y \geq 1$
$y = \cot x$	$x \neq n\pi$	all reals

Problem Set 2.3

A   *In Problems 1–6 give each functional value from memory.*

1. **a.** $\cos 0$ **b.** $\sec \frac{3\pi}{2}$ **c.** $\tan \pi$ **d.** $\cot \pi$ **e.** $\csc \frac{3\pi}{2}$ **f.** $\sin 0$

2. **a.** $\cot \frac{\pi}{2}$ **b.** $\cos \frac{\pi}{6}$ **c.** $\sec \pi$ **d.** $\tan \frac{\pi}{2}$ **e.** $\sin \frac{\pi}{6}$ **f.** $\csc \pi$

3.   a. $\cot\frac{\pi}{3}$ b. $\csc\frac{\pi}{6}$ c. $\cos\frac{\pi}{4}$ d. $\sin\frac{\pi}{4}$ e. $\tan\frac{\pi}{3}$ f. $\sec\frac{\pi}{2}$

4.   a. $\csc\frac{\pi}{4}$ b. $\cot\frac{\pi}{4}$ c. $\sin\frac{\pi}{2}$ d. $\cos\frac{\pi}{2}$ e. $\sec\frac{\pi}{3}$ f. $\tan\frac{\pi}{4}$

5.   a. $\cot\frac{\pi}{6}$ b. $\sin\pi$ c. $\csc\frac{\pi}{3}$ d. $\sec\frac{\pi}{4}$ e. $\cos\pi$ f. $\tan\frac{\pi}{6}$

6.   a. $\sin\frac{3\pi}{2}$ b. $\csc\frac{\pi}{4}$ c. $\sec\frac{\pi}{6}$ d. $\tan 2\pi$ e. $\tan 0$ f. $\cos\frac{3\pi}{2}$

7.   Draw a quick sketch of $y = \sec x$ from memory.

8.   Draw a quick sketch of $y = \csc x$ from memory.

9.   Draw a quick sketch of $y = \cot x$ from memory.

Graph the functions in each of Problems 10–30.

10. $y = 2\sec x$ 11. $y = \sec 2x$ 12. $y = 2\sec x + 1$

13. $y = 2\sec(x + 1)$ 14. $y = 2\csc x$ 15. $y = \csc 2x$

16. $y = \csc 2x + 1$ 17. $y = \csc(2x + 1)$ 18. $y = 2\sec(x + \frac{\pi}{3})$

19. $y - 2 = 2\sec(x + \frac{2\pi}{3})$ 20. $y = \csc(2x - \frac{\pi}{2})$ 21. $y - 1 = \csc(2x - \frac{\pi}{2})$

22. $y = 2\cot x$ 23. $y = \cot 2x$ 24. $y = 2\cot x - 1$

25. $y + 1 = 2\cot(x - \frac{\pi}{6})$ 26. $y = 2\cot(x - 1)$ 27. $y = \sec(x - 2) + 1$

28. $y = 2\csc(x - 1) - 2$ 29. $y = 2\cot(x + 1) - 2$ 30. $y = \frac{1}{2}\csc(x - 2) - 1$

*2.4 ADDITION OF ORDINATES

From radio modulation systems and sounds of musical instruments, to the quality of a singer's voice and the pattern of a human heartbeat, the periodic nature of a sine curve plays a role. In this chapter we have considered, among other things, variations of the sine curve, but many of the more common periodic phenomena require the idea of graphing the sum of trigonometric functions.

Adding Ordinates

We can graph the sum of two curves by **adding ordinates**. That is, we will add second components, point by point, until we have plotted enough points to see the shape of the curve that is the sum. For example, the sum of y_1 and y_2 at x_1 is found by adding the length of segment y_1 to the point (x_1, y_2) as shown in Figure 2.27 (page 83).

_____

* Optional section.

Historical Note

The French mathematician Joseph Fourier (1768–1830) was studying problems of heat flow and used trigonometry to describe what he saw. In more advanced mathematics, a series is defined as a sum of terms of a sequence and if these terms are trigonometric, then the series is called a **Fourier series**. The following is an example of a Fourier series:

$$y = \cos x + \tfrac{1}{3}\cos 3x + \tfrac{1}{5}\cos 5x + \cdots$$

As you can see, addition of ordinates can be used with Fourier series. The main characteristic of a Fourier series is that it can be used to represent functions that ordinarily are represented by different expressions in different parts of some interval. Fourier series are used today in studying heat flow, electrical fields, and sound waves.

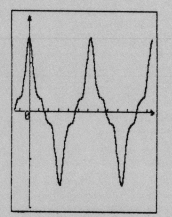

Graph of the function
$y = \cos x + \tfrac{1}{3}\cos 3x + \tfrac{1}{5}\cos 5x$
on a Casio fx-7000G

For the graph shown above the following choices were made for the domain and range values:

Xmin: −1.57
  max: 12.566371
  scl: 1
Ymin: −2
  max: 2
  scl: .5

FIGURE 2.27 Finding
$(x_1, y_1 + y_2)$

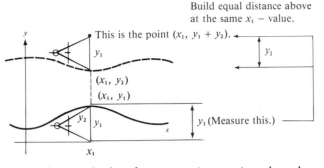

If either, or both, of y_1 or y_2 is negative, then the addition is performed by considering y_1 and y_2 as directed distances. This process is shown for several points in Figure 2.28.

FIGURE 2.28 Adding ordinates

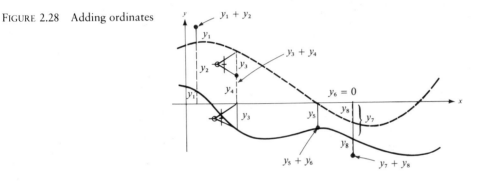

Example 1

Graph $y = \dfrac{x}{2} + \cos x$.

Solution

Let $y_1 = \dfrac{x}{2}$ and $y_2 = \cos x$, and write $y = y_1 + y_2$. First graph y_1 (the dotted line in Figure 2.29a). Next, graph y_2 (the solid black curve in Figure 2.29a). Choose some x value, and, using your eye or a ruler, measure y_1 and y_2 and then plot $y_1 + y_2$. Do this by counting or measuring *squares* on your graph; do *not* think in term of units. Repeat this procedure for several points, as shown by the color dots in Figure 2.29a. Finally, connect the dots for $y = y_1 + y_2$, as shown in Figure 2.29b.

FIGURE 2.29 Graph of
$y = \frac{x}{2} + \cos x$

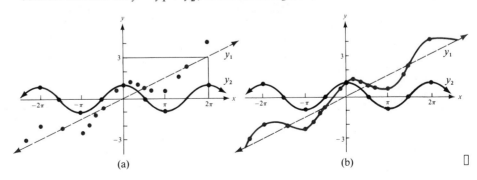

(a)   (b)

Example 2

Graph $y = 2\cos 3x + 3\sin 2x$ by adding ordinates.

Solution

Let $y = y_1 + y_2$, where $y_1 = 2\cos 3x$ and $y_2 = 3\sin 2x$. First graph $y_1 = 2\cos 3x$ and $y_2 = 3\sin 2x$ on the same axes, as shown in Figure 2.30. Next, select a particular x value—say x_1—and graphically find $(x_1, y_1 + y_2)$ by adding ordinates. Do this for several points, as shown, and draw a smooth curve (in color) to represent $y = y_1 + y_2$.

FIGURE 2.30　Graph of $y = 2\cos 3x + 3\sin 2x$ by adding ordinates

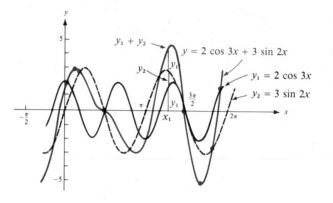

Graph of the function $y = 2\cos 3x + 3\sin 2x$ on a Casio fx-7000G

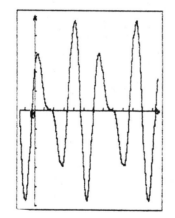

For the graph shown above the following choices were made for the domain and range values:

Xmin: -1.57
　max: 12.566371
　scl: 1

Ymin: -5
　max: 5
　scl: 1

Graphs of Other Trigonometric Curves

Example 3

Graph $y = \dfrac{x}{2} \sin x$.

Solution

This function is a product where

$$y_1 = \frac{x}{2}$$

$$y_2 = \sin x$$

and

$$y = y_1 y_2$$

The procedure for finding a product is similar to that for finding sums: using your eye or a ruler, count squares to find y_1 and y_2 and then plot the *product* $y_1 y_2$. If you focus on values of x for which either y_1 or y_2 is 0 or 1, you will find this process to be particularly easy, as shown in Figure 2.31.

FIGURE 2.31   Graph of $y = \frac{x}{2} \sin x$

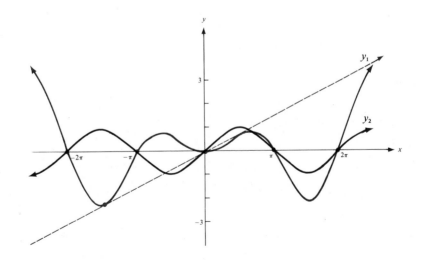

If you have two engines or motors that are running side by side at almost the same speed, then you can hear the noise as a series of beats. These beats are the result of the peaks of the addition of ordinates of the tones of the separate engines. For example, if $y_1 = \sin 7x$ and $y_2 = \sin 8x$, then the result of $y = y_1 + y_2$ is a series of beats as shown in Figure 2.32 (page 86).

$y_1 = \sin 7x$

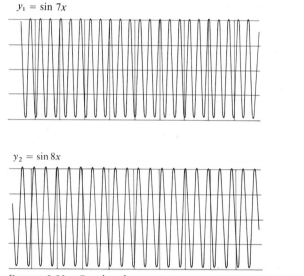

$y = y_1 + y_2$

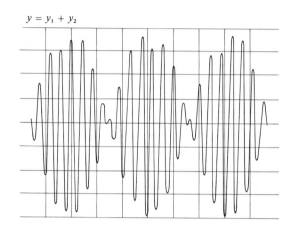

$y_2 = \sin 8x$

FIGURE 2.32   Graphs of $y_1 = \sin 7x$, $y_2 = \sin 8x$, and $y = \sin 7x + \sin 8x$ showing the beats in noise

In radio, modulation is the modification of the amplitude, period (frequency), and starting point (phase) of an electrical wave in accordance with a modulating signal wave. The common systems are amplitude modulation (as in AM radio) and frequency modulation (as in FM radio). Example 4 illustrates a function that can occur in radio transmission that involves multiplication of ordinates.

Example 4

Graph $y = (3 \sin x)(\cos 6x)$ from 0 to 2π.

Solution

Graph this curve by plotting points as shown in Figure 2.33.

FIGURE 2.33   Graph of $y = (3 \sin x)(\cos 6x)$ by plotting points

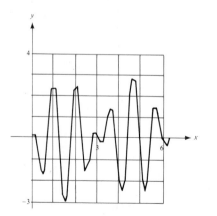

Graph of the function
$y = (3 \sin x)(\cos 6x)$ on a Casio
fx-7000G

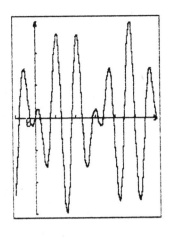

For the graph shown above the following choices were made for the domain and range values:

Xmin: -1
max: 6
scl: 1

Ymin: -3
max: 3
scl: 1

Problem Set 2.4

A   *Graph the functions given in Problems 1–24 by adding ordinates.*

1. $y = \frac{x}{2} + \sin x$
2. $y = \frac{x}{3} + \sin x$
3. $y = \frac{x}{3} + \cos x$
4. $y = \frac{2x}{3} + \cos x$
5. $y = \frac{2x}{3} + \sin x$
6. $y = x + \sin x$
7. $y = x + \cos x$
8. $y = x + \cos 2x$
9. $y = 2x + \cos 2x$

B
10. $y = \sin x + \cos x$
11. $y = 2 \sin x + \cos x$
12. $y = \sin x + 2 \cos x$
13. $y = \sin 2x + \cos 2x$
14. $y = 2 \sin 2x + \cos 2x$
15. $y = \sin 2x + 2 \cos 2x$
16. $y = 2 \sin \frac{x}{2} + \cos x$
17. $y = 2 \sin x + \cos \frac{x}{2}$
18. $y = 3 \sin \frac{x}{3} + \cos \frac{x}{3}$
19. $y = x \sin x$
20. $y = x \cos x$
21. $y = \frac{x}{2} \cos x$
22. $y = -\frac{x}{2} \cos x$
23. $y = -\frac{x}{2} \sin x$
24. $y = -x \cos x$

25. *Business.* The sales of rototillers are very seasonal. A new business can describe its sales by $y = .02x + |\sin \frac{x\pi}{6}|$, where x is the number of months since going into business on January 1, 1990. Plot the sales graph for the first three years.

26. *Business forecasting.* Using the information from Problem 25, plot the graph for rototiller sales for the year 1993.

27. *Physics.* Two light waves are said to be *destructive* if the amplitude of the sum of the waves is zero at every point. Show that $y = \sin x$ and $y = \cos(x + \frac{\pi}{2})$ combine to form destructive interference.

28. *Oceanography.* The sun–moon tide curves are shown here.* During a new moon, the sun and moon tidal bulges are centered at the same longitude, so their effects are added to produce maximum high tides and minimum low tides. This produces maximum tidal range.

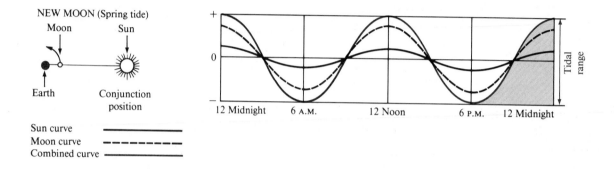

Write possible equations for the sun curve, moon curve, and for the combined curve. Assume the tidal range is 10 ft and the period is 12.

29. *Oceanography.* The sun–moon tide curves are shown below.* During the third quarter (neap tide) the sun and moon tidal bulges are at right angles to each other. Thus, the sun tide reduces the effect of the moon tide.

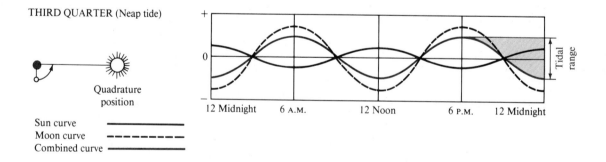

Write possible equations for the sun curve, moon curve, and for the combined curve. Assume the tidal range is 4 ft and the period is 12.

---

* From *Introductory Oceanography*, 5th edition, by H. V. Thurman, p. 253. Reprinted with permission of Merrill, an imprint of Macmillan Publishing Company. Copyright ©1988 Merrill Publishing Company, Columbus, Ohio.

C   *Sketch the curves in Problems 30–38 from 0 to 2π by plotting points or by using a graphics calculator.*

30. $y = x \sin x$ **31.** $y = x^2 \cos x$ **32.** $y = \frac{1}{x} \cos x$

33. $y = \dfrac{\sin x}{1 + x^2}$ **34.** $y = (2 \sin x)(\cos 6x)$ **35.** $y = (3 \cos x)(\sin 6x)$

36. $y = (4 \sin x)(\cos 8x)$ **37.** $y = \cos^2 x$ **38.** $y^3 = \sin x$

39. *Physics.* Graph the sound wave given by $y = .04 \sin 200\pi x + .06 \sin 400\pi x$ for $0 \le x \le \frac{1}{100}$.

40. Graph $y = \sin 7x + \sin 9x$, and notice the effect of beats.

41. *Electrical engineering.* The current I from two power sources is given by

$$I = 3 \sin 240\pi t + 6 \cos 120\pi t$$

where t is time in seconds. Sketch the graph of the current for $0 \le t \le \frac{1}{30}$.

2.5 SUMMARY AND REVIEW

	OBJECTIVES	PAGES/EXAMPLES
2.1 GRAPHS OF COSINE, SINE, AND TANGENT FUNCTIONS	1. Graph the cosine, sine, and tangent curves from memory by using a frame.	pp. 58, 59, 62; Problems 6–8.
	2. Graph the standard cosine, sine, and tangent curves by building a frame from a given starting point.	pp. 58, 60, 62; Examples 1–3; Problems 9–20.
	3. Graph trigonometric curves by plotting points.	p. 56; Problems 21–30.
2.2 GENERAL COSINE, SINE, AND TANGENT CURVES	4. Know the standard forms of the cosine, sine, and tangent curves.	p. 71
	5. Graph one period of the general cosine, sine, and tangent curves by using a frame.	p. 71; Examples 2–5; Problems 1–21.
	6. Graph the general cosine, sine, and tangent curves.	p. 71; Examples 6–8; Problems 22–36.
2.3 GRAPHS OF SECANT, COSECANT, AND COTANGENT FUNCTIONS	7. Graph the secant, cosecant, and cotangent curves from memory.	pp. 76–78; Problems 7–9.
	8. Graph the general secant, cosecant, and cotangent curves.	p. 79; Examples 1–3; Problems 10–30.
2.4 ADDITION OF ORDINATES (OPTIONAL)	9. Graph trigonometric curves by adding ordinates.	p. 81; Examples 1–2; Problems 1–18.
	10. Graph trigonometric curves by multiplying ordinates.	pp. 85–88; Examples 3–4; Problems 19–24, 30–38.

Terms

Adding ordinates [2.4]   Cotangent curve [2.3]   Secant curve [2.3]
Amplitude [2.2]   Frame [2.1, 2.2]   Sine curve [2.1]
Asymptote [2.1, 2.2]   Period [2.1, 2.2]   Tangent curve [2.1]
Cosecant curve [2.3]   Phase shift [2.2]   Translation [2.2]
Cosine curve [2.1]

Chapter 2 Review of Objectives

The problem numbers correspond to the objectives listed in Section 2.5.

1. From memory draw a quick sketch of the following curves.
   a. $y = \cos x$ b. $y = \sin x$ c. $y = \tan x$
2. Plot the given point, and draw a frame using the given point as a starting point. Draw one period of the requested curve.
   a. $(\frac{\pi}{6}, -\frac{3}{2})$; sine curve   b. $(-\frac{3\pi}{2}, 1)$; cosine curve
   c. $(-\frac{\pi}{3}, 2)$; tangent curve   d. $(-1, 1)$; sine curve
3. Graph the given curves by plotting points.
   a. $y = \sin x + \cos x$ b. $y = \dfrac{\tan x}{x}$
4. Fill in the blanks.

   $$\text{Standard forms:} \quad y - k = a \cos b(x - h)$$
   $$y - k = a \sin b(x - h)$$
   $$y - k = a \tan b(x - h)$$

   The origin is translated to _____ a. _____ .
   The height of the frame is _____ b. _____ .
   The length of the frame is _____ c. _____ for cosine, _____ d. _____ for sine, and _____ e. _____ for tangent.
   The *a* value for the sine curve is called its _____ f. _____ , and the length of the frame is called its _____ g. _____ .
5. Graph one period of the given curves.
   a. $y - 2 = \sin(x - \frac{\pi}{6})$ b. $y = \cos(x + \frac{\pi}{4})$ c. $y = \frac{1}{2}\tan\frac{1}{2}x$
   d. $y = \tan(x - \frac{2\pi}{3}) - 2$ e. $y = 2\cos\frac{2}{3}x$
6. Graph the given curves.
   a. $y = 2\cos(2x + \frac{\pi}{3})$ b. $y = 3\sin\frac{\pi x}{4}$
   c. $y + 3 = 3\tan\frac{1}{2}(x - \frac{\pi}{3})$ d. $y + 1 = \cos(x - \frac{\pi}{3})$
   e. $y = \frac{3}{2}\tan\frac{3}{2}x$
7. From memory draw a quick sketch of the following curves.
   a. $y = \sec x$ b. $y = \csc x$ c. $y = \cot x$
8. Graph the given curves.
   a. $y - 1 = \sec(\frac{1}{2}x - \frac{1}{4})$ b. $y + 2 = \cot(x - \frac{\pi}{3})$
   c. $y + 1 = \csc 2(x - \frac{\pi}{4})$ d. $y - \frac{3}{2} = \cot(x - 1)$
9. Graph the given curves by adding ordinates.
   a. $y = \frac{3x}{4} + \cos x$ b. $y = \sin 3x + 2\cos x$

10. Graph the given curves by multiplying ordinates.
    a. $y = 2x \sin x$ b. $y = \frac{2x}{3} \cos 2x$

Miscellaneous Problems

Graph each of the curves given in Problems 1–30.

1. $y = 5 \cos x$
2. $y = 5 \sin x$
3. $y = 5 \tan x$
4. $y = \sin 5x$
5. $y = \tan 5x$
6. $y = \cos 5x$
7. $y = \tan 5x - 3$
8. $y = \cos 5x - 1$
9. $y = \sin 5x + 1$
10. $y = 5 \cos x - 3$
11. $y = 5 \sin x - 3$
12. $y = 5 \tan x - 3$
13. $y = 5 \sin(x - 3)$
14. $y = 5 \tan \frac{1}{2}(x - 3)$
15. $y = 5 \cos(x - 3)$
16. $y = \tan(5x - 3)$
17. $y = \cos(5x - 3)$
18. $y = \sin(5x - 3)$
19. $y - 2 = \cos(3x - 1)$
20. $y + 1 = \sin(2x + 1)$
21. $y + \frac{1}{2} = \tan(4x - 1)$
22. $y + 1 = \sin(2x - \pi)$
23. $y - \frac{\pi}{3} = \cos(3x + 2)$
24. $y - 2 = \tan(2x + \frac{\pi}{2})$
25. $y = 2 \sec 3x$
26. $y = \sqrt{2} \csc 2x$
27. $y = \pi \cot x$
28. $y - 1 = \sec(x - \frac{\pi}{6})$
29. $y + 1 = \csc(x - \frac{\pi}{4})$
30. $y - 1 = \cot(x + \frac{\pi}{6})$

APPLICATION FOR FURTHER STUDY

COMPUTER GRAPHING OF TRIGONOMETRIC CURVES

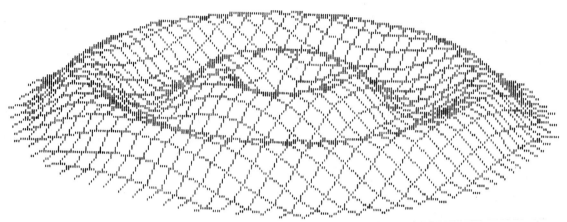

Graph of $z = \sin^2(\sqrt{x^2 + y^2})$. Courtesy of Carroll Wilde.

Getting Started

Begin by graphing the standard cosine, sine, and tangent curves. Most programs require parentheses following the function you are graphing; the input will look like this:

```
Y = COS(X);   Y = SIN(X);   Y = TAN(X)
```

Remember, a computer graphs a function simply by plotting at a very high speed a large number of points, giving the illusion that it is drawing a smooth curve. This effect is analogous to the well-known illusion that the individual frames of a moving picture give of motion.

Changes in Amplitude

Experiment 1

Graph $y = \sin x$. Without erasing this curve, graph each of the following curves:

a. $y = 3 \sin x$ **b.** $y = \frac{1}{2} \sin x$ **c.** $y = \pi \sin x$.

What is the effect of a numerical coefficient on a sine curve? What will $y = a \sin x$ look like?

Solution

a. $y = 3 \sin x$. Many programs require an asterisk for multiplication between the numerical coefficient and the trigonometric function, as shown here:

```
Y=3*SIN(X)
```

This graph is shown in Figure 2.34a on page 93.
b. $y = \frac{1}{2} \sin x$. *Input:* `Y=(1/2)*SIN(X)`
This graph is shown in Figure 2.34b.

92

c. $y = \pi \sin x$. *Input:* `Y=3.1416*SIN(X)`

Notice that an approximate value of π must be input. The graph is shown in Figure 2.34c.

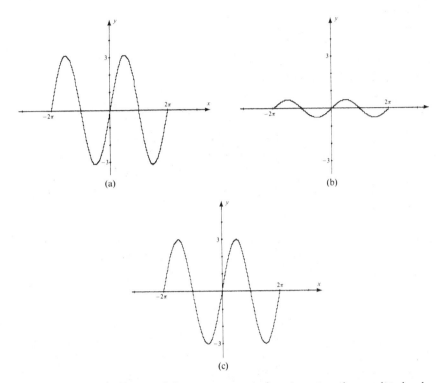

(a)

(b)

(c)

The numerical coefficient of the trigonometric function gives the amplitude; the graph of $y = a \sin x$ is a standard sine curve with an amplitude of $|a|$ rather than of 1. □

COMPUTER APPLICATIONS

There are many commercially available computer programs to help you do trigonometric graphing. One such package is called TRIGPAK by John Mowbray. It requires an IBM-compatible computer. This program can be used to graph complicated curves; it can also serve as an instructional aid in teaching you how to graph the curves in the first place. If you use TRIGPAK, you will need to make some menu choices before you start. When you start up TRIGPAK you will see a *Main Menu*. At the main menu you will be given several choices; select O (oh) and return. The main menu will change to an *Operations Menu*. Once again select choice O (oh) and you will be given several forms of trigonometric curves. Select choice 3; this is the choice that is keyed into the framing technique presented in this book. You will then be able to explore using the terminology used in this chapter.

Changes in Period

Graph $y = \cos x$. Without erasing this curve, graph

a. $y = \cos 2x$ **b.** $y = \cos \frac{1}{2}x$ **c.** $y = \cos \pi x$

What is the effect of a numerical coefficient on the angle of a cosine curve? What will $y = \cos bx$ look like?

Solution

a. $y = \cos 2x$. *Input:* Y=COS(2*X)
The graph is shown in Figure 2.35a.
b. $y = \cos \frac{1}{2}x$. *Input:* Y=COS((1/2)*X)
The graph is shown in Figure 2.35b.
c. $y = \cos \pi x$. *Input:* Y=COS(3.1416*X)
The graph is shown in Figure 2.35c.

FIGURE 2.35 Graphs showing changes in period

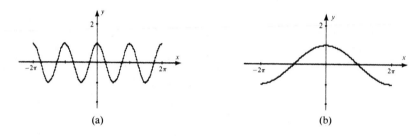

(a)   (b)

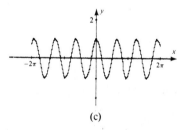

(c)

The numerical coefficient of the angle of a cosine curve changes the period of the cosine curve. The graph of $y = \cos bx$ is a standard cosine curve with a period of $\frac{2\pi}{b}$ instead of 2π. ▯

Phase Displacement

Graph $y = \tan x$. Without erasing this curve, graph

a. $y = \tan(x - \frac{\pi}{2})$ **b.** $y = \tan(x + \frac{\pi}{2})$
c. $y = \tan(x + 1)$ **d.** $y = \tan(x - 2)$

What is the effect of a number subtracted from the angle of a tangent curve? What will $y = \tan(x - h)$ look like?

Solution

a. $y = \tan(x - \frac{\pi}{2})$. *Input:* Y=TAN(X-3.1416/2)
   The graph is shown in Figure 2.36a.
b. $y = \tan(x + \frac{\pi}{2})$. *Input:* Y=TAN(X+3.1416/2)
   The graph is shown in Figure 2.36b.
c. $y = \tan(x + 1)$. *Input:* Y=TAN(X+1)
   The graph is shown in Figure 2.36c.
d. $y = \tan(x - 2)$. *Input:* Y=TAN(X-2)
   The graph is shown in Figure 2.36d.

FIGURE 2.36 Graphs showing changes in phase displacement

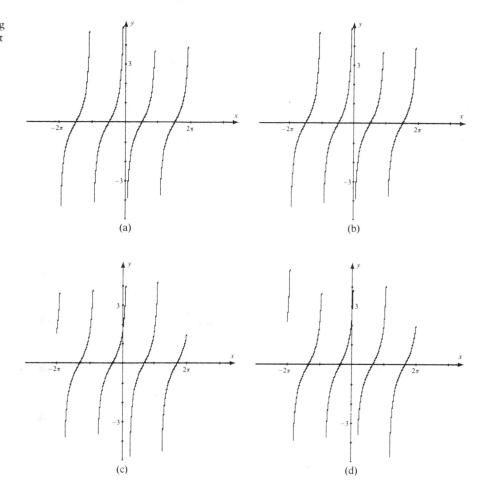

The number subtracted from the angle is the amount that the graph is displaced from the standard tangent curve. □

95

Translation

Experiment 4

Graph $y = \sin x$. Without erasing this curve, graph

a. $y + 1 = \sin(x - \frac{\pi}{2})$ **b.** $y - 1 = \sin(x - 2)$ **c.** $y - 2 = \sin(x + 1)$

What will $y - k = \sin(x - h)$ look like?

Solution

a. $y + 1 = \sin(x - \frac{\pi}{2})$. *Input:* `Y=SIN(X-3.1416/2)-1`
The graph is shown in Figure 2.37a.
b. $y - 1 = \sin(x - 2)$. *Input:* `Y=SIN(X-2)+1`
The graph is shown in Figure 2.37b.
c. $y - 2 = \sin(x + 1)$. *Input:* `Y=SIN(X+1)+2`
The graph is shown in Figure 2.37c.

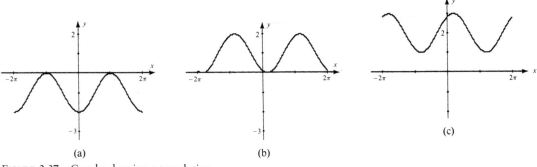

(a)   (b)   (c)

FIGURE 2.37   Graphs showing a translation

The curve $y - k = \sin(x - h)$ is a standard sine curve that has been translated to the point (h, k).   ⬚

Graphing General Trigonometric Curves

After using a computer to learn the shapes of the standard trigonometric curves and the effects of amplitude, period, displacement changes, and translations, you can then use it to help you graph some very complicated trigonometric curves.

Example 1

Graph $y = \dfrac{10 \sin x}{x}$.

Solution

Input: `Y=10*SIN(X)/X`
The resulting graph is shown in Figure 2.38.

FIGURE 2.38 Graph of
$$y = \frac{10 \sin x}{x}$$

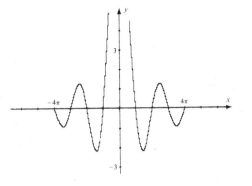

Example 2

Graph $y = \sin^2 x$, $y = \cos^2 x$, and $y = \sin^2 x + \cos^2 x$ on the same grid.

Solution

Input: `Y=(SIN(X))↑2,  Y=(COS(X))↑2,` and
`Y=(SIN(X))↑2+(COS(X))↑2`
Graph these curves as shown in Figure 2.39.

FIGURE 2.39

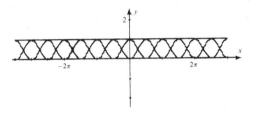

Example 3

Graph $y = \sin x$, $y = \sin x + \frac{\sin 3x}{3}$, and $y = \sin x + \frac{\sin 3x}{3} + \frac{\sin 5x}{5}$.

Solution

Input: `Y=SIN(X),  Y=SIN(X)+SIN(3X)/3,` and
`Y=SIN(X)+SIN(3X)/3+SIN(5X)/5`
Graph these curves as shown in Figure 2.40.

FIGURE 2.40

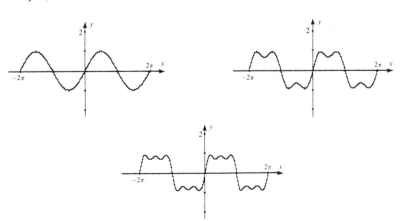

Joseph Fourier used combinations of the trigonometric functions, which today are used in advanced mathematics in the study of electrical circuits, sound, and heat flow, to mention just a few applications. Example 3 shows the first three terms of a *Fourier series*:

$$\sin x + \frac{\sin 3x}{3} + \frac{\sin 5x}{5} + \frac{\sin 7x}{7} + \cdots$$

The wave displayed on an oscilloscope for the sound of a tuning fork vibrating at f cycles per second with amplitude a is given by

$$y = a \sin 2\pi f x$$

Now, for a vibrating string (as in a piano) of length L, tension T, and mass m per unit length, the value for f is

$$f = \frac{1}{2L} \sqrt{\frac{T}{m}}$$

For example, the frequency of middle C on a properly tuned piano is 261.626. Using this value for f, and an amplitude of 8, we obtain the equation $y = 8 \sin 2\pi f x$. The first overtone for this note is $y = 4 \sin 4\pi f x$, producing the sound wave described by the equation

$$y = 8 \sin 2\pi f x + 4 \sin 4\pi f x$$

where $f = 261.626$. This sound wave can be graphed by inputting

```
Y=8*SIN(2*3.1416*261.626*X)+4*SIN(4*3.1416*261.626*X)
```

The graph is shown in Figure 2.41.

FIGURE 2.41   Graph of a sound wave

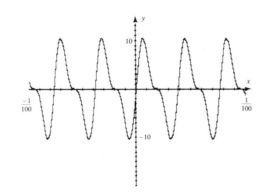

Problems for Further Study: Computer Graphing of Trigonometric Curves

1. Repeat Experiment 1 for $y = \cos x$.
2. Repeat Experiment 1 for $y = \tan x$.
3. Repeat Experiment 2 for $y = \sin x$.
4. Repeat Experiment 2 for $y = \tan x$.
5. Repeat Experiment 3 for $y = \sin x$.
6. Repeat Experiment 3 for $y = \cos x$.
7. Repeat Experiment 4 for $y = \tan x$.
8. Repeat Experiment 4 for $y = \cos x$.

Graph the curves in Problems 9–20 by using a computer. You can either use a printout of the graph or you can copy what is on the screen onto your own graph paper.

9. $y = \dfrac{10 \cos x}{x}$

10. $y = \dfrac{\tan x}{5x}$

11. $y = \dfrac{x}{\cos x}$

12. $y = \dfrac{x}{\sin x}$

13. $y = \dfrac{x}{\tan x}$

14. $y = \dfrac{3 \sin x}{\cos x}$

15. $y = 2 \sin 3x + 3 \sin 2x$

16. $y = \cos 9x + \sin 11x$

17. $y = 3 \sin x - 2 \cos \pi x$

18. $y = \sin x + \cos 2x - \sin 3x$

19. $y = \cot^2 x - \csc^2 x$

20. $y = \sec^2 x - \tan^2 x$

21. Graph the following curves on the same grid.

    a. $y = \cos x$ b. $y = \cos x + \dfrac{\cos 3x}{3}$ c. $y = \cos x + \dfrac{\cos 3x}{3} + \dfrac{\cos 5x}{5}$

22. Graph the following curves on the same grid.

    a. $y = \sin x$ b. $y = \sin x + \dfrac{\sin 2x}{2}$ c. $y = \sin x + \dfrac{\sin 2x}{2} + \dfrac{\sin 3x}{3}$

23. *Music.* Graph the sound wave for $f = 294$.
24. *Music.* Graph the sound wave for $f = 294$, along with its first overtone.
25. *Electrical.* The current I for a two-power-source circuit is

$$y = 2 \sin 100\pi t + 4 \cos 120\pi t$$

Graph this current from $-.2$ to $.2$.

CUMULATIVE REVIEW FOR CHAPTERS 1 AND 2

SAMPLE TEST 1: OBJECTIVE FORM

1. State the generalized definition of the trigonometric functions.

2. Fill in the blanks.

FUNCTION		COFUNCTION		RECIPROCAL
cosine	a. _____		b. _____	
tangent	c. _____		d. _____	
secant	e. _____		f. _____	
cosecant	g. _____		h. _____	
sine	i. _____		j. _____	

3. Using the definition of the trigonometric functions, find their values for an angle α whose terminal side passes through $(\sqrt{7}, 3)$.

4. The westernmost city in the continental United States is Ferndale, California, and it is located at 124°W longitude, 41°N latitude. How far (to the nearest 100 miles) is Ferndale from the equator if you assume that the earth's radius is approximately 3,950 miles?

5. Complete the following table of exact values.

angle in degrees	$-45°$	a.	b.	210°	90°
angle in radians	c.	π	$\frac{-4\pi}{3}$	d.	e.
sine	f.	g.	h.	i.	j.
cosine	k.	l.	m.	n.	o.
tangent	p.	q.	r.	s.	t.

6. From memory, draw a quick sketch of each of the following. Label the axes and scale.
   a. $y = \sin x$ b. $y = \cot x$ c. $y = \csc x$

7. Graph one period of $y = 2\cos 3x$.

8. Graph one period of $y = \frac{1}{2}\tan\frac{3}{2}x$.

9. Graph $y - 3 = \sin(x - \frac{\pi}{3})$.

10. Graph $y + 2 = \cos(3x + 2\pi)$.

SAMPLE TEST 2: FILL IN AND MULTIPLE CHOICE

Fill in the blanks in Problems 1–5 with the word or words that make the statements complete and correct.

1. Let θ be an angle in standard position with a point $P(x, y)$ on the terminal side a distance of r from the origin $(r \neq 0)$. Then

   $\sin \theta =$ _____ ;

   $\sec \theta =$ _____ ;

   $\cot \theta =$ _____ .

2. The angle with measure 2 is in Quadrant(s) _____ .

3. An angle rotated in a clockwise direction is _____ .

4. In Quadrant II, the _____ and _____ are negative.

5. The period of the standard sine curve is _____ .

*Choose the **best** answer from the choices given.*

6. The reciprocal of secant is
   A. cosecant   B. sine
   C. cosine   D. cotangent
   E. none of these

7. csc 15.8° correct to four decimal places is
   A. 3.6727   B. .9622
   C. 1.0393   D. -10.8806
   E. none of these

8. If θ is an angle in standard position and P_θ is a point on a unit circle on the terminal side of θ, then the

coordinates of θ are
A. impossible to find B. $(0, 1)$
C. $(\sin\theta, \cos\theta)$ D. $(\cos\theta, \sin\theta)$
E. none of these

9. To calculate $\dfrac{25}{\tan 38°}$ use the following calculator sequence

ALGEBRAIC LOGIC

A. $\boxed{25}\ \boxed{\div}\ \boxed{\tan}\ \boxed{38}\ \boxed{=}$

B. $\boxed{25}\ \boxed{\div}\ \boxed{\tan}\ \boxed{38}$

C. $\boxed{25}\ \boxed{\div}\ \boxed{\tan}\ \boxed{38}\ \boxed{=}$

D. $\boxed{25}\ \boxed{\times}\ \boxed{38}\ \boxed{\tan}\ \boxed{1/x}\ \boxed{=}$

E. none of these

or

RPN LOGIC

A. $\boxed{25}\ \boxed{ENTER}\ \boxed{\tan}\ \boxed{38}\ \boxed{\div}$

B. $\boxed{25}\ \boxed{\div}\ \boxed{\tan}\ \boxed{38}$

C. $\boxed{25}\ \boxed{ENTER}\ \boxed{\tan}\ \boxed{38}\ \boxed{=}$

D. $\boxed{25}\ \boxed{ENTER}\ \boxed{38}\ \boxed{\tan}\ \boxed{1/x}\ \boxed{\times}$

E. none of these

10. If $\theta = -3$, then its reference angle is
A. 3 B. -3
C. $\pi - 3$ D. $3 - \pi$
E. none of these

11. The length of an arc with a central angle of $40°$ and radius of 1 is
A. less than 1 B. equal to 1
C. between 1 and 40 D. equal to 40
E. none of these

12. The size of θ is about
A. $\frac{\pi}{6}$
B. $\frac{1}{2}$
C. -6
D. $-\frac{5\pi}{4}$
E. none of these

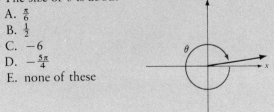

13. An angle with measure equal to $\frac{3\pi}{4}$ is
A. $45°$ B. $-45°$
C. $135°$ D. all are equal
E. none of these

14. $\cos\frac{5\pi}{6}$ is
A. $\frac{1}{2}$ B. $\frac{\sqrt{3}}{2}$
C. $\frac{\sqrt{2}}{2}$ D. $\frac{\sqrt{3}}{3}$
E. none of these

15. $\tan\left(-\frac{3\pi}{4}\right)$ is
A. $-\frac{\sqrt{3}}{3}$ B. -1
C. $-\sqrt{3}$ D. $\sqrt{3}$
E. none of these

16. The length of the frame for $y = 3\tan 2x$ is
A. $\frac{\pi}{2}$ B. π
C. $\frac{2\pi}{3}$ D. 2π
E. none of these

17. The graph of the cosecant curve passes through the point
A. $(0, 0)$ B. $(\frac{\pi}{2}, 1)$
C. $(\frac{3\pi}{2}, 1)$ D. all of the above
E. none of the above

18. The center of the frame for
$$y - 1 = 2\tan\left(x - \frac{\pi}{3}\right)\text{ is}$$
A. $(1, 2)$ B. $(1, \frac{\pi}{3})$
C. $(\frac{\pi}{3}, 1)$ D. $(\frac{\pi}{3}, 2)$
E. none of these

19. When graphing the curve
$$y + 1 = 3\sin\left(x + \frac{\pi}{3}\right)$$
the point (h, k) is
A. $(-1, \frac{\pi}{3})$ B. $(\frac{\pi}{3}, -1)$
C. $(0, 3)$ D. $(-\frac{\pi}{3}, 1)$
E. none of these

20. The top right-hand corner of the frame for
$$y = 2\tan\left(x + \frac{3\pi}{4}\right)\text{ is}$$
A. $(-\frac{\pi}{4}, 0)$ B. $(\frac{\pi}{4}, 0)$
C. $(\frac{\pi}{4}, 2)$ D. $(-\frac{\pi}{4}, 2)$
E. none of these

CHAPTER 3

IDENTITIES

CHAPTER OVERVIEW

This chapter focuses on building a new skill, one that will be needed in subsequent courses. It is called *proving identities*, but what it means is showing that a given open equation is true for all values in the domain. The first three sections of this chapter are designed to give you confidence in proving identities, and then the remaining sections develop those trigonometric identities you will need for your subsequent work in mathematics. The identities you need are listed on the inside back cover of the book for your easy reference. You should be familiar with these identities, but, at the same time, you should keep them in proper perspective. Do not become overwhelmed with the long list of derived identities; instead, approach each one as just a new identity to prove. After each is derived, you can add it to your list for future reference.

3.1 PROVING IDENTITIES

Fundamental Identities

In Chapter 1 we considered eight fundamental identities that are used to simplify and change the form of a great variety of trigonometric expressions. We will repeat them here for your easy reference.

FUNDAMENTAL
IDENTITIES

> **Reciprocal Identities**
>
> **1.** $\sec\theta = \dfrac{1}{\cos\theta}$ **2.** $\csc\theta = \dfrac{1}{\sin\theta}$ **3.** $\cot\theta = \dfrac{1}{\tan\theta}$
>
> **Ratio Identities**
>
> **4.** $\tan\theta = \dfrac{\sin\theta}{\cos\theta}$ **5.** $\cot\theta = \dfrac{\cos\theta}{\sin\theta}$
>
> **Pythagorean Identities**
>
> **6.** $\cos^2\theta + \sin^2\theta = 1$ **7.** $1 + \tan^2\theta = \sec^2\theta$ **8.** $\cot^2\theta + 1 = \csc^2\theta$

In addition to these fundamental identities, you will need to remember from arithmetic and algebra the proper procedure for simplifying expressions, especially the concepts of common denominators and complex fractions. You will combine this algebraic simplification with a knowledge of the fundamental identities, as shown by the next example.

Example 1

Simplify $\dfrac{\sin\theta + \dfrac{\cos^2\theta}{\sin\theta}}{\dfrac{\cos\theta}{\sin\theta}}$ and leave your answer in terms of sine and cosine functions only.

Solution

$$\frac{\sin\theta + \dfrac{\cos^2\theta}{\sin\theta}}{\dfrac{\cos\theta}{\sin\theta}} = \frac{\dfrac{\sin^2\theta + \cos^2\theta}{\sin\theta}}{\dfrac{\cos\theta}{\sin\theta}} \qquad \textit{Common denominator}$$

$$= \frac{\sin^2\theta + \cos^2\theta}{\sin\theta} \cdot \frac{\sin\theta}{\cos\theta} \qquad \textit{Dividing rational expressions}$$

$$= \frac{1}{\cos\theta} \qquad \textit{Substitution}$$

Procedure for Proving Trigonometric Equations Are Identities

In this section you will learn how to prove that a given expression is an identity. Remember, an identity is an open equation that is true for all replacements for the variable that are in the domain. The procedure for proving identities is not the same as the procedure for solving an equation. Recall that in Chapter 1 we wrote

$$\sec\theta = \frac{1}{\cos\theta} \quad \text{as} \quad \sec\theta\cos\theta = 1$$

by multiplying both sides by $\cos\theta$. We could do this because *we had previously proved that it was an identity*. If you are trying to prove that an equation is an identity you CANNOT perform operations on both sides of the equation because you do not know both sides are equal (indeed, that is what you are proving).

Suppose you are given a trigonometric equation such as

$$\tan\theta + \cot\theta = \sec\theta\csc\theta$$

and are asked to show that it is an identity. This equation has a left side, L, and a right side, R, and you must prove that $L = R$. In order to do this, there are three ways of proceeding:

1. Reduce the left-hand side to the right-hand side by using algebra and the fundamental identities.
2. Reduce the right-hand side to the left-hand side.
3. Reduce both sides **independently** to the same expression.

When you are asked to prove an identity, we mean prove that the given equation is an identity. We will illustrate these techniques with several examples.

Example 2

Prove that $\tan\theta + \cot\theta = \sec\theta\csc\theta$ is an identity.

Solution

Method I: Reduce the left-hand side to the right-hand side.

$$\tan\theta + \cot\theta = \frac{\sin\theta}{\cos\theta} + \frac{\cos\theta}{\sin\theta}$$

$$= \frac{\sin^2\theta}{\cos\theta\sin\theta} + \frac{\cos^2\theta}{\cos\theta\sin\theta}$$

$$= \frac{\sin^2\theta + \cos^2\theta}{\cos\theta\sin\theta}$$

$$= \frac{1}{\cos\theta\sin\theta}$$

$$= \frac{1}{\cos\theta} \cdot \frac{1}{\sin\theta}$$

$$= \sec\theta\csc\theta$$

In proving identities, it is often advantageous to change the trigonometric functions to cosines and sines.

When you obtain an expression that is the same as the other side of the equation, you are finished with the proof. Thus, $\tan\theta + \cot\theta = \sec\theta\csc\theta$.

Method II: Reduce the right-hand side to the left-hand side.

$$\sec\theta\csc\theta = \frac{1}{\cos\theta} \cdot \frac{1}{\sin\theta}$$

$$= \frac{1}{\cos\theta\sin\theta}$$

$$= \frac{\sin^2\theta + \cos^2\theta}{\cos\theta\sin\theta}$$

$$= \frac{\sin^2\theta}{\cos\theta\sin\theta} + \frac{\cos^2\theta}{\cos\theta\sin\theta}$$

$$= \frac{\sin\theta}{\cos\theta} + \frac{\cos\theta}{\sin\theta}$$

$$= \tan\theta + \cot\theta$$

You would, of course, choose only one of these three methods.

Thus, $\tan\theta + \cot\theta = \sec\theta\csc\theta$.

Method III: Reduce both sides to a form that is the same. Be sure to work with each side separately; do not treat this as an equation and add expressions to both sides or multiply both sides by the same expression.

$$\tan\theta + \cot\theta = \frac{\sin\theta}{\cos\theta} + \frac{\cos\theta}{\sin\theta} \qquad\qquad \sec\theta\csc\theta = \frac{1}{\cos\theta} \cdot \frac{1}{\sin\theta}$$

$$= \frac{\sin^2\theta + \cos^2\theta}{\cos\theta\sin\theta} \qquad\qquad\qquad = \frac{1}{\cos\theta\sin\theta}$$

$$= \frac{1}{\cos\theta\sin\theta}$$

These are identical.

Thus, $\tan\theta + \cos\theta = \sec\theta\csc\theta$.

Of course, we would pick only one of these methods when proving a particular identity. Usually it is easier to begin with the more complicated side and try to reduce it to the simpler side. If both sides seem equally complicated, you might change all the functions to sines and cosines and then simplify.

Example 3

Prove that $\dfrac{1}{1 + \cos \theta} + \dfrac{1}{1 - \cos \theta} = 2 \csc^2 \theta$ is an identity.

Solution

$$\frac{1}{1 + \cos \theta} + \frac{1}{1 - \cos \theta} = \frac{(1 - \cos \theta) + (1 + \cos \theta)}{(1 + \cos \theta)(1 - \cos \theta)} \quad \textit{Begin with the more}$$
$$\textit{complicated side.}$$

$$= \frac{2}{1 - \cos^2 \theta}$$

$$= \frac{2}{\sin^2 \theta}$$

$$= 2 \csc^2 \theta \qquad \qquad \square$$

Example 4

Prove that $\sec 4\theta + \cos 4\theta = 2 \sec 4\theta - \tan 4\theta \sin 4\theta$ is an identity.

Solution

$$\sec 4\theta + \cos 4\theta = \frac{1}{\cos 4\theta} + \cos 4\theta \qquad \textit{When both sides seem equally}$$
$$\textit{complicated, a good procedure to}$$
$$= \frac{1 + \cos^2 4\theta}{\cos 4\theta} \qquad \qquad \textit{follow is to change everything to}$$
$$\textit{cosines and sines and then}$$
$$\textit{simplify.}$$

$$= \frac{1 + (1 - \sin^2 4\theta)}{\cos 4\theta}$$

$$= \frac{2 - \sin^2 4\theta}{\cos 4\theta} \qquad \qquad \textit{Notice that the angle here is } 4\theta.$$
$$\textit{The identity is proved in terms}$$
$$= \frac{2}{\cos 4\theta} - \frac{\sin^2 4\theta}{\cos 4\theta} \qquad \textit{of } 4\theta.$$

$$= 2 \cdot \frac{1}{\cos 4\theta} - \frac{\sin 4\theta}{\cos 4\theta} \cdot \frac{\sin 4\theta}{1}$$

$$= 2 \sec 4\theta - \tan 4\theta \sin 4\theta \qquad \qquad \square$$

Example 5

Prove that $\dfrac{\sec 2\lambda + \cot 2\lambda}{\sec 2\lambda} = 1 + \csc 2\lambda - \sin 2\lambda$ is an identity.

Solution

Begin with the left-hand side. When working with a fraction consisting of a single term as a denominator, it is often helpful to separate the fraction into the sum of several fractions.

$$\frac{\sec 2\lambda + \cot 2\lambda}{\sec 2\lambda} = \frac{\sec 2\lambda}{\sec 2\lambda} + \frac{\cot 2\lambda}{\sec 2\lambda}$$

$$= 1 + \cot 2\lambda \cdot \frac{1}{\sec 2\lambda}$$

$$= 1 + \frac{\cos 2\lambda}{\sin 2\lambda} \cdot \cos 2\lambda$$

$$= 1 + \frac{\cos^2 2\lambda}{\sin 2\lambda}$$

$$= 1 + \frac{1 - \sin^2 2\lambda}{\sin 2\lambda}$$

$$= 1 + \frac{1}{\sin 2\lambda} - \frac{\sin^2 2\lambda}{\sin 2\lambda}$$

$$= 1 + \csc 2\lambda - \sin 2\lambda$$

We will consider some additional simplifications of trigonometric identities in the next section.

Problem Set 3.1

A   *Simplify each of the expressions in Problems 1–6 using only cosines, sines, and the fundamental identities.*

1. $\dfrac{1 - \cos^2\theta}{\sin\theta}$

2. $\dfrac{1 - \sin^2\theta}{\cos\theta}$

3. $\dfrac{\dfrac{\sin\theta}{\cos\theta} + \dfrac{\cos\theta}{\sin\theta}}{\dfrac{1}{\sin\theta\cos\theta}}$

4. $\sin\theta + \dfrac{\cos^2\theta}{\sin\theta}$

5. $\dfrac{\cos\theta + \dfrac{\sin^2\theta}{\cos\theta}}{\sin\theta}$

6. $\dfrac{\dfrac{\cos^4\theta}{\sin^2\theta} + \cos^2\theta}{\dfrac{\cos^2\theta}{\sin^2\theta}}$

Reduce each of the expressions in Problems 7–12 to involve only sines and cosines, and then simplify.

7. $\sin\theta + \cot\theta$

8. $\sec\theta + \tan\theta$

9. $\dfrac{\tan\theta + \cot\theta}{\sec\theta\csc\theta}$

10. $\dfrac{\sec\theta + \csc\theta}{\tan\theta\cot\theta}$

11. $\sec^2\theta + \tan^2\theta$

12. $\csc^2\theta + \cot^2\theta$

Prove that the equations in Problems 13–24 are identities.

13. $\tan \theta = \sin \theta \sec \theta$

14. $\cot \theta = \cos \theta \csc \theta$

15. $\sec \theta = \dfrac{\cos \theta}{1 - \sin^2\theta}$

16. $\csc \theta = \dfrac{\sin \theta}{1 - \cos^2\theta}$

17. $\tan 2\theta \csc 2\theta = \sec 2\theta$

18. $\cos 2\theta = \sin 2\theta \cot 2\theta$

19. $\tan \theta \cos \theta = \sin \theta$

20. $\cot \theta \sin \theta = \cos \theta$

21. $\dfrac{\sin \theta \csc \theta}{\tan \theta} = \cot \theta$

22. $\dfrac{\cot \theta \tan \theta}{\sin \theta} = \csc \theta$

23. $\dfrac{\cot t}{\sin^2 t + \cos^2 t} = \dfrac{1}{\tan t}$

24. $\dfrac{\sec^2 t}{1 + \tan^2 t} = 1$

B   *Prove that the equations in Problems 25–31 are identities.*

25. $\csc \theta + \sin \theta = 2 \csc \theta - \cot \theta \cos \theta$

26. $\csc \theta - \sin \theta = \cos^2\theta \csc \theta$

27. $(\csc 2\theta - \sin 2\theta)(\csc 2\theta + \sin 2\theta) = \csc^2 2\theta (1 - \sin^4 2\theta)$

28. $\dfrac{\tan 2\theta + \cot 2\theta}{\sec 2\theta} = \csc 2\theta$

29. $\dfrac{\tan 3\theta + \cot 3\theta}{\csc 3\theta} = \sec 3\theta$

30. $\dfrac{\sec \lambda + \tan^2\lambda}{\sec \lambda} = 1 + \sec \lambda - \cos \lambda$

31. $\dfrac{\sin 2\lambda}{\tan 2\lambda} + \dfrac{\cos 2\lambda}{\cot 2\lambda} = \cos 2\lambda + \sin 2\lambda$

C   *Prove that the equations in Problems 32–41 are identities.*

32. $(\tan 2\beta - \cot 2\beta)^2 = \tan^2 2\beta (2 - \csc^2 2\beta)^2$

33. $(1 - \sin^2\alpha)(1 + \tan^2\alpha) = 1$

34. $(1 - \cos^2 2\alpha)(1 + \cot^2 2\alpha) = 1$

35. $\dfrac{\cos^2\gamma + \tan^2\gamma - 1}{\sin^2\gamma} = \tan^2\gamma$

36. $\cot^2\gamma = \dfrac{\sin^2\gamma + \cot^2\gamma - 1}{\cos^2\gamma}$

37. $\dfrac{\sec^2\gamma + \tan^2\gamma - 1}{\csc^2\gamma} = 2 \tan^2\gamma - 2 \sin^2\gamma$

38. $(\cos \alpha - \cos \beta)^2 + (\sin \alpha - \sin \beta)^2 = 2 - 2(\cos \alpha \cos \beta + \sin \alpha \sin \beta)$

39. $(\sec \alpha + \sec \beta)^2 - (\tan \alpha - \tan \beta)^2 = 2 + 2(\sec \alpha \sec \beta + \tan \alpha \tan \beta)$

40. $\tan A + \cot B = (\sin A \sin B + \cos A \cos B) \sec A \csc B$

41. $\sec A + \csc B = (\cos A \sin^2 B + \sin B \cos^2 A) \sec^2 A \csc^2 B$

3.2   PROVING IDENTITIES (CONTINUED)

There are many "tricks of the trade" that can be used in proving identities. Some of these "tricks" are developed in the following examples. Since $a + b$ is the **conjugate** of $a - b$, the procedure illustrated in Example 1 is sometimes called **multiplying by the conjugate.**

Example 1   Prove that $\dfrac{\cos \theta}{1 - \sin \theta} = \dfrac{1 + \sin \theta}{\cos \theta}$ is an identity.

Solution

Sometimes, when there is a binomial in the numerator or denominator, the identity can be proved by multiplying the numerator and denominator by the conjugate of the binomial. That is, multiply both the numerator and the denominator of the left side by $1 + \sin\theta$. (This does not change the value of the expression because you are multiplying by 1.)

$$\frac{\cos\theta}{1-\sin\theta} = \frac{\cos\theta}{1-\sin\theta}\cdot\frac{1+\sin\theta}{1+\sin\theta}$$

$$= \frac{\cos\theta(1+\sin\theta)}{1-\sin^2\theta}$$

$$= \frac{\cos\theta(1+\sin\theta)}{\cos^2\theta}$$

$$= \frac{1+\sin\theta}{\cos\theta}$$

You could also have proved this identity by multiplying the numerator and denominator of the right-hand side by $(1-\sin\theta)$. ◻

Example 2

Prove that $\dfrac{\sec^2 2\theta - \tan^2 2\theta}{\tan 2\theta + \sec 2\theta} = \dfrac{\cos 2\theta}{1+\sin 2\theta}$ is an identity.

Solution

Sometimes the identity can be proved by factoring and reducing.

$$\frac{\sec^2 2\theta - \tan^2 2\theta}{\sec 2\theta + \tan 2\theta} = \frac{(\sec 2\theta + \tan 2\theta)(\sec 2\theta - \tan 2\theta)}{\sec 2\theta + \tan 2\theta}$$

Notice that the angle here is 2θ. The identity is proved in terms of 2θ.

$$= \sec 2\theta - \tan 2\theta$$

$$= \frac{1}{\cos 2\theta} - \frac{\sin 2\theta}{\cos 2\theta}$$

$$= \frac{1-\sin 2\theta}{\cos 2\theta}$$

$$= \frac{1-\sin 2\theta}{\cos 2\theta}\cdot\frac{1+\sin 2\theta}{1+\sin 2\theta}$$

$$= \frac{1-\sin^2 2\theta}{\cos 2\theta(1+\sin 2\theta)}$$

$$= \frac{\cos^2 2\theta}{\cos 2\theta(1+\sin 2\theta)}$$

$$= \frac{\cos 2\theta}{1+\sin 2\theta}$$ ◻

Example 3

Prove that $\dfrac{-2\sin\theta\cos\theta}{1-\sin\theta-\cos\theta} = 1+\sin\theta+\cos\theta$ is an identity.

Solution

Sometimes, when there is a fraction on one side, the identity can be proved by multiplying the other side by 1, where 1 is written so that the desired denominator is obtained. Thus, for this example,

$$1+\sin\theta+\cos\theta = (1+\sin\theta+\cos\theta)\cdot\frac{1-\sin\theta-\cos\theta}{1-\sin\theta-\cos\theta}$$

$$= \frac{(1+\sin\theta+\cos\theta)(1-\sin\theta-\cos\theta)}{1-\sin\theta-\cos\theta}$$

$$= \frac{1-\sin\theta-\cos\theta+\sin\theta-\sin^2\theta-\sin\theta\cos\theta+\cos\theta-\cos\theta\sin\theta-\cos^2\theta}{1-\sin\theta-\cos\theta}$$

$$= \frac{1-(\sin^2\theta+\cos^2\theta)-2\sin\theta\cos\theta}{1-\sin\theta-\cos\theta} = \frac{-2\sin\theta\cos\theta}{1-\sin\theta-\cos\theta} \qquad \square$$

In summary, there is no one best way to proceed in proving identities. However, the following hints should help.

HINTS FOR PROVING
TRIGONOMETRIC
IDENTITIES

Tricks of the Trade

1. If one side contains one function only, write all the trigonometric functions on the other side in terms of that function.
2. If the denominator of a fraction consists of only one term, break up the fraction into the sum of several fractions.
3. Simplify by combining fractions.
4. Factoring is sometimes helpful.
5. Change all trigonometric functions to sines and cosines and simplify.
6. Multiply numerator and denominator by the conjugate of either the numerator or the denominator.
7. If there are squares of functions, look for alternate forms of the Pythagorean identities.
8. Avoid the introduction of radicals.

Problem Set 3.2

A   *Prove that the equations in Problems 1–16 are identities.*

1. $\sin\theta = \sin^3\theta + \cos^2\theta\sin\theta$

2. $\sec\theta = \sec\theta\sin^2\theta + \cos\theta$

3. $\tan\theta = \cot\theta\tan^2\theta$

4. $\dfrac{\sin\theta\cos\theta + \sin^2\theta}{\sin\theta} = \cos\theta + \sin\theta$

5. $\tan^2\theta - \sin^2\theta = \tan^2\theta \sin^2\theta$

6. $\cot^2\theta \cos^2\theta = \cot^2\theta - \cos^2\theta$

7. $\tan A + \cot A = \sec A \csc A$

8. $\cot A = \csc A \sec A - \tan A$

9. $\sin x + \cos x = \dfrac{\sec x + \csc x}{\csc x \sec x}$

10. $\dfrac{\cos\gamma + \tan\gamma \sin\gamma}{\sec\gamma} = 1$

11. $\dfrac{1 - \sec^2 t}{\sec^2 t} = -\sin^2 t$

12. $\dfrac{1 + \cot^2 t}{\cot^2 t} = \sec^2 t$

13. $(\sec\theta - \cos\theta)^2 = \tan^2\theta - \sin^2\theta$

14. $\dfrac{\sin\theta}{\csc\theta} + \dfrac{\cos\theta}{\sec\theta} = 1$

15. $1 - \sin 2\theta = \dfrac{1 - \sin^2 2\theta}{1 + \sin 2\theta}$

16. $\dfrac{1 - \tan^2 3\theta}{1 - \tan 3\theta} = 1 + \tan 3\theta$

B *Prove that the equations in Problems 17–31 are identities.*

17. $\csc 3\beta - \cos 3\beta \cot 3\beta = \sin 3\beta$

18. $\dfrac{1 + \cot^2 A}{1 + \tan^2 A} = \cot^2 A$

19. $\dfrac{\sin^2\beta - \cos^2\beta}{\sin\beta + \cos\beta} = \sin\beta - \cos\beta$

20. $\dfrac{\tan^2\gamma - \cot^2\gamma}{\tan\gamma + \cot\gamma} = \tan\gamma - \cot\gamma$

21. $\tan^2 2\gamma + \sin^2 2\gamma + \cos^2 2\gamma = \sec^2 2\gamma$

22. $\cot^2 C + \cos^2 C + \sin^2 C = \csc^2 C$

23. $\dfrac{\tan\theta + \cot\theta}{\sec\theta \csc\theta} = 1$

24. $\dfrac{\tan\theta - \cot\theta}{\sec\theta \csc\theta} = \sin^2\theta - \cos^2\theta$

25. $\dfrac{1}{\sin\theta + \cos\theta} + \dfrac{1}{\sin\theta - \cos\theta} = \dfrac{\sin\theta}{\sin^2\theta - \frac{1}{2}}$

26. $\dfrac{1}{\sec\theta + \tan\theta} + \dfrac{1}{\sec\theta - \tan\theta} = 2\sec\theta$

27. $\dfrac{1 + \tan B}{1 - \tan B} = \dfrac{\sec^2 B + 2\tan B}{2 - \sec^2 B}$

28. $(\cot x + \csc x)^2 = \dfrac{\sec x + 1}{\sec x - 1}$

29. $\dfrac{\sin^3 x - \cos^3 x}{\sin x - \cos x} = 1 + \sin x \cos x$

30. $\dfrac{\tan^3 t - \cot^3 t}{\tan t - \cot t} = \sec^2 t + \cot^2 t$

31. $\sqrt{(3\cos\theta - 4\sin\theta)^2 + (3\sin\theta + 4\cos\theta)^2} = 5$

C *Prove that the equations in Problems 32–37 are identities.*

32. $\sin\theta + \cos\theta + 1 = \dfrac{2\sin\theta \cos\theta}{\sin\theta + \cos\theta - 1}$

33. $\dfrac{2\tan^2\theta + 2\tan\theta \sec\theta}{\tan\theta + \sec\theta - 1} = \tan\theta + \sec\theta + 1$

34. $\dfrac{\csc\theta + 1}{\csc\theta - 1} - \dfrac{\sec\theta - \tan\theta}{\sec\theta + \tan\theta} = 4\tan\theta \sec\theta$

35. $\dfrac{\cos\theta + \sin\theta}{\cos\theta - \sin\theta} + \dfrac{\cot\theta - 1}{\cot\theta + 1} = \dfrac{-2}{\sin^2\theta - \cos^2\theta}$

36. $\dfrac{\cos\theta + 1}{\cos\theta - 1} + \dfrac{1 - \sec\theta}{1 + \sec\theta} = -2\cot^2\theta - 2\csc^2\theta$

37. $\dfrac{\sin\theta}{1 - \cos\theta} + \dfrac{\cos\theta}{1 - \sin\theta} = (1 + \sin\theta + \cos\theta)(\sec\theta \csc\theta)$

*3.3   DISPROVING IDENTITIES

As you worked the problems in the last two sections, you may have come upon an equation that you could not prove to be an identity and in desperation said "This can't be done!" There is a very simple way to prove that a given equation is not an identity. If you can find *one replacement* of the variable for which the functions are defined that will make the equation false, then you have proved that the equation is not an identity. When you prove that an equation is not an identity, we say you have *disproved an identity*.

Example 1

A common mistake for trigonometry students is to write $\sin 2\theta$ as $2 \sin \theta$. Show that

$$\sin 2\theta = 2 \sin \theta$$

is not an identity.

Solution

Let $\theta = 30°$.

LEFT SIDE:

RIGHT SIDE:

$$\sin 2\theta = \sin 2(30°)$$

$$2 \sin 30° = 2\left(\frac{1}{2}\right)$$

$$= \sin 60°$$

$$= 1$$

$$= \frac{\sqrt{3}}{2}$$

Thus, $\sin 2\theta = 2 \sin \theta$ is not an identity; it is disproved by a counterexample.   □

Example 2

Is $\cos^2 x - \sin^2 x = \sin x$ an identity?

Solution

Begin by assuming it is an identity and try to prove it as shown in the last section.

$$\cos^2 x - \sin^2 x = (1 - \sin^2 x) - \sin^2 x$$

$$= 1 - 2 \sin^2 x$$

If you suspect that it might not be an identity, you can try to prove that it is not an identity by counterexample. Choose some value for the variable, say $x = 30°$.

LEFT SIDE:

RIGHT SIDE:

$$\cos^2 30° - \sin^2 30° = \left(\frac{\sqrt{3}}{2}\right)^2 - \left(\frac{1}{2}\right)^2$$

$$\sin 30° = \frac{1}{2}$$

$$= \frac{3}{4} - \frac{1}{4} = \frac{1}{2}$$

---

* Optional section.

For $x = 30°$, $\cos^2 x - \sin^2 x = \sin x$, so you have neither proved nor disproved that the equation is an identity. Choose some other value for the variable, say $x = 60°$.

LEFT SIDE: RIGHT SIDE:

$$\cos^2 60° - \sin^2 60° = \left(\frac{1}{2}\right)^2 - \left(\frac{\sqrt{3}}{2}\right)^2 \qquad \sin 60° = \frac{\sqrt{3}}{2}$$

$$= \frac{1}{4} - \frac{3}{4}$$

$$= -\frac{1}{2}$$

For $x = 60°$, $\cos^2 x - \sin^2 x \neq \sin x$, so the statement is not an identity; it is disproved by a counterexample. □

As you can see from Example 2, if you happen to pick a value that satisfies the equation, you have proved nothing. For this reason, it is often worthwhile to use an unusual value when checking for a counterexample, which is easy if you use a calculator. For Example 2, suppose we pick $x = 17°$. Evaluate both the left and right sides of the equation using a calculator.

For the left side, evaluate $\cos^2 17° - \sin^2 17°$ by calculator:

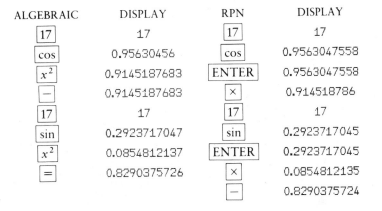

ALGEBRAIC	DISPLAY	RPN	DISPLAY
17	17	17	17
cos	0.95630456	cos	0.9563047558
x^2	0.9145187683	ENTER	0.9563047558
−	0.9145187683	×	0.914518786
17	17	17	17
sin	0.2923717047	sin	0.2923717045
x^2	0.0854812137	ENTER	0.2923717045
=	0.8290375726	×	0.0854812135
		−	0.8290375724

For the right side, $\sin 17° = 0.2923717047$

Therefore, $\cos^2 x - \sin^2 x \neq \sin x$.

"The exception proves the rule" is a saying that is, of course, completely false. In fact, *any* exception disproves the rule, and this is what counterexamples are all about. However, you should not pick values for counterexamples that cause any of the functions to be undefined. For example, when working with a tangent function, do not pick the angle equal to $\frac{\pi}{2}$ or to $\frac{3\pi}{2}$. Even the fundamental identities do not have meaning for undefined values.

Problem Set 3.3

A

Disprove each identity in Problems 1–10 by finding a counterexample. Be sure you do not choose a value that makes any of the functions undefined.

1. $\sec^2 x - 1 = \sqrt{3}\,\tan x$

2. $\sec^2 x - 1 = \tan x$

3. $2\cos 2x \sin 2x = \sin 2x$

4. $2\cos 2x \sin 2x = \cos 2x$

5. $\sin x = \cos x$

6. $\tan x = \cot x$

7. $\cos^2 x - 3\sin x + 3 = 0$

8. $\sin^2 x + \cos x = 0$

9. $\tan^2 x = \sec x$

10. $2\sin^2 x - 2\cos^2 x = 1$

B

Prove or disprove that each of the equations in Problems 11–20 is an identity.

11. $\sec^2\theta\cos\theta + 2\tan\theta\cos\theta = \sec^2\theta\cos\theta + 2\sin\theta$

12. $\sin^2\theta\cos\theta + 2\tan\theta\sin\theta = \sin^2\theta\cos\theta + 2\cos\theta$

13. $\sin\theta = \cot\theta\sec\theta - \cot\theta\cos\theta$

14. $\cos\lambda = \tan\lambda\csc\lambda - \tan\lambda\sin\lambda$

15. $\tan\lambda + \cot\lambda = \sec\lambda\csc\lambda$

16. $\dfrac{\sin\alpha}{\tan\alpha} + \dfrac{\cos\alpha}{\cot\alpha} = \cos\alpha + \sin\alpha$

17. $\dfrac{\cot\alpha + \sin^2\alpha\sec\alpha}{\sin\alpha} = \sec\alpha\sin\alpha$

18. $\dfrac{\tan\beta + \cot\beta}{\sec\beta\csc\beta} = 1$

19. $\sin\beta + \cot\beta\csc\beta = \sec\beta$

20. $\dfrac{\cos^4\gamma + \cos^2\gamma\sin^2\gamma}{\sin^2\gamma\cot^2\gamma} = \cos^2\gamma$

C

Prove or disprove each of the identities in Problems 21–31.

21. $\dfrac{1 + \tan\gamma}{1 - \tan\gamma} = \dfrac{\sec^2\gamma\cos\gamma + 2\sin\gamma}{2\cos\gamma - \sec\gamma}$

22. $\csc\theta + \cot\theta = \dfrac{1}{\csc\theta - \cot\theta}$

23. $\sec^2\lambda - \csc^2\lambda = (2\sin^2\lambda - 1)(\sec^2\lambda + \csc^2\lambda)$

24. $2\csc A = 2\csc A - \cot A\cos A + \cos^2 A\csc A$

25. $\sec^2 2\lambda + \csc^2 2\lambda = \csc^2 2\lambda\sec^2 2\lambda$

26. $\dfrac{\tan\theta}{\cot\theta} - \dfrac{\cot\theta}{\tan\theta} = \sec^2\theta - \csc^2\theta$

27. $\dfrac{\cos^4\theta - \sin^4\theta}{(\cos^2\theta - \sin^2\theta)^2} = \dfrac{\cos\theta}{\cos\theta + \sin\theta} + \dfrac{\sin\theta}{\cos\theta - \sin\theta}$

28. $\dfrac{\cos^2\theta - \cos\theta\csc\theta}{\cos^2\theta\csc\theta - \cos\theta\csc^2\theta} = \sin\theta$

29. $\dfrac{\tan^2\theta - 2\tan\theta}{2\tan\theta - 4} = \dfrac{1}{2}\tan\theta$

30. $\dfrac{\cos\theta + \cos^2\theta}{\cos\theta + 1} = \dfrac{\cos\theta\sin\theta + \cos^2\theta}{\sin\theta + \cos\theta}$

31. $\dfrac{\csc\theta + 1}{\cot^2\theta + \csc\theta + 1} = \dfrac{\sin^2\theta + \sin\theta\cos\theta}{\sin\theta + \cos\theta}$

3.4   COFUNCTION AND OPPOSITE-ANGLE IDENTITIES

When proving identities, it is sometimes necessary to simplify the functional value of the sum or difference of two angles. That is, if α and β represent any two angles, we know that in general

$$\cos(\alpha - \beta) \neq \cos\alpha - \cos\beta$$

For example, if $\alpha = 60°$ and $\beta = 30°$,

$$\cos(60° - 30°) = \cos 30° \quad \text{while} \quad \cos 60° - \cos 30° = \frac{1}{2} - \frac{\sqrt{3}}{2}$$

$$= \frac{\sqrt{3}}{2} \leftarrow \textit{These are not the same} \rightarrow = \frac{1 - \sqrt{3}}{2}$$

Thus,

$$\cos(60° - 30°) \neq \cos 60° - \sin 30°$$

In fact, we will now show that

$$\cos(\alpha - \beta) = \cos\alpha\cos\beta + \sin\alpha\sin\beta$$

This result not only will enable us to find the cosine of the difference of two angles, but also will provide the cornerstone upon which we will build a great many additional identities.

Distance Formula

To prove $\cos(\alpha - \beta) = \cos\alpha\cos\beta + \sin\alpha\sin\beta$ we will need to use the **distance formula** from algebra.

DISTANCE FORMULA

> If $P_1(x_1, y_1)$ and $P_2(x_2, y_2)$ are any two points, then the distance d between P_1 and P_2 is
>
> $$d = \sqrt{(x_2 - x_1)^2 + (y_2 - y_1)^2}$$

The proof of this formula follows directly from the Pythagorean theorem and is found in most algebra books.

Example 1   Find the distance between $(1, -2)$ and $(-3, 4)$.

Solution
$$d = \sqrt{(-3 - 1)^2 + (4 + 2)^2}$$
$$= \sqrt{16 + 36}$$
$$= \sqrt{52}$$
$$= 2\sqrt{13}$$

Example 2   Find the distance between $(\cos \theta, \sin \theta)$ and $(1, 0)$.

Solution

$$d = \sqrt{(1 - \cos \theta)^2 + (0 - \sin \theta)^2}$$
$$= \sqrt{1 - 2\cos \theta + \cos^2 \theta + \sin^2 \theta}$$
$$= \sqrt{1 - 2\cos \theta + 1}$$
$$= \sqrt{2 - 2\cos \theta}$$

□

Chord Length

Example 2 is used to find the length of any chord (in a unit circle) whose corresponding arc subtends or is intercepted by a central angle θ, where θ is in standard position. Let A be the point $(1, 0)$ and P the point on the intersection of the terminal side of angle θ and the unit circle (see Figure 3.1). This means (from the definition of cosine and sine) that the coordinates of P are $(\cos \theta, \sin \theta)$. The length of the chord AP, denoted by $|AP|$, is the number found in Example 2.

$P(\cos \theta, \sin \theta)$

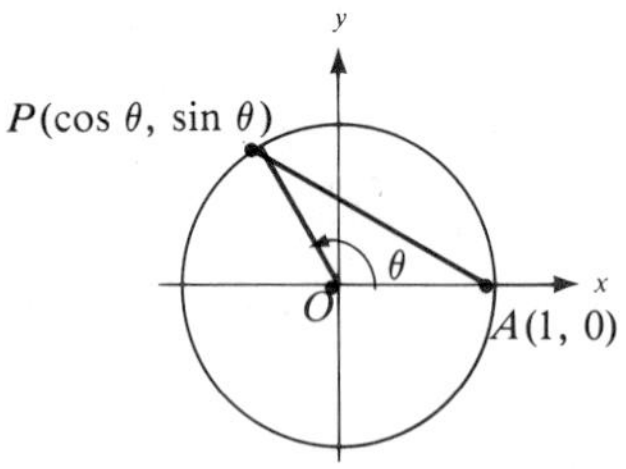

FIGURE 3.1   Length of a chord subtended by an angle θ

CHORD LENGTH

> The length of chord AP is
>
> $$|AP| = \sqrt{2 - 2\cos \theta}$$
>
> where θ is the central angle between OA and OP.

Example 3   Find $|P_\alpha P_\beta|$, where P_α and P_β are points on a unit circle determined by the angles α and β, respectively.

Solution   Since $P_\alpha P_\beta$ forms a chord in a unit circle and since the central angle θ is $\alpha - \beta$ (see Figure 3.2),

$$|P_\alpha P_\beta| = \sqrt{2 - 2\cos(\alpha - \beta)}$$

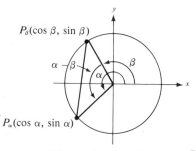

$P_\beta(\cos \beta, \sin \beta)$

$P_\alpha(\cos \alpha, \sin \alpha)$

FIGURE 3.2   Distance between P_α and P_β □

Suppose we rework Example 3 using the distance formula instead of the chord length formula:

$$|P_\alpha P_\beta| = \sqrt{(\cos \beta - \cos \alpha)^2 + (\sin \beta - \sin \alpha)^2}$$
$$= \sqrt{\cos^2 \beta - 2\cos \alpha \cos \beta + \cos^2 \alpha + \sin^2 \beta - 2\sin \alpha \sin \beta + \sin^2 \alpha}$$
$$= \sqrt{(\cos^2 \beta + \sin^2 \beta) + (\cos^2 \alpha + \sin^2 \alpha) - 2(\cos \alpha \cos \beta + \sin \alpha \sin \beta)}$$
$$= \sqrt{2 - 2(\cos \alpha \cos \beta + \sin \alpha \sin \beta)}$$

This result and the result from Example 3 look very different, yet they both represent the same distance. If we equate these expressions (since they are equivalent), we obtain a very useful result (after simplifying):

$$\sqrt{2 - 2\cos(\alpha - \beta)} = \sqrt{2 - 2(\cos\alpha\cos\beta + \sin\alpha\sin\beta)}$$
$$2 - 2\cos(\alpha - \beta) = 2 - 2(\cos\alpha\cos\beta + \sin\alpha\sin\beta)$$
$$-2\cos(\alpha - \beta) = -2(\cos\alpha\cos\beta + \sin\alpha\sin\beta)$$
$$\cos(\alpha - \beta) = \cos\alpha\cos\beta + \sin\alpha\sin\beta$$

We can now use this identity in a variety of different contexts and as a basis for deriving other identities in this chapter.

Cofunction Identities

We will begin numbering the identities with 9 to remind you that the first eight identities are the fundamental identities. These identities 9–11 are called the **cofunction identities** because they relate the sine and cosine as well as the tangent and cotangent.

COFUNCTION IDENTITIES

> For any real number (or angle) θ,
>
> 9. $\cos(\frac{\pi}{2} - \theta) = \sin\theta$ 10. $\sin(\frac{\pi}{2} - \theta) = \cos\theta$ 11. $\tan(\frac{\pi}{2} - \theta) = \cot\theta$

Proof of Identity 9: Let $\alpha = \frac{\pi}{2}$ and $\beta = \theta$ in the identity $\cos(\alpha - \beta) = \cos\alpha\cos\beta + \sin\alpha\sin\beta$.

$$\cos(\tfrac{\pi}{2} - \theta) = \cos\tfrac{\pi}{2}\cos\theta + \sin\tfrac{\pi}{2}\sin\theta$$
$$= 0 \cdot \cos\theta + 1 \cdot \sin\theta$$
$$\therefore \cos(\tfrac{\pi}{2} - \theta) = \sin\theta$$

The *proof of Identity* 10 depends on Identity 9. For any θ,

$$\cos\theta = \cos[\tfrac{\pi}{2} - (\tfrac{\pi}{2} - \theta)]$$
$$= \sin(\tfrac{\pi}{2} - \theta) \qquad \textit{This is Identity 9}$$
$$\therefore \sin(\tfrac{\pi}{2} - \theta) = \cos\theta$$

Identities involving the tangent are usually proved after proving similar identities for cosine and sine. The fundamental identity

$$\tan\theta = \frac{\sin\theta}{\cos\theta}$$

is applied first, allowing you then to use the appropriate identities for cosine and sine.

This process is illustrated with the *proof of Identity* 11:

$$\tan\left(\tfrac{\pi}{2} - \theta\right) = \frac{\sin\left(\tfrac{\pi}{2} - \theta\right)}{\cos\left(\tfrac{\pi}{2} - \theta\right)}$$

$$= \frac{\cos\theta}{\sin\theta}$$

$$= \cot\theta$$

The cofunction identities allow us to change a trigonometric function to the cofunction of its complement.

Example 4

Write each function in terms of its cofunction.

a. $\sin 28°$ **b.** $\cos 43°$ **c.** $\cot 9°$ **d.** $\sin\frac{\pi}{6}$

Solution

a. $\sin 28° = \cos(90° - 28°)$ **b.** $\cos 43° = \sin(90° - 43°)$

$\qquad = \cos 62°$ $\qquad\qquad = \sin 47°$

c. $\cot 9° = \tan(90° - 9°)$ **d.** $\sin\frac{\pi}{6} = \cos\left(\frac{\pi}{2} - \frac{\pi}{6}\right)$

$\qquad = \tan 81°$ $\qquad\qquad = \cos\frac{2\pi}{6}$

$\qquad\qquad\qquad\qquad\qquad\qquad = \cos\frac{\pi}{3}$ □

Opposite Angle Identities

Suppose the given angle in Example 4 is larger than 90°; for example,

$$\cos 125° = \sin(90° - 125°)$$

$$= \sin(-35°)$$

This result can be further simplified using the following **opposite-angle identities**.

OPPOSITE-ANGLE
IDENTITIES

For any real number (or angle) θ,
12. $\cos(-\theta) = \cos\theta$ **13.** $\sin(-\theta) = -\sin\theta$ **14.** $\tan(-\theta) = -\tan\theta$

Proof of Identity 12: Let $\alpha = 0$ and $\beta = \theta$ in the identity $\cos(\alpha - \beta) = \cos\alpha\cos\beta + \sin\alpha\sin\beta$.

$$\cos(0 - \theta) = \cos 0\cos\theta + \sin 0\sin\theta$$

$$= 1 \cdot \cos\theta + 0 \cdot \sin\theta$$

$$= \cos\theta$$

But, if you simplify directly,

$$\cos(0 - \theta) = \cos(-\theta)$$

$$\therefore \cos(-\theta) = \cos\theta$$

Proof of Identity 13:   Let $\alpha = \frac{\pi}{2}$ and $\beta = -\theta$.

$$\cos(\tfrac{\pi}{2} + \theta) = \cos[\tfrac{\pi}{2} - (-\theta)]$$
$$= \cos\tfrac{\pi}{2}\cos(-\theta) + \sin\tfrac{\pi}{2}\sin(-\theta)$$
$$= 0 \cdot \cos(-\theta) + 1 \cdot \sin(-\theta)$$
$$= \sin(-\theta)$$

Also,

$$\cos(\tfrac{\pi}{2} + \theta) = \cos(\theta + \tfrac{\pi}{2})$$
$$= \cos[\theta - (-\tfrac{\pi}{2})]$$
$$= \cos\theta\cos(-\tfrac{\pi}{2}) + \sin\theta\sin(-\tfrac{\pi}{2})$$
$$= \cos\theta \cdot \ \ 0 \ \ \ \ + \sin\theta \cdot (-1)$$
$$= -\sin\theta$$
$$\therefore \ \sin(-\theta) = -\sin\theta$$

Proof of Identity 14:

$$\tan(-\theta) = \frac{\sin(-\theta)}{\cos(-\theta)}$$
$$= \frac{-\sin\theta}{\cos\theta}$$
$$= -\tan\theta$$

Example 5

Write each as a function of a positive angle or number.

a. $\cos(-19°)$ **b.** $\sin(-19°)$ **c.** $\tan(-2)$

Solution

a. $\cos(-19°) = \cos 19°$ **b.** $\sin(-19°) = -\sin 19°$

c. $\tan(-2) = -\tan 2$

Example 6

Write the given functions in terms of their cofunctions.

a. $\cos 125°$ **b.** $\sin 102°$ **c.** $\cot 2.5$

Solution

a. $\cos 125° = \sin(90° - 125°)$ *You might recall we worked problems like this in*
$$= \sin(-35°)$$ *Chapter 1 using reference angles and the reduction*
$$= -\sin 35°$$ *principle. Note that 35° is the reference angle for*
(−35°) and sine is negative in Quadrant II.

b. $\sin 102° = \cos(90° - 102°)$ *Can you verify this using the reduction principle*
$$= \cos(-12°)$$ *from Chapter 1?*
$$= \cos 12°$$

c. $\cot 2.5 = \tan(\tfrac{\pi}{2} - 2.5)$
$$\approx \tan(-.9292)$$
$$= -\tan .9292$$

You may also need to use opposite-angle identities together with other identities. For example, you know from algebra that

$$a - b \quad \text{and} \quad b - a$$

are opposites. This means that $a - b = -(b - a)$. In trigonometry you often see angles like $\frac{\pi}{2} - \theta$ and want to write $\theta - \frac{\pi}{2}$. This means that $\frac{\pi}{2} - \theta = -(\theta - \frac{\pi}{2})$. In particular,

$$\cos(\tfrac{\pi}{2} - \theta) = \cos[-(\theta - \tfrac{\pi}{2})]$$
$$= \cos(\theta - \tfrac{\pi}{2}) \qquad \textit{By Identity 12}$$

Example 7

Write the given functions using the opposite-angle identities.

a. $\sin(\frac{\pi}{2} - \theta)$ **b.** $\tan(\frac{\pi}{2} - \theta)$ **c.** $\cos(\pi - \theta)$

Solution

a. $\sin(\frac{\pi}{2} - \theta) = \sin[-(\theta - \frac{\pi}{2})]$ **b.** $\tan(\frac{\pi}{2} - \theta) = \tan[-(\theta - \frac{\pi}{2})]$
                   $= -\sin(\theta - \frac{\pi}{2})$ $= -\tan(\theta - \frac{\pi}{2})$

c. $\cos(\pi - \theta) = \cos[-(\theta - \pi)]$
                $= \cos(\theta - \pi)$ □

Graphing with Opposites

The procedure for graphing $y = -\cos x$ is identical to the procedure for graphing $y = \cos x$, except that, after the frame is drawn, the endpoints are at the bottom of the frame instead of at the top, and the midpoint is at the top, as shown in Figure 3.3.

FIGURE 3.3   One period of $y = -\cos x$

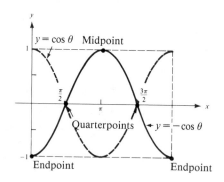

Example 8

Graph $y = \sin(-x)$.

Solution

First use an opposite-angle identity, if necessary: $\sin(-x) = -\sin x$. To graph $y = -\sin x$, build the frame as before; the endpoints and midpoints are the same, but the quarterpoints are reversed, as shown in Figure 3.4 on page 121.

FIGURE 3.4   One period of
$y = -\sin x$

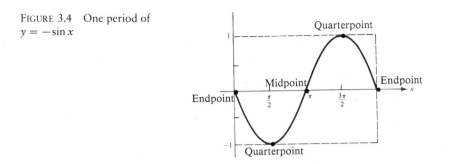

Problem Set 3.4

A   *Use the distance formula to find the distance between the points given in Problems 1–6.*

1. $(6, 2)$ and $(1, 10)$ **2.** $(-2, 3)$ and $(5, 10)$

3. $(1, 6)$ and $(5, -8)$ **4.** (α, θ) and (β, ω)

5. $(\cos \alpha, \sin \alpha)$ and $(0, 1)$ **6.** $(\cos \theta, \sin \theta)$ and $(\cos \gamma, \sin \gamma)$

Write each function in Problems 7–12 in terms of its cofunction by using the cofunction identities.

7. $\cos 15°$ **8.** $\sin 38°$ **9.** $\cot 41°$

10. $\sin \frac{\pi}{6}$ **11.** $\cos \frac{5\pi}{6}$ **12.** $\cot \frac{2\pi}{3}$

Write the functions in Problem 13–18 as functions of positive angles.

13. $\cos(-18°)$ **14.** $\tan(-49°)$ **15.** $\sin(-41°)$

16. $\sin(-31°)$ **17.** $\cos(-39°)$ **18.** $\tan(-24°)$

B   **19.**   Write each as a function of $\theta - \frac{\pi}{3}$.
                    a. $\cos(\frac{\pi}{3} - \theta)$ **b.** $\sin(\frac{\pi}{3} - \theta)$ **c.** $\tan(\frac{\pi}{3} - \theta)$

           20.   Write each as a function of $(\theta - \frac{2\pi}{3})$.
                    a. $\cos(\frac{2\pi}{3} - \theta)$ **b.** $\sin(\frac{2\pi}{3} - \theta)$ **c.** $\tan(\frac{2\pi}{3} - \theta)$

           21.   Write each as a function of $\alpha - \beta$.
                    a. $\cos(\beta - \alpha)$ **b.** $\sin(\beta - \alpha)$ **c.** $\tan(\beta - \alpha)$

Use the opposite-angle identities, if necessary, to graph the functions in Problems 22–30.

22. $y = -2\cos x$ **23.** $y = -3\sin x$ **24.** $y = \tan(-x)$

25. $y = \cos(-2x)$ **26.** $y = \sin(-3x)$ **27.** $y - 1 = \sin(2 - x)$

28. $y - 2 = \cos(\pi - x)$ **29.** $y - 3 = \sin(\pi - x)$ **30.** $y + 1 = \tan(\frac{\pi}{6} - x)$

C   *Prove that the equations in Problems 31–32 are identities.*

           31. $\cot(-\theta) = -\cot \theta$ **32.** $\sec(-\theta) = \sec \theta$

3.5 ADDITION LAWS

The identities of this section are based on the cosine identity, derived at the beginning of the last section, and are called the **addition laws**. Since subtraction can easily be written as a sum, the designation *addition laws* refers to both addition and subtraction.

ADDITION LAWS

> **15.** $\cos(\alpha + \beta) = \cos\alpha\cos\beta - \sin\alpha\sin\beta$
>
> **16.** $\cos(\alpha - \beta) = \cos\alpha\cos\beta + \sin\alpha\sin\beta$
>
> **17.** $\sin(\alpha + \beta) = \sin\alpha\cos\beta + \cos\alpha\sin\beta$
>
> **18.** $\sin(\alpha - \beta) = \sin\alpha\cos\beta - \cos\alpha\sin\beta$
>
> **19.** $\tan(\alpha + \beta) = \dfrac{\tan\alpha + \tan\beta}{1 - \tan\alpha\tan\beta}$
>
> **20.** $\tan(\alpha - \beta) = \dfrac{\tan\alpha - \tan\beta}{1 + \tan\alpha\tan\beta}$

Example 1

Write $\cos\left(\frac{2\pi}{3} + \theta\right)$ as a function of θ only.

Solution

Use Identity 15:

$$\cos\left(\tfrac{2\pi}{3} + \theta\right) = \cos\tfrac{2\pi}{3}\cos\theta - \sin\tfrac{2\pi}{3}\sin\theta$$
$$= \left(-\tfrac{1}{2}\right)\cos\theta - \left(\tfrac{\sqrt{3}}{2}\right)\sin\theta \qquad \textit{Substitute exact values where possible.}$$
$$= -\tfrac{1}{2}(\cos\theta + \sqrt{3}\sin\theta) \qquad \textit{Simplify.} \qquad\qquad □$$

Example 2

Evaluate

$$\frac{\tan 18° - \tan 40°}{1 + \tan 18°\tan 40°}$$

using Appendix D Table or using a calculator.

Solution

You can do a lot of arithmetic, or use Identity 20.

$$\frac{\tan 18° - \tan 40°}{1 + \tan 18°\tan 40°} = \tan(18° - 40°)$$
$$= \tan(-22°)$$
$$= -\tan 22° \qquad \textit{Identity 14 from Section 3.4}$$
$$\approx -.4040 \qquad \textit{Use Appendix D Table or a calculator.} \qquad □$$

Ptolemy lived in Alexandria and worked at the museum there. Along with Hipparchus and Menelaus, he invented trigonometry in order to build a quantitative astronomy that could be used to predict the paths and positions of the heavenly bodies and to aid in telling time, calendar-reckoning, navigating, and studying geography. In his book *Syntaxis Mathematica*, he derived many of the identities of this chapter. Although his purposes were more related to solving triangles (see Chapter 5), today a major focus of trigonometry is on the relationships of the functions themselves, which is what we have been doing in this chapter.

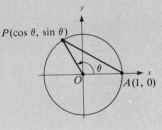

Example 3

Find the exact value of $\cos 345°$.

Solution

$$\cos 345° = \cos 15°$$

First use the reduction principle, if appropriate.

$$= \cos(45° - 30°)$$

Next, write the angle in terms of a sum or difference of angles whose exact values you know; in this case, $15° = 45° - 30°$.

$$= \cos 45° \cos 30° + \sin 45° \sin 30°$$

Apply Identity 16.

$$= \frac{\sqrt{2}}{2} \cdot \frac{\sqrt{3}}{2} + \frac{\sqrt{2}}{2} \cdot \frac{1}{2}$$

Use exact values.

$$= \frac{\sqrt{6}}{4} + \frac{\sqrt{2}}{4}$$

Simplify.

$$= \frac{\sqrt{6} + \sqrt{2}}{4}$$

We will conclude this section by proving some of the addition laws; others are left as exercises.

Proof of Identity 15:

$$\cos(\alpha + \beta) = \cos[\alpha - (-\beta)]$$
$$= \cos \alpha \cos(-\beta) + \sin \alpha \sin(-\beta)$$
$$= \cos \alpha \cos \beta - \sin \alpha \sin \beta \qquad \textit{Use Identities 12 and 13.}$$

Identity 16 is the main cosine identity proved at the beginning of Section 3.4. It is numbered and included here for the sake of completeness.

Proof of Identity 17:

$$\sin(\alpha + \beta) = \cos[\tfrac{\pi}{2} - (\alpha + \beta)]$$
$$= \cos[(\tfrac{\pi}{2} - \alpha) - \beta]$$
$$= \cos(\tfrac{\pi}{2} - \alpha) \cos \beta + \sin(\tfrac{\pi}{2} - \alpha) \sin \beta$$
$$= \sin \alpha \cos \beta + \cos \alpha \sin \beta$$

Proof of Identity 18:   Replace β by $-\beta$ in Identity 17; the details are left as an exercise.

Proof of Identity 19:

$$\tan(\alpha + \beta) = \frac{\sin(\alpha + \beta)}{\cos(\alpha + \beta)}$$

$$= \frac{\sin \alpha \cos \beta + \cos \alpha \sin \beta}{\cos \alpha \cos \beta - \sin \alpha \sin \beta}$$

$$= \frac{\sin \alpha \cos \beta + \cos \alpha \sin \beta}{\cos \alpha \cos \beta - \sin \alpha \sin \beta} \cdot \frac{\dfrac{1}{\cos \alpha \cos \beta}}{\dfrac{1}{\cos \alpha \cos \beta}} \qquad \textit{Multiply by 1.}$$

$$= \frac{\dfrac{\sin \alpha \cos \beta}{\cos \alpha \cos \beta} + \dfrac{\cos \alpha \sin \beta}{\cos \alpha \cos \beta}}{\dfrac{\cos \alpha \cos \beta}{\cos \alpha \cos \beta} - \dfrac{\sin \alpha \sin \beta}{\cos \alpha \cos \beta}}$$

$$= \frac{\tan \alpha + \tan \beta}{1 - \tan \alpha \tan \beta}$$

The *proof of Identity* 20 is left as an exercise.

Problem Set 3.5

A   *Change each of the expressions in Problems 1–6 to functions of θ only.*

1.   **a.** $\cos(30° + \theta)$ **b.** $\sin(\theta - 45°)$ 2.   **a.** $\tan(45° + \theta)$ **b.** $\tan(\theta - 45°)$

3.   **a.** $\cos(\tfrac{\pi}{3} - \theta)$ **b.** $\cos(\theta - \tfrac{\pi}{4})$ 4.   **a.** $\sin(\tfrac{2\pi}{3} + \theta)$ **b.** $\sin(\theta - \tfrac{\pi}{6})$

5. **a.** $\cos(\theta + \theta)$ **b.** $\sin(\theta + \theta)$ 6. **a.** $\tan(\theta + \theta)$ **b.** $\cot(\theta + \theta)$

Evaluate each of the expressions in Problems 7–16. You may use Appendix D Table or a calculator.

7. $\sin 158° \cos 92° - \cos 158° \sin 92°$

8. $\cos 114° \cos 85° + \sin 114° \sin 85°$

9. $\cos 30° \cos 48° - \sin 30° \sin 48°$

10. $\sin 18° \cos 23° + \cos 18° \sin 23°$

11. $\sin 125° \cos 85° - \cos 125° \sin 85°$

12. $\cos 215° \cos 145° - \sin 215° \sin 145°$

13. $\dfrac{\tan 32° + \tan 18°}{1 - \tan 32° \tan 18°}$

14. $\dfrac{\tan 59° - \tan 25°}{1 + \tan 59° \tan 25°}$

15. $\dfrac{\tan 120° - \tan 89°}{1 + \tan 120° \tan 89°}$

16. $\dfrac{\tan 245° + \tan 210°}{1 - \tan 245° \tan 210°}$

B   *Using the identities of this section, find the exact values of the sine, cosine, and tangent of each of the angles given in Problems 17–22.*

17. $-15°$ 18. $195°$ 19. $75°$ 20. $165°$ 21. $345°$ 22. $105°$

Prove the identities in Problems 23–30.

23. $\sin(\alpha - \beta) = \sin \alpha \cos \beta - \cos \alpha \sin \beta$

24. $\tan(\alpha - \beta) = \dfrac{\tan \alpha - \tan \beta}{1 + \tan \alpha \tan \beta}$

25. $\cot(\alpha + \beta) = \dfrac{\cot \alpha \cot \beta - 1}{\cot \beta + \cot \alpha}$

26. $\cot(\alpha - \beta) = \dfrac{\cot \alpha \cot \beta + 1}{\cot \beta - \cot \alpha}$

27. $\dfrac{\cos 5\theta}{\sin \theta} - \dfrac{\sin 5\theta}{\cos \theta} = \dfrac{\cos 6\theta}{\sin \theta \cos \theta}$

28. $\dfrac{\sin 6\theta}{\sin 3\theta} - \dfrac{\cos 6\theta}{\cos 3\theta} = \sec 3\theta$

29. $\sin(\alpha + \beta) \cos \beta - \cos(\alpha + \beta) \sin \beta = \sin \alpha$

30. $\cos(\alpha - \beta) \cos \beta - \sin(\alpha - \beta) \sin \beta = \cos \alpha$

C   *Prove the identities in Problems 31–47.*

31. $\sin(\frac{3\pi}{2} - \theta) = -\cos \theta$

32. $\cos(\frac{3\pi}{2} - \theta) = -\sin \theta$

33. $\sin(\frac{3\pi}{2} + \theta) = -\cos \theta$

34. $\cos(\frac{3\pi}{2} + \theta) = \sin \theta$

35. $\sin(\pi - \theta) = \sin \theta$

36. $\cos(\pi - \theta) = -\cos \theta$

37. $\tan(\pi - \theta) = -\tan \theta$

38. $\sin(\pi + \theta) = -\sin \theta$

39. $\cos(\pi + \theta) = -\cos \theta$

40. $\tan(\pi + \theta) = \tan \theta$

41. $\sec(\frac{\pi}{2} - \theta) = \csc \theta$

42. $\cot(\frac{\pi}{2} - \theta) = \tan \theta$

43. $\csc(\frac{\pi}{2} - \theta) = \sec \theta$

44. $\dfrac{\sin(\theta + h) - \sin \theta}{h} = \cos \theta \left(\dfrac{\sin h}{h} \right) - \sin \theta \left(\dfrac{1 - \cos h}{h} \right)$

45. $\dfrac{\cos(\theta + h) - \cos \theta}{h} = -\sin \theta \left(\dfrac{\sin h}{h} \right) - \cos \theta \left(\dfrac{1 - \cos h}{h} \right)$

46. $\sin(\alpha + \beta + \gamma) = \sin \alpha \cos \beta \cos \gamma + \cos \alpha \sin \beta \cos \gamma + \cos \alpha \cos \beta \sin \gamma - \sin \alpha \sin \beta \sin \gamma$

47. $\cos(\alpha + \beta + \gamma) = \cos \alpha \cos \beta \cos \gamma - \cos \alpha \sin \beta \sin \gamma - \sin \alpha \cos \beta \sin \gamma - \sin \alpha \sin \beta \cos \gamma$

3.6   DOUBLE-ANGLE AND HALF-ANGLE IDENTITIES

Double-Angle Identities

Two additional, special cases of the addition laws of Section 3.5 are now considered. The first is that of the **double-angle identities**.

DOUBLE-ANGLE
IDENTITIES

> **21.** $\cos 2\theta = \cos^2\theta - \sin^2\theta$ **22.** $\sin 2\theta = 2 \sin\theta \cos\theta$
>
> $\quad\quad\quad = 2\cos^2\theta - 1$ **23.** $\tan 2\theta = \dfrac{2\tan\theta}{1 - \tan^2\theta}$
>
> $\quad\quad\quad = 1 - 2\sin^2\theta$

To prove these identities, use the addition laws where $\alpha = \theta$ and $\beta = \theta$.

$$\cos 2\theta = \cos(\theta + \theta) = \cos\theta\cos\theta - \sin\theta\sin\theta$$
$$= \cos^2\theta - \sin^2\theta$$

$$\sin 2\theta = \sin(\theta + \theta) = \sin\theta\cos\theta + \cos\theta\sin\theta$$
$$= 2\sin\theta\cos\theta$$

$$\tan 2\theta = \tan(\theta + \theta) = \frac{\tan\theta + \tan\theta}{1 - \tan\theta\tan\theta}$$
$$= \frac{2\tan\theta}{1 - \tan^2\theta}$$

(You did this in Problems 5–6 in Problem Set 3.5.)

Example 1 $\cos 100x = \cos(2 \cdot 50x) = \cos^2 50x - \sin^2 50x$ ⬚

Example 2 $\sin 120° = \sin 2(60°) = 2\sin 60° \cos 60°$ ⬚

Example 3 $\cos 3\theta = \cos(2\theta + \theta) = \cos 2\theta \cos\theta - \sin 2\theta \sin\theta$
$$= (\cos^2\theta - \sin^2\theta)\cos\theta - (2\sin\theta\cos\theta)\sin\theta$$
$$= \cos^3\theta - \sin^2\theta\cos\theta - 2\sin^2\theta\cos\theta$$
$$= \cos^3\theta - 3\sin^2\theta\cos\theta$$
$$= \cos^3\theta - 3(1 - \cos^2\theta)\cos\theta$$
$$= \cos^3\theta - 3\cos\theta + 3\cos^3\theta$$
$$= 4\cos^3\theta - 3\cos\theta$$ ⬚

Example 4

Evaluate $\dfrac{2\tan\frac{\pi}{16}}{1-\tan^2\frac{\pi}{16}}$.

Solution

Notice that this is the right-hand side of Identity 23, so it is the same as $\tan\left(2\cdot\frac{\pi}{16}\right)=\tan\frac{\pi}{8}$. Now use tables or a calculator to find $\tan\frac{\pi}{8}\approx.4142$.   □

Example 5

If $\cos\theta=\frac{3}{5}$ and θ is in Quadrant IV, find $\cos 2\theta$, $\sin 2\theta$, and $\tan 2\theta$.

Solution

Since $\cos 2\theta=2\cos^2\theta-1$,

$$\cos 2\theta=2\left(\frac{3}{5}\right)^2-1$$

$$=2\left(\frac{9}{25}\right)-1$$

$$=-\frac{7}{25}$$

For the other functions of 2θ, you need to know $\sin\theta$. Begin with the fundamental identity relating cosine and sine.

$$\sin^2\theta=1-\cos^2\theta$$

$$\sin\theta=-\sqrt{1-\cos^2\theta}\quad\textit{Negative since the sine is negative in Quadrant IV}$$

$$=-\sqrt{1-\left(\frac{3}{5}\right)^2}$$

$$=-\sqrt{1-\frac{9}{25}}$$

$$=-\sqrt{\frac{16}{25}}$$

$$=-\frac{4}{5}$$

Now, $\sin 2\theta=2\sin\theta\cos\theta$

$$=2\left(-\frac{4}{5}\right)\left(\frac{3}{5}\right)$$

$$=-\frac{24}{25}$$

Finally, $\tan 2\theta=\dfrac{\sin 2\theta}{\cos 2\theta}$

$$=\frac{-\frac{24}{25}}{-\frac{7}{25}}=\frac{24}{7}$$ □

Half-Angle Identities

Sometimes, as in Example 3, we want to write $\cos 2\theta$ in terms of cosines, and at other times we want to write it in terms of sines. Thus,

$$\begin{aligned}
\cos 2\theta &= \cos^2\theta - \sin^2\theta \\
&= \cos^2\theta - (1 - \cos^2\theta) \\
&= 2\cos^2\theta - 1
\end{aligned}$$

and

$$\begin{aligned}
\cos 2\theta &= \cos^2\theta - \sin^2\theta \\
&= (1 - \sin^2\theta) - \sin^2\theta \\
&= 1 - 2\sin^2\theta
\end{aligned}$$

These last two identities lead us to the second important special case of the addition laws, called the **half-angle identities**. We wish to solve $\cos 2\alpha = 2\cos^2\alpha - 1$ for $\cos^2\alpha$.

$$\begin{aligned}
2\cos^2\alpha - 1 &= \cos 2\alpha \\
2\cos^2\alpha &= 1 + \cos 2\alpha \\
\cos^2\alpha &= \frac{1 + \cos 2\alpha}{2}
\end{aligned}$$

Now, if $\alpha = \frac{1}{2}\theta$, then $2\alpha = \theta$ and

$$\cos^2 \tfrac{1}{2}\theta = \frac{1 + \cos\theta}{2}$$

If $\frac{1}{2}\theta$ is in Quadrant I or IV, then

$$\cos \tfrac{1}{2}\theta = \sqrt{\frac{1 + \cos\theta}{2}}$$

If $\frac{1}{2}\theta$ is in Quadrant II or III, then

$$\cos \tfrac{1}{2}\theta = -\sqrt{\frac{1 + \cos\theta}{2}}$$

These results are summarized by writing

$$\cos \tfrac{1}{2}\theta = \pm\sqrt{\frac{1 + \cos\theta}{2}}$$

However, *you must be careful*. The sign $+$ or $-$ is chosen according to which quadrant $\frac{1}{2}\theta$ is in. The formula requires either $+$ or $-$, but not both. This use of $\pm$ is different from the use of $\pm$ in algebra. For example, when using $\pm$ in the quadratic formula, we are indicating *two* possible correct roots—one plus, one minus. In this trigonometric identity we will obtain *one* correct sign, plus or minus, depending on the quadrant of $\frac{1}{2}\theta$.*

For the sine, solve $\cos 2\alpha = 1 - 2\sin^2\alpha$ for $\sin^2\alpha$.

$$\cos 2\alpha = 1 - 2\sin^2\alpha$$

$$2\sin^2\alpha = 1 - \cos 2\alpha$$

$$\sin^2\alpha = \frac{1 - \cos 2\alpha}{2}$$

Replace $\alpha = \frac{1}{2}\theta$, and

$$\sin^2 \tfrac{1}{2}\theta = \frac{1 - \cos\theta}{2}$$

or

$$\sin \tfrac{1}{2}\theta = \pm\sqrt{\frac{1 - \cos\theta}{2}}$$

where the sign depends on the quadrant of $\frac{1}{2}\theta$. If $\frac{1}{2}\theta$ is in Quadrant I or II, you use $+$; if it is in Quadrant III or IV, you use $-$.

Finally, to find the half-angle identity for the tangent, write

$$\tan \tfrac{1}{2}\theta = \frac{\sin\tfrac{1}{2}\theta}{\cos\tfrac{1}{2}\theta}$$

$$= \frac{\pm\sqrt{\dfrac{1 - \cos\theta}{2}}}{\pm\sqrt{\dfrac{1 + \cos\theta}{2}}}$$

$$= \pm\sqrt{\frac{1 - \cos\theta}{1 + \cos\theta}}$$

$$= \pm\sqrt{\frac{1 - \cos\theta}{1 + \cos\theta} \cdot \frac{1 - \cos\theta}{1 - \cos\theta}}$$

$$= \pm\sqrt{\frac{(1 - \cos\theta)^2}{\sin^2\theta}} \qquad \textit{Remember that } 1 - \cos^2\theta = \sin^2\theta.$$

$$= \frac{1 - \cos\theta}{\sin\theta} \qquad \begin{array}{l}\textit{Notice that } 1 - \cos\theta \textit{ is positive. Also, since } \tan\tfrac{1}{2}\theta \\ \textit{and } \sin\theta \textit{ have the same sign regardless of the} \\ \textit{quadrant of } \theta, \textit{ the desired result follows.}\end{array}$$

You can also show that $\tan \tfrac{1}{2}\theta = \dfrac{\sin\theta}{1 + \cos\theta}$.

---

* You should also remember from algebra that $\frac{1}{2}\theta$ and $\dfrac{\theta}{2}$ mean the same thing, so $\cos\frac{1}{2}\theta$ and $\cos\dfrac{\theta}{2}$ are the same thing, but $\dfrac{\cos\theta}{2}$ is different.

HALF-ANGLE IDENTITIES

24. $\cos\frac{1}{2}\theta = \pm\sqrt{\dfrac{1 + \cos\theta}{2}}$ 26. $\tan\frac{1}{2}\theta = \dfrac{1 - \cos\theta}{\sin\theta}$

25. $\sin\frac{1}{2}\theta = \pm\sqrt{\dfrac{1 - \cos\theta}{2}}$ $= \dfrac{\sin\theta}{1 + \cos\theta}$

Example 6

Find the exact value of $\cos\frac{9\pi}{8}$.

Solution

$$\cos\frac{9\pi}{8} = \cos\left(\frac{1}{2}\cdot\frac{9\pi}{4}\right) = -\sqrt{\frac{1 + \cos\frac{9\pi}{4}}{2}}$$

$$= -\sqrt{\frac{1 + \cos\frac{\pi}{4}}{2}}$$

Choose a negative sign, since $\frac{9\pi}{8}$ is in Quadrant III and the cosine is negative in this quadrant.

$$= -\sqrt{\frac{1 + \frac{\sqrt{2}}{2}}{2}}$$

$$= -\sqrt{\frac{2 + \sqrt{2}}{4}}$$

$$= -\frac{1}{2}\sqrt{2 + \sqrt{2}}$$

Example 7

If $\cot 2\theta = \frac{3}{4}$, find $\cos\theta$, $\sin\theta$, and $\tan\theta$, where both θ and 2θ are in Quadrant I.

Solution

You need to find $\cos 2\theta$ so that you can use it in the half-angle identities. To do this, first find $\tan 2\theta$:

$$\tan 2\theta = \frac{1}{\cot 2\theta} = \frac{4}{3}$$

Next, find $\sec 2\theta$:

$$\sec 2\theta = \pm\sqrt{1 + \tan^2 2\theta}$$

$$= \sqrt{1 + \frac{16}{9}} \quad \text{It is positive because } 2\theta \text{ is in Quadrant I.}$$

$$= \frac{5}{3}$$

Finally, $\cos 2\theta$ is the reciprocal of $\sec 2\theta$:

$$\cos 2\theta = \frac{3}{5}$$

Next use the half-angle identities:

$$\cos \theta = \pm \sqrt{\frac{1 + \cos 2\theta}{2}} \qquad \text{and} \qquad \sin \theta = \pm \sqrt{\frac{1 - \cos 2\theta}{2}}$$

Do you see that θ is one-half of 2θ in these formulas?

$$\cos \theta = + \sqrt{\frac{1 + \frac{3}{5}}{2}} \qquad\qquad \sin \theta = + \sqrt{\frac{1 - \frac{3}{5}}{2}}$$

Positive value chosen because θ is in Quadrant I

$$\cos \theta = \frac{2}{\sqrt{5}} \qquad\qquad\qquad \sin \theta = \frac{1}{\sqrt{5}}$$

$$\tan \theta = \frac{\sin \theta}{\cos \theta}$$

$$= \frac{1/\sqrt{5}}{2/\sqrt{5}}$$

$$= \frac{1}{2}$$

Example 8   Prove that $\sin \theta = \dfrac{2 \tan \frac{1}{2}\theta}{1 + \tan^2 \frac{1}{2}\theta}$.

Solution   When proving identities involving functions of different angles, you should write all the trigonometric functions in the problems as functions of a single angle.

$$\frac{2 \tan \frac{1}{2}\theta}{1 + \tan^2 \frac{1}{2}\theta} = \frac{2 \dfrac{\sin \frac{1}{2}\theta}{\cos \frac{1}{2}\theta}}{\sec^2 \frac{1}{2}\theta}$$

$$= 2 \frac{\sin \frac{1}{2}\theta}{\cos \frac{1}{2}\theta} \cdot \cos^2 \frac{1}{2}\theta$$

$$= 2 \sin \tfrac{1}{2}\theta \cos \tfrac{1}{2}\theta$$

$$= \sin \theta \qquad \text{From Identity 22 (double-angle identity)}$$

Problem Set 3.6

A   *Use the double-angle or half-angle identities to evaluate each of Problems 1–9 using exact values.*

1. $2 \cos^2 22.5° - 1$

2. $\dfrac{2 \tan \frac{\pi}{8}}{1 - \tan^2 \frac{\pi}{8}}$

3. $\sqrt{\dfrac{1 - \cos 60°}{2}}$

4. $\cos^2 15° - \sin^2 15°$ **5.** $1 - 2\sin^2 90°$ **6.** $-\sqrt{\dfrac{1 - \cos 420°}{2}}$

7. $\sin 22.5°$ **8.** $\cos \frac{\pi}{8}$ **9.** $\tan 22.5°$

In each of Problems 10–15, find the exact values of cosine, sine, and tangent of 2θ where $0 \le \theta < 2\pi$.

10. $\sin \theta = \frac{3}{5}$; θ in Quadrant I   **11.** $\sin \theta = \frac{5}{13}$; θ in Quadrant II

12. $\tan \theta = -\frac{5}{12}$; θ in Quadrant IV   **13.** $\tan \theta = -\frac{3}{4}$; θ in Quadrant II

14. $\cos \theta = \frac{5}{9}$; θ in Quadrant I   **15.** $\cos \theta = -\frac{5}{13}$; θ in Quadrant III

In each of Problems 16–21, find the exact values of cosine, sine, and tangent of $\frac{1}{2}\theta$ where $0 \le \theta < 2\pi$.

16. $\sin \theta = \frac{3}{5}$; θ in Quadrant I   **17.** $\sin \theta = \frac{5}{13}$; θ in Quadrant II

18. $\tan \theta = -\frac{5}{12}$; θ in Quadrant IV   **19.** $\tan \theta = -\frac{3}{4}$; θ in Quadrant II

20. $\cos \theta = \frac{5}{9}$; θ in Quadrant I   **21.** $\cos \theta = -\frac{5}{13}$; θ in Quadrant III

B

In each of Problems 22–27, find $\cos \theta$, $\sin \theta$, and $\tan \theta$ when θ is in Quadrant I $(0° < \theta < 90°)$ and $\cot 2\theta$ is given.

22. $\cot 2\theta = -\frac{3}{4}$ **23.** $\cot 2\theta = 0$ **24.** $\cot 2\theta = \frac{1}{\sqrt{3}}$

25. $\cot 2\theta = -\frac{1}{\sqrt{3}}$ **26.** $\cot 2\theta = -\frac{4}{3}$ **27.** $\cot 2\theta = \frac{4}{3}$

28. *Aviation.* An airplane flying faster than the speed of sound is said to have a speed greater than Mach 1. The Mach number is the ratio of the speed of the plane to the speed of sound and is denoted by M. When a plane flies faster than the speed of sound, a sonic boom is heard, created by sound waves that form a cone with a vertex angle θ, as shown in Figure 3.5.

FIGURE 3.5   Pattern of sound waves creating a sonic boom

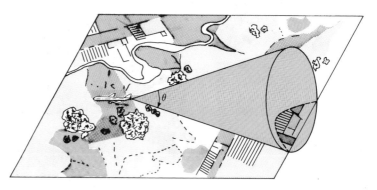

It can be shown that, if $M > 1$, then

$$\sin \frac{\theta}{2} = \frac{1}{M}$$

a. If $\theta = \frac{\pi}{6}$, find the Mach number to the nearest tenth.

b. Find the exact Mach number for part a.

29. *Navigation.* If a boat is moving at a constant rate that is faster than the water waves it produces, then the boat sends out waves in the shape of a cone with a vertex angle θ, as shown in Figure 3.6.

FIGURE 3.6   Pattern of bow waves

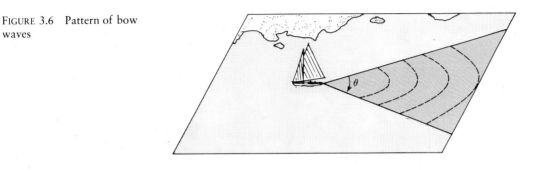

If r represents the ratio of the speed of the boat to the speed of the wave and if $r > 1$, then

$$\sin \frac{\theta}{2} = \frac{1}{r}$$

a. If $\theta = \frac{\pi}{4}$, find r to the nearest tenth.
b. Find the exact value of r for part a.

Prove each of the identities in Problems 30–38.

30. $\sin \alpha = 2 \sin \frac{\alpha}{2} \cos \frac{\alpha}{2}$

31. $\cos 4\theta = \cos^2 2\theta - \sin^2 2\theta$

32. $\sin 2\theta = \dfrac{2 \tan \theta}{1 + \tan^2 \theta}$

33. $\tan \frac{3}{2}\beta = \dfrac{2 \tan \frac{3\beta}{4}}{1 - \tan^2 \frac{3\beta}{4}}$

34. $\tan \frac{1}{2}\theta = \dfrac{1 - \cos \theta}{\sin \theta}$

35. $\tan \frac{1}{2}\theta = \dfrac{\sin \theta}{1 + \cos \theta}$

36. $\sin 2\theta \sec \theta = 2 \sin \theta$

37. $\sin 2\theta \tan \theta = 2 \sin^2 \theta$

38. $\cos^4 \theta - \sin^4 \theta = \cos 2\theta$

C   *Prove each of the identities in Problems 39–44.*

39. $\sec 2\theta - \tan 2\theta = \dfrac{1 - \tan \theta}{1 + \tan \theta}$

40. $\sin 3\theta = 3 \sin \theta - 4 \sin^3 \theta$

41. $\tan \frac{B}{2} = \csc B - \cot B$

42. $\sin 4\theta = 4 \sin \theta \cos \theta - 8 \sin^3 \theta \cos \theta$

43. $\frac{1}{2} \cot x - \frac{1}{2} \tan x = \cot 2x$

44. $\cos^4 \theta = \frac{1}{8}(3 + 4 \cos 2\theta + \cos 4\theta)$

3.7   PRODUCT AND SUM IDENTITIES

It is sometimes convenient, or even necessary, to write a trigonometric sum as a product or a product as a sum. To do so, we again turn to the identities for the sum and difference of two angles.

Product Identities

Add and subtract the following pair of identities.

$$\cos \alpha \cos \beta + \sin \alpha \sin \beta = \cos(\alpha - \beta)$$
$$\cos \alpha \cos \beta - \sin \alpha \sin \beta = \cos(\alpha + \beta)$$

Adding:　　　　$2 \cos \alpha \cos \beta = \cos(\alpha - \beta) + \cos(\alpha + \beta)$

Subtracting:　　$2 \sin \alpha \sin \beta = \cos(\alpha - \beta) - \cos(\alpha + \beta)$

Also:

$$\sin \alpha \cos \beta + \cos \alpha \sin \beta = \sin(\alpha + \beta)$$
$$\sin \alpha \cos \beta - \cos \alpha \sin \beta = \sin(\alpha - \beta)$$

Adding:　　　　$2 \sin \alpha \cos \beta = \sin(\alpha + \beta) + \sin(\alpha - \beta)$

Subtracting:　　$2 \cos \alpha \sin \beta = \sin(\alpha + \beta) - \sin(\alpha - \beta)$

These identities are called the **product identities**.

PRODUCT IDENTITIES

> 27.　$2 \cos \alpha \cos \beta = \cos(\alpha - \beta) + \cos(\alpha + \beta)$
>
> 28.　$2 \sin \alpha \sin \beta = \cos(\alpha - \beta) - \cos(\alpha + \beta)$
>
> 29.　$2 \sin \alpha \cos \beta = \sin(\alpha + \beta) + \sin(\alpha - \beta)$
>
> 30.　$2 \cos \alpha \sin \beta = \sin(\alpha + \beta) - \sin(\alpha - \beta)$

Example 1

Write $2 \sin 3 \sin 1$ as the sum of two functions.

Solution

Use Identity 28, where $\alpha = 3$ and $\beta = 1$.

$$2 \sin 3 \sin 1 = \cos(3 - 1) - \cos(3 + 1)$$
$$= \cos 2 - \cos 4$$

Example 2

Write $\sin 40° \cos 12°$ as the sum of two functions.

Solution

Use Identity 29, where $\alpha = 40°$ and $\beta = 12°$.

$$2 \sin 40° \cos 12° = \sin(40° + 12°) + \sin(40° - 12°)$$
$$= \sin 52° + \sin 28°$$

But what about the coefficient 2? Since you know that the preceding is an *equation* that is true, you can divide both sides by 2 to obtain

$$\sin 40° \cos 12° = \tfrac{1}{2}(\sin 52° + \sin 28°)$$

Sum Identities

Identities 27–30 will change a product to a sum. They can also be used to change a sum to a product. To do this, put them in a more useful form for this purpose by letting

$x = \alpha + \beta$ and $y = \alpha - \beta$. To rewrite Identity 27 in sum form, substitute x for $\alpha + \beta$ and y for $\alpha - \beta$:

$$2 \cos \alpha \cos \beta = \cos(\alpha - \beta) + \cos(\alpha + \beta)$$
$$= \cos y + \cos x$$
$$= \cos x + \cos y$$

Now, you need to eliminate α and β:

$$\begin{cases} x = \alpha + \beta \\ y = \alpha - \beta \end{cases}$$
Add: $x + y = 2\alpha$
$$\frac{x + y}{2} = \alpha$$

and   Subtract:

$$\begin{cases} x = \alpha + \beta \\ y = \alpha - \beta \end{cases}$$
$x - y = 2\beta$
$$\frac{x - y}{2} = \beta$$

Thus,

$$2 \cos\left(\frac{x + y}{2}\right) \cos\left(\frac{x - y}{2}\right) = \cos x + \cos y$$

This is called a **sum identity**. If you make the same substitutions into Identities 28, 29, and 30, you will obtain the other sum identities.

SUM IDENTITIES

31. $\cos x + \cos y = 2 \cos\left(\dfrac{x + y}{2}\right) \cos\left(\dfrac{x - y}{2}\right)$

32. $\cos x - \cos y = -2 \sin\left(\dfrac{x + y}{2}\right) \sin\left(\dfrac{x - y}{2}\right)$

33. $\sin x + \sin y = 2 \sin\left(\dfrac{x + y}{2}\right) \cos\left(\dfrac{x - y}{2}\right)$

34. $\sin x - \sin y = 2 \sin\left(\dfrac{x - y}{2}\right) \cos\left(\dfrac{x + y}{2}\right)$

Example 3

Solution

Write $\sin 35° + \sin 27°$ as a product.

$x = 35°$, $y = 27°$, and

$$\frac{x + y}{2} = \frac{35° + 27°}{2} = 31°$$

$$\frac{x - y}{2} = 4°$$

Therefore, $\sin 35° + \sin 27° = 2 \sin 31° \cos 4°$.

You sometimes will use these product and sum identities to prove other identities.

Example 4   Prove $\dfrac{\sin 7\gamma + \sin 5\gamma}{\cos 7\gamma - \cos 7\gamma} = -\cot \gamma.$

Solution

$$\frac{\sin 7\gamma + \sin 5\gamma}{\cos 7\gamma - \cos 5\gamma} = \frac{2 \sin\left(\dfrac{7\gamma + 5\gamma}{2}\right)\cos\left(\dfrac{7\gamma - 5\gamma}{2}\right)}{-2 \sin\left(\dfrac{7\gamma + 5\gamma}{2}\right)\sin\left(\dfrac{7\gamma - 5\gamma}{2}\right)}$$

$$= \frac{2 \sin 6\gamma \cos \gamma}{-2 \sin 6\gamma \sin \gamma}$$

$$= -\frac{\cos \gamma}{\sin \gamma}$$

$$= -\cot \gamma$$ ⬚

Problem Set 3.7

A   *Write each of the expressions in Problems 1–12 as the sum or difference of two functions.*

1. $2 \cos 75° \cos 35°$ **2.** $2 \cos 46° \cos 18°$ **3.** $2 \sin 35° \sin 24°$

4. $2 \sin 53° \cos 24°$ **5.** $\sin 70° \sin 88°$ **6.** $\cos 53° \cos 70°$

7. $\sin 41° \cos 19°$ **8.** $\cos 115° \sin 200°$ **9.** $\sin 225° \sin 300°$

10. $\sin 2\theta \sin 5\theta$ **11.** $\cos \theta \cos 3\theta$ **12.** $\cos 3\theta \sin 2\theta$

Write each of the expressions in Problems 13–24 as a product of two functions.

13. $\sin 43° + \sin 63°$ **14.** $\sin 22° - \sin 6°$ **15.** $\cos 81° - \cos 79°$

16. $\cos 78° + \cos 25°$ **17.** $\sin 215° + \sin 300°$ **18.** $\cos 25° - \cos 100°$

19. $\sin x - \sin 2x$ **20.** $\cos 5x - \cos 3x$ **21.** $\sin x + \sin 2x$

22. $\cos 3\theta + \cos 2\theta$ **23.** $\cos 5y + \cos 9y$ **24.** $\sin 6z - \sin 9z$

B   *Prove each of the identities given in Problems 25–32.*

25. $\cos x - \cos y = -2 \sin\left(\dfrac{x + y}{2}\right)\sin\left(\dfrac{x - y}{2}\right)$ **26.** $\sin x + \sin y = 2 \sin\left(\dfrac{x + y}{2}\right)\cos\left(\dfrac{x - y}{2}\right)$

27. $\sin x - \sin y = 2 \sin\left(\dfrac{x - y}{2}\right)\cos\left(\dfrac{x + y}{2}\right)$ **28.** $\left(\sin \dfrac{\theta}{2} + \cos \dfrac{\theta}{2}\right)^2 = 1 + \sin \theta$

29. $\cos^2 \dfrac{\theta}{2} - \sin^2 \dfrac{\theta}{2} = \cos \theta$ **30.** $\dfrac{\sin 5\theta + \sin 3\theta}{\cos 5\theta + \cos 3\theta} = \tan 4\theta$

31. $\dfrac{\cos 3\theta - \cos \theta}{\sin \theta - \sin 3\theta} = \tan 2\theta$ **32.** $\dfrac{\cos 5w + \cos w}{\cos w - \cos 5w} = \dfrac{\cot 2w}{\tan 3w}$

C   **33.** Here is an alternate proof for the formula $\cos(\alpha + \beta) = \cos\alpha\cos\beta - \sin\alpha\sin\beta$.

   a. *Case* i: Let $\alpha = 0$ and $\beta = 0$. Prove the given identity.

   b. *Case* ii: Let $\alpha \neq 0$ and $\beta \neq 0$. Let α be in standard position and β be any angle drawn so that its initial side is along the terminal side of α (see Figure 3.7).

FIGURE 3.7   Figure for proof of sum formula

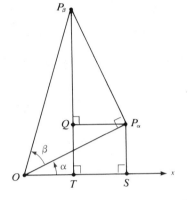

Let P_α be an arbitrary point on the terminal side of α. Let P_β be the point on the terminal side of β so that $P_\alpha P_\beta$ is perpendicular to OP_α. Draw perpendiculars $P_\alpha S$ and $P_\beta T$ to the x-axis. Draw $P_\alpha Q$ perpendicular to $P_\beta T$. Thus,

$$\cos(\alpha + \beta) = \frac{OT}{OP_\beta}$$

Now show

$$\cos(\alpha + \beta) = \frac{OS}{OP_\alpha} \cdot \frac{OP_\alpha}{OP_\beta} - \frac{QP_\alpha}{P_\alpha P_\beta} \cdot \frac{P_\alpha P_\beta}{OP_\beta}$$

and therefore

$$\cos(\alpha + \beta) = \cos\alpha\cos\beta - \sin\alpha\sin\beta$$

3.8   Summary and Review

	OBJECTIVES	PAGES/EXAMPLES
3.1 Proving Identities	**1.** Simplify trigonometric expressions.	p. 103; Examples 2–5; Problems 13–31.
	2. Prove identities using the fundamental identities.	p. 104; Example 1; Problems 1–12.
3.2 Proving Identities (Continued)	**3.** Prove identities by breaking up the fraction into a sum of fractions.	p. 110; Problems 4, 9–12.

---

* Optional section.

22. Given $\cot 2\theta$, find $\cos\theta$, $\sin\theta$, and $\tan\theta$.   p. 130; Example 7; Problems 22–27.

23. Prove identities using the half-angle and double-angle identities.   p. 131; Example 8; Problems 30–38.

3.7 PRODUCT AND SUM IDENTITIES

24. Use the product identities to write a product as a sum.   p. 134; Examples 1 and 2; Problems 1–12.

25. Use the sum identities to write a sum as a product.   p. 134; Example 3; Problems 13–24.

26. Prove identities using the identities of this chapter.   p. 136; Example 4; Problems 25–32.

Terms

Addition laws [3.5]
Chord length [3.4]
Cofunction identities [3.4]
*Counterexample [3.3]
Distance formula [3.4]

Double-angle identities [3.6]
Half-angle identities [3.6]
Identity [3.1]

Opposite-angle identities [3.4]
Product identities [3.7]
Sum identities [3.7]

Chapter 3 Review of Objectives

The problem numbers correspond to the objectives listed in Section 3.8. (Objectives 1 and 2 apply to Problems 1–8.) Prove the identities in Problems 1–8.

1. $\dfrac{\csc^2\alpha}{1 + \cot^2\alpha} = 1$

2. $2\sec x = \dfrac{\sec^2 x + \tan^2 x + 1}{\sec x}$

3. $\tan\beta + \sec\beta = \dfrac{1 + \csc\beta}{\cos\beta\csc\beta}$

4. $\dfrac{1}{\sin\theta + \cos\theta} + \dfrac{1}{\sin\theta - \cos\theta} = \dfrac{2\sin\theta}{\sin^4\theta - \cos^4\theta}$

5. $\dfrac{\sin^2\theta - \cos^2\theta}{\sin\theta + \cos\theta} = \sin\theta - \cos\theta$

6. $\dfrac{\cos\theta}{\sec\theta} - \dfrac{\sin\theta}{\cos\theta} = \dfrac{\cos\theta\cot\theta - \tan\theta}{\csc\theta}$

7. $\dfrac{1 - \cos x}{1 + \cos x} = \dfrac{2 - 2\cos x - \sin^2 x}{\sin^2 x}$

8. $\dfrac{1 + \tan^2\theta}{\csc\theta} = \sec\theta\tan\theta$

---

* Optional section.

*9. Prove that the given equation is not an identity: $\tan\theta + \csc\theta = \cot\theta$.

*10. Prove or disprove that the given equations are identities.

    a. $(\sin t + \cos t)^2 = 1$ **b.** $\cos t \tan t \csc t \sec t = 1$

11. Suppose the circle given in Figure 3.8 is a unit circle. Find the distance between P_α and P_β.

12. Suppose the circle given in Figure 3.8 is a unit circle. Find the length of chord AP_θ. This is the length of any chord in a unit circle subtended by a central angle θ.

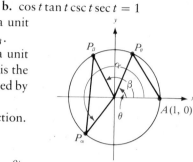

13. Write each function in terms of its cofunction.

    a. $\sin 38°$ **b.** $\tan\frac{\pi}{8}$ **c.** $\cos 2$

14. Write each as a function of $(\theta - \frac{\pi}{6})$.

    a. $\cos(\frac{\pi}{6} - \theta)$ **b.** $\sin(\frac{\pi}{6} - \theta)$ **c.** $\tan(\frac{\pi}{6} - \theta)$

FIGURE 3.8

15. Graph $y - 1 = 2\sin(\frac{\pi}{6} - x)$.

16. Use exact values and the addition laws to write $\cos(\theta - 30°)$ as a function of $\cos\theta$ and $\sin\theta$.

17. Find the exact value of $\sin 105°$.

18. Prove $\cos(\alpha - \beta) = \cos\alpha\cos\beta + \sin\alpha\sin\beta$.

19. Evaluate

$$\frac{2\tan\frac{\pi}{6}}{1 - \tan^2\frac{\pi}{6}}$$

using exact values.

20. If $\cos\theta = \frac{4}{5}$ and θ is in Quadrant IV, find the exact values of $\cos 2\theta$, $\sin 2\theta$, and $\tan 2\theta$.

21. Evaluate

$$-\sqrt{\frac{1 + \cos 240°}{2}}$$

using exact values.

22. If $\cot 2\theta = -\frac{4}{3}$, find the exact values of $\cos\theta$, $\sin\theta$, and $\tan\theta$ where θ is in Quadrant I.

23. If $\cos 2\alpha = 2\cos^2\alpha - 1$, then show

$$\cos\tfrac{1}{2}\theta = \pm\sqrt{\frac{1 + \cos\theta}{2}}$$

where $\theta = 2\alpha$.

24. Write $\sin 3\theta\cos\theta$ as a sum.

25. Write $\sin(x + h) - \sin x$ as a product.

26. Prove $\dfrac{\sin 5\theta + \sin 3\theta}{\cos 5\theta - \cos 3\theta} = -\cot\theta$.

---

* Optional section.

Miscellaneous Problems

Prove that each of the equations in Problems 1–38 is an identity.

1. $\dfrac{\sec^2 u}{\tan^2 u} = \csc^2 u$

2. $\dfrac{\sec u}{\tan^2 u} = \cot u \csc u$

3. $\sec^2 \gamma + 2 \tan \gamma = \dfrac{\sec^2 \gamma \cos \gamma + 2 \sin \gamma}{\cos \gamma}$

4. $\sec \gamma + \tan \gamma = \dfrac{\csc \gamma + 1}{\csc \gamma \cos \gamma}$

5. $\tan 3\theta \sin 3\theta = \sec 3\theta - \cos 3\theta$

6. $\csc 3\theta - \sin 3\theta = \cot 3\theta \cos 3\theta$

7. $1 + \sin^2 \lambda = 2 - \cos^2 \lambda$

8. $2 - \sin^2 3\lambda = 1 + \cos^2 3\lambda$

9. $\dfrac{\sin \alpha}{\tan \alpha} + \dfrac{\cos \alpha}{\cot \alpha} = \cos \alpha + \sin \alpha$

10. $\dfrac{1}{1 + \cos 2\alpha} + \dfrac{1}{1 - \cos 2\alpha} = 2 \csc^2 2\alpha$

11. $\sec \beta + \cos \beta = \dfrac{2 - \sin^2 \beta}{\cos \beta}$

12. $2 \sin^2 3\beta - 1 = 1 - 2 \cos^2 3\beta$

13. $(\tan 5\beta - 1)(\tan 5\beta + 1) = \sec^2 5\beta - 2$

14. $\tan^2 7\alpha = (\sec 7\alpha - 1)(\sec 7\alpha + 1)$

15. $(\sin \gamma - \cos \gamma)^2 = 1 - 2 \sin \gamma \cos \gamma$

16. $(\sin \gamma + \cos \gamma)(\sin \gamma - \cos \gamma) = 2 \sin^2 \gamma - 1$

17. $\dfrac{\sin \gamma}{\cos \gamma} + \cot \gamma = \sec \gamma \csc \gamma$

18. $\tan 3\gamma + \cot 3\gamma = \sec 3\gamma \csc 3\gamma$

19. $(\sin A \cos A \cos B + \sin B \cos B \cos A) \sec A \sec B = \sin A + \sin B$

20. $(\cos A \cos B \tan A + \sin A \sin B \cot B) \csc A \sec B = 2$

21. $\sin \lambda = \dfrac{\sin^2 \lambda + \sin \lambda \cos \lambda + \sin \lambda}{\sin \lambda + \cos \lambda + 1}$

22. $\dfrac{1 + \cot 2\lambda \sec 2\lambda}{\tan 2\lambda + \sec 2\lambda} = \cot 2\lambda$

23. $\sin 2\alpha \cos 2\alpha (\tan 2\alpha + \cot 2\alpha) = 1$

24. $(\sin \beta - \cos \beta)^2 + (\sin \beta + \cos \beta)^2 = 2$

25. $\dfrac{1 - \cos \theta}{1 + \cos \theta} = \left(\dfrac{1 - \cos \theta}{\sin \theta}\right)^2$

26. $\dfrac{(\sec^2\gamma + \tan^2\gamma)^2}{\sec^4\gamma - \tan^4\gamma} = 1 + 2\tan^2\gamma$

27. $\dfrac{(\cos^2\gamma - \sin^2\gamma)^2}{\cos^4\gamma - \sin^4\gamma} = 2\cos^2\gamma - 1$

28. $(\sec 2\theta + \csc 2\theta)^2 = \dfrac{1 + 2\sin 2\theta \cos 2\theta}{\cos^2 2\theta \sin^2 2\theta}$

29. $\dfrac{1}{\sec\theta + \tan\theta} = \sec\theta - \tan\theta$

30. $\dfrac{1 + \tan^3\theta}{1 + \tan\theta} = \sec^2\theta - \tan\theta$

31. $\dfrac{1 - \sec^3\theta}{1 - \sec\theta} = \tan^2\theta + \sec\theta + 2$

32. $\csc(-\theta) = -\csc\theta$

33. $\dfrac{\cot 3\theta - \sin 3\theta}{\sin 3\theta} = \dfrac{\cos^2 3\theta + \cos 3\theta - 1}{\sin^2 3\theta}$

34. $\cot\beta - \sin\beta = (\cos^2\beta + \cos\beta - 1)\csc\beta$

35. $\dfrac{\tan(\alpha + \beta) - \tan\beta}{1 + \tan(\alpha + \beta)\tan\beta} = \tan\alpha$

36. $\dfrac{\tan\alpha - \tan(\alpha - \beta)}{1 + \tan\alpha\tan(\alpha - \beta)} = \tan\beta$

37. $\tan(\alpha + \beta + \gamma) = \dfrac{\tan\alpha + \tan\beta + \tan\gamma - \tan\alpha\tan\beta\tan\gamma}{1 - \tan\beta\tan\gamma - \tan\alpha\tan\gamma - \tan\alpha\tan\beta}$

38. $\cot(\alpha + \beta + \gamma) = \dfrac{\cot\alpha\cot\beta\cot\gamma - \cot\alpha - \cot\beta - \cot\gamma}{\cot\beta\cot\gamma + \cot\alpha\cot\gamma + \cot\alpha\cot\beta - 1}$

Find the exact value of each expression in Problems 39–48.

39. $\cos 345°$

40. $\sin(\theta - 60°)$ in terms of $\sin\theta$ and $\cos\theta$

41. $\sin 50° \cos 20° - \cos 50° \sin 20°$

42. $2\cos^2 15° - 1$

43. $\dfrac{\tan 15° + \tan 30°}{1 - \tan 15° \tan 30°}$

44. $\dfrac{\tan 120° - \tan 60°}{1 + \tan 120° \tan 60°}$

45. $\dfrac{2\tan 30°}{1 - \tan^2 30°}$

46. $\dfrac{1 - \cos 120°}{\sin 120°}$

47. $\sqrt{\dfrac{1 - \cos 270°}{2}}$

48. $-\sqrt{\dfrac{1 + \cos 270°}{2}}$

49. If $\sin\theta = \frac{4}{5}$ and θ is in Quadrant II, find the exact value of $\cos 2\theta$.

50. If $\cot 2\theta = \frac{4}{3}$, find the exact value of $\cos\theta$, $\theta > 90°$.

APPLICATION FOR FURTHER STUDY

SUNRISE, SUNSET, AND LINEAR INTERPOLATION

COURTESY OF SARA HUNSAKER

The most obvious example of a periodic phenomenon is the rotation of the earth, which produces our day/night cycles. Suppose we consult a table that gives the times of sunrise and sunset. These times depend on the latitude of the location, as illustrated by Table 3.1 on page 145, which gives the times of sunrise and sunset for selected cities. If the sunrise and sunset for latitude 35°N are plotted as shown in Figure 3.9, you can see that they are definitely periodic, although not quite sine or cosine curves. However, if you plot the length of daylight, you will find that the result is nearly a sine curve (see Problem 7 in this section).

If you wish to use Table 3.1 for your own town's latitude and it is not listed, you can use a procedure called **linear interpolation**. Your local Chamber of Commerce will probably be able to tell you your town's latitude. The latitude for Santa Rosa, California, is about 38°, which falls between 35° and 40° on the table (see Figure 3.10).

ANGLES   TIMES

35°   7:08   *Known, from Table 3.1*

38° x *Unknown; this is the latitude of the town for which the times of sunrise and sunset are not available.*

40°   7:22   *Known, from Table 3.1*

Use the known information and a proportion to find x:

$$5\left[3\left[\begin{array}{cc} 35° & 7:08 \\ 38° & x \end{array}\right]? \atop 40° \quad 7:22 \right]14$$

FIGURE 3.9   Sunrise and sunset
times for latitude 35°N

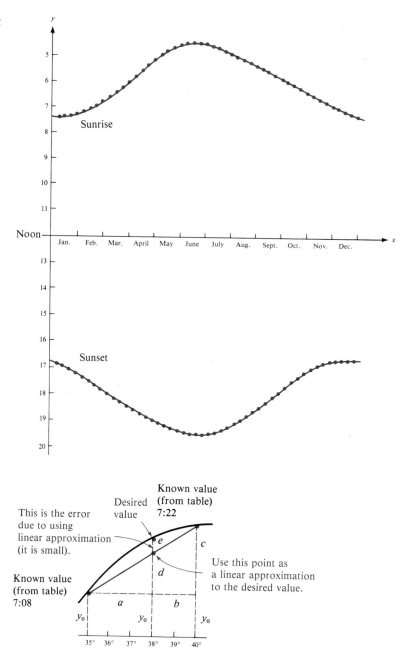

FIGURE 3.10   Linear
interpolation

TABLE 3.1 Times of Sunrise and Sunset for Various Latitudes

Date	Time of Sunrise						Time of Sunset					
	20°N LATITUDE (HAWAII)	30°N LATITUDE (NEW ORLEANS)	35°N LATITUDE (ALBUQUERQUE)	40°N LATITUDE (PHILADELPHIA)	45°N LATITUDE (MINNEAPOLIS)	60°N LATITUDE (ALASKA)	20°N LATITUDE (HAWAII)	30°N LATITUDE (NEW ORLEANS)	35°N LATITUDE (ALBUQUERQUE)	40°N LATITUDE (PHILADELPHIA)	45°N LATITUDE (MINNEAPOLIS)	60°N LATITUDE (ALASKA)
	h m	h m	h m	h m	h m	h m	h m	h m	h m	h m	h m	h m
Jan. 1	6 35	6 56	7 08	7 22	7 38	9 03	17 31	17 10	16 58	16 44	16 28	15 03
Jan. 15	6 38	6 57	7 08	7 20	7 35	8 48	17 41	17 22	17 12	16 59	16 44	15 31
Jan. 30	6 36	6 52	7 01	7 11	7 23	8 19	17 51	17 35	17 27	17 17	17 05	16 09
Feb. 14	6 30	6 41	6 48	6 55	7 03	7 42	17 59	17 48	17 42	17 34	17 26	16 48
Mar. 1	6 20	6 26	6 30	6 34	6 39	6 59	18 05	17 59	17 56	17 52	17 47	17 27
Mar. 16	6 08	6 09	6 10	6 11	6 11	6 15	18 10	18 09	18 08	18 08	18 07	18 04
Mar. 31	5 55	5 51	5 49	5 46	5 43	5 29	18 14	18 18	18 20	18 23	18 26	18 41
Apr. 15	5 42	5 34	5 28	5 23	5 16	4 44	18 18	18 27	18 32	18 38	18 45	19 18
Apr. 30	5 32	5 18	5 11	5 02	4 51	4 01	18 23	18 37	18 44	18 53	19 04	19 55
May 15	5 24	5 07	4 57	4 45	4 31	3 23	18 29	18 46	18 56	19 08	19 22	20 31
May 30	5 20	5 00	4 48	4 34	4 18	2 53	18 35	18 55	19 07	19 21	19 38	21 04
June 14	5 20	4 58	4 45	4 30	4 13	2 37	18 40	19 02	19 15	19 30	19 48	21 24
June 29	5 23	5 02	4 49	4 34	4 16	2 40	18 43	19 05	19 18	19 33	19 51	21 26
July 14	5 29	5 08	4 56	4 43	4 26	3 01	18 43	19 03	19 15	19 29	19 45	21 09
July 29	5 34	5 17	5 07	4 55	4 41	3 33	18 39	18 56	19 06	19 17	19 31	20 38
Aug. 13	5 39	5 26	5 18	5 09	4 59	4 09	18 30	18 43	18 51	19 00	19 10	19 59
Aug. 28	5 43	5 35	5 29	5 24	5 17	4 45	18 19	18 27	18 33	18 38	18 45	19 16
Sept. 12	5 47	5 43	5 40	5 38	5 35	5 20	18 06	18 10	18 12	18 14	18 17	18 31
Sept. 27	5 50	5 51	5 51	5 52	5 53	5 55	17 52	17 51	17 50	17 49	17 49	17 45
Oct. 12	5 54	6 00	6 03	6 07	6 11	6 31	17 39	17 33	17 29	17 26	17 21	17 01
Oct. 22	5 57	6 06	6 12	6 18	6 25	6 56	17 32	17 22	17 17	17 11	17 04	16 32
Nov. 6	6 04	6 18	6 26	6 35	6 45	7 35	17 23	17 10	17 02	16 52	16 42	15 52
Nov. 21	6 12	6 30	6 40	6 52	7 05	8 12	17 19	17 02	16 51	16 40	16 26	15 19
Dec. 6	6 22	6 42	6 54	7 07	7 23	8 44	17 20	17 00	16 48	16 35	16 19	14 58
Dec. 21	6 30	6 52	7 04	7 18	7 35	9 02	17 26	17 05	16 52	16 38	16 21	14 54

The data in this table are courtesy of the U.S. Naval Observatory. This table of times of sunrise and sunset may be used in any year of the 20th century with an error not exceeding two minutes and generally less than one minute. The times are fairly accurate anywhere in the vicinity of the stated latitude, with an error of less than one minute for each 9 miles from the given latitude.

Since $38°$ is $\frac{3}{5}$ of the way between $35°$ and $40°$, the sunrise time will be $\frac{3}{5}$ of the way between 7:08 and 7:22:

$$\frac{3}{5} = \frac{x}{14}$$

$$x = \frac{3}{5}(14)$$

$= 8.4$ *To the nearest minute, this is* :08.

The sunrise time is about $7:08 + :08 = 7:16$.

Problems for Further Study: Sunrise, Sunset, and Linear Interpolation

Use linear interpolation and Table 3.1 to approximate the times requested in each of Problems 1–6.

1. Sunset for Santa Rosa (lat. 38°N) on January 1.
2. Sunrise for Tampa, Florida (lat. 28°N), on June 29.
3. Sunset for Tampa, Florida (lat. 28°N), on June 29.
4. Sunset for Winnipeg, Canada (lat. 50°N), on October 12.
5. Sunrise for Winnipeg, Canada (lat. 50°N), on October 12.
6. Sunrise for Juneau, Alaska (lat. 58°N), on March 1.
7. Plot the length of daylight in Chattanooga, Tennessee (lat. 35°N).
8. Plot the time of sunrise and sunset for Seward, Alaska (lat. 60°N).
9. Plot the length of daylight for Seward, Alaska (lat. 60°N).
10. *Astronomy.* Plot the sunrise and sunset for your town's latitude by drawing a graph similar to that in Figure 3.9.
11. *Astronomy.* Plot the length of daylight for your town from the data in Problem 10.

CHAPTER 4

TRIGONOMETRIC EQUATIONS
AND INVERSE FUNCTIONS

CHAPTER OVERVIEW

In Chapter 3 we introduced the concept of a trigonometric identity. Suppose you are given an open equation with one or more trigonometric functions. The equation is either true for all permissible replacements of the variable or it is not. That is, it is or is not an identity. If it is not an identity, then it is simply called a **trigonometric equation,** and in this chapter you will learn how to solve trigonometric equations. Remember, **to solve an equation** means to find all replacements for the variable that make the equation true.

4.1   INTRODUCTION TO TRIGONOMETRIC EQUATIONS

Consider the trigonometric equation

$$\cos \theta = \tfrac{1}{2}$$

It is a conditional equation, since some values of θ make it true and some make it false. For example,

$\theta = 40°$ is not a solution since $\cos 40° \neq \tfrac{1}{2}$.

$\theta = 420°$ is a solution since $\cos 420° = \tfrac{1}{2}$.

If you are given

$$\cos \theta = x$$

and wish to solve for θ, you must find the set of all values, or replacements, of θ that make the equation true. In other words, you want to find *the angle* or real number θ whose cosine is x. A notation has been invented to denote this idea:

$$\theta = \cos^{-1} x \quad \text{or} \quad \theta = \arccos x$$

CALCULATOR GRAPHICS

You can use a calculator with graphics capability to approximate solutions to equations. For example, in the text we are looking for the solution of the equation $\cos x = \frac{1}{2}$. On the Casio fx-7000G you can first graph $y = \cos x$ and then graph the function $y = \frac{1}{2}$. This is shown at the right. The solution of the equation $\cos x = \frac{1}{2}$ is found using the trace function. Press

Shift   Trace

and watch the cursor move along the line $y = .5$; the displays when it is closest to the points of intersection are

$X = -64.14893617$

$X = 56.4893617$

$X = 297.7659574$

$X = 418.4042553$

$X = 659.6808511$

On the other hand, if you graph $y = \cos x$ and then trace along the cosine curve you obtain the following (x, y) values:

$(-64.14898617°, .436633193)$

$(65.106938298°, .420937624)$

$(297.7659574°, .4658609794)$

$(418.4042553°, .5239276476)$

$(659.6808511°, .495168343)$

Since $y = .5$, you can see from the above components that there is a great deal of approximation. Do not be misled about a calculator's accuracy because it shows you 10 or 12 decimal places. You could redraw these curves using the factor key to obtain better accuracy. However, a picture such as the one shown here can help to give you a good intuitive idea about what you are looking for when solving an equation.

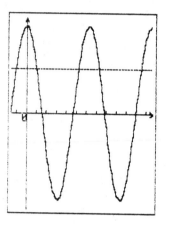

Graph of the function $y = \cos x$ and $y = \frac{1}{2}$ on a Casio fx-7000G.

For the graph shown above the following choices were made for the domain and range values (we are working in degree mode for this problem):

Xmin: -90

  max: 720

  scl: 30

Ymin: -1.1

  max: 1.1

  scl: .1

First Graph $y = \cos x$; then

  Graph $y = .5$

INVERSE
TRIGONOMETRIC
RELATIONS

$\theta = \arccos x$ and $\theta = \cos^{-1} x$ are equivalent to $\cos \theta = x$.
  θ is the angle whose cosine is equal to x.

$\theta = \arcsin x$ and $\theta = \sin^{-1} x$ are equivalent to $\sin \theta = x$.
  θ is the angle whose sine is equal to x.

$\theta = \arctan x$ and $\theta = \tan^{-1} x$ are equivalent to $\tan \theta = x$.
  θ is the angle whose tangent is equal to x.

$\theta = \text{arcsec } x$ and $\theta = \sec^{-1} x$ are equivalent to $\sec \theta = x$.
  θ is the angle whose secant is equal to x.

$\theta = \text{arccsc } x$ and $\theta = \csc^{-1} x$ are equivalent to $\csc \theta = x$.
  θ is the angle whose cosecant is equal to x.

$\theta = \text{arccot } x$ and $\theta = \cot^{-1} x$ are equivalent to $\cot \theta = x$.
  θ is the angle whose cotangent is equal to x.

WARNING!

$$(\cos x)^{-1} = \frac{1}{\cos x} \text{ but}$$

$$\cos^{-1} x \neq \frac{1}{\cos x}$$

Remember, $\cos^{-1} x$ means inverse cosine, and does NOT represent the reciprocal of cosine.

The expression $\cos^{-1} x$ is pronounced "inverse cosine," and the other functions are similarly pronounced. Note the warning in the margin; the -1 notation is NOT an exponent, but is used as part of the notation for the inverse.

We use the inverse notation to solve trigonometric equations, but first we need to practice using this notation. We talk about **evaluating** an inverse trigonometric relation. It means the same thing as it does when referring to evaluating trigonometric functions—namely, to find the value. To do this we can use a table of exact values, tables, or calculators.

Solving Equations with Exact Values

Example 1

Solve $\sin \theta = \frac{1}{2}$.

Solution

Use the definition to write $\sin \theta = \frac{1}{2}$ as $\theta = \sin^{-1} \frac{1}{2}$. We have solved for θ, but have not yet evaluated (i.e., found the numerical value). To evaluate, notice that this means that θ is the angle whose sine is $\frac{1}{2}$. Since this is an exact value we know $\theta = 30°$ or $\frac{\pi}{6}$, but we also need to remember that sine is positive (notice that $\frac{1}{2}$ is a positive number and $\sin \theta$ is equal to this positive number) in Quadrants I and II. Thus, we use our table of exact values to find the angle in Quadrant I and then the reduction principle to put the angle in Quadrant II: *What is the angle in Quadrant* II *whose reference angle is* 30°? The answer is 150°. You can check your work by finding $\sin 30° = \sin 150° = \frac{1}{2}$. Do these values form the complete solution set for θ? The answer is no; since we know that sine is periodic (with period 360°) we can add multiples of 360° to each of these values:

$$\theta = 30° + 360°k \quad \text{and} \quad \theta = 150° + 360°k$$

where k is any integer.

The positive angles less than one revolution are called the **primary angles**.

Example 2

Solve $2 \cos \theta = -1$.

Solution

Algebraically, solve first for the function $\cos \theta$ by dividing both sides by 2:

$$\cos \theta = -\tfrac{1}{2}$$

Next, use the definition of inverse cosine to solve for θ: $\theta = \cos^{-1}(-\tfrac{1}{2})$. Compare this with Example 1 and you see it is very similar except that here the function is negative (notice $-\tfrac{1}{2}$ is a negative number and $\cos \theta$ is equal to this negative number). To evaluate we use our knowledge of exact values. In the first quadrant, $\cos 60° = \tfrac{1}{2}$ (from the table of exact values) and cosine is negative in Quadrants II and III so we ask the following question: *What are the angles in Quadrants* II *and* III *whose reference angle is* 60°? We find these values (120° and 240°), and then add multiples of 360° to each:

$$\theta = 120° + 360°k \quad \text{and} \quad \theta = 240° + 360°k$$

where k is any integer.*   ☐

Even though we will sometimes solve trigonometric equations using degree measure (as in Examples 1 and 2), we will more commonly solve them using real numbers (radian measure). In fact, we agree from now on to give all answers using real numbers unless degree answers are specifically requested.

Example 3

Solve $\sec \theta = -2$.

Solution

First, solve for θ: $\theta = \sec^{-1}(-2)$. We recognize this as an exact value and in the first quadrant $\sec \tfrac{\pi}{3} = 2$. In this problem we see the secant is negative and therefore must be in Quadrants II and III. *What angles in Quadrants* II *and* III *have a reference angle of* $\tfrac{\pi}{3}$?

$$\theta = \tfrac{2\pi}{3} + 2\pi k \quad \text{and} \quad \theta = \tfrac{4\pi}{3} + 2\pi k$$

Notice that 2π has been added to the angles in the second and third quadrants.   ☐

Solving Equations by Table or by Calculator

In all the above examples, the angle could be found by a knowledge of exact values. Not all equations, however, will involve exact values. For those that do not, we will use Appendix D Table or a calculator.

Example 4

Solve $\sin \theta = -.4695$.

Solution

First solve: $\theta = \sin^{-1}(-.4695)$. By Appendix D Table: Use .4695 to find the reference angle, θ'; the value of sine that comes the closest to .4695 is .49, so $\theta' \approx .49$. Now use this reference angle to find the primary angles in Quadrants III and IV;

---

* In these problems it will be assumed that k is any integer. You can use this in your answers without saying "k is any integer" on each problem.

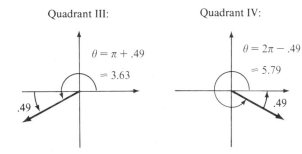

and finally add multiples of 2π:

$\theta = 3.63 + 6.28k$ (Quadrant III; $\pi + \theta'$)

$\theta = 5.79 + 6.28k$ (Quadrant IV; $2\pi - \theta'$)

By calculator, we will first find the reference angle by using .4695:

PRESS: .4695 INV sin*   DISPLAY: 0.4887244

Use this reference angle ($\theta' \approx .4887244$) to place it in Quadrants III and IV (it is a good idea to store the displayed number and then use it for each quadrant):

$\theta = 3.6303 + 2\pi k$ and $\theta = 5.7945 + 2\pi k$ ⬜

You might ask why we use reference angles when finding the angle using a calculator. Why not simply input the negative value as follows:

.4695 +/− INV sin   DISPLAY: −0.4887244

This answer (to four places) is $-.4887$, is negative, and is not equal to one of the primary angles 3.6303 or 5.7945; it is, however, in the list of solutions for Example 4:

$5.7945 + 2\pi(-1) \approx -.4887$

This negative angle, however, does not make it easy to find the value in Quadrant III. We will discuss this more completely in Section 4.3, but you should note that the calculator does not give solutions to equations, but values for what are called the **inverse trigonometric functions**. These functions will be defined in Section 4.3, and your calculators have been preprogrammed to give the values of these functions. In the first quadrant, however, the values of the trigonometric functions give us the reference angles that we need to solve trigonometric equations.

---

* Many calculators will use the second-function key instead of INV.

Example 5

Solve $\cos 2\theta = 1$.

Solution

Notice the difference between this equation and the one given in Example 2.
First, solve for θ:

$$2\theta = \cos^{-1} 1$$
$$\theta = \tfrac{1}{2}\cos^{-1} 1$$

Next, evaluate this expression: since $\cos 0 = 1$ we see that $\cos^{-1} 1 = 0 + 2\pi k$; thus,

$$\theta = \tfrac{1}{2}(0 + 2\pi k) = \pi k$$ ▯

Algebraic Solution for the Angle

In Examples 1–5, we solved some simple trigonometric equations. The next step is to practice the algebraic steps necessary to put them into a form that can be easily evaluated. You can see from Example 5 that this process can look complicated. You need to distinguish the function from the angle. Both of these must be distinguished from the unknown. Consider

$$\underbrace{\cos(2\overset{\downarrow}{x} + 1)}= 0$$

The unknown is x.
The angle is $2x + 1$.
The function is cosine.

The steps in solving a trigonometric equation are now given.

PROCEDURE FOR SOLVING TRIGONOMETRIC EQUATIONS

1. Solve for a single trigonometric function. You may use identities, factoring, or the quadratic formula.

2. Solve for the angle. You will use the definition of the inverse trigonometric functions for this step.

3. Solve for the unknown.

Example 6

Solve for the trigonometric function first; next name the angle; finally, solve for the unknown. **Do not evaluate.**

a. $5\sin(\theta + 1) = \tfrac{1}{2}$ **b.** $\sin 5(\theta + 1) = \tfrac{1}{2}$
c. $(3\cos\theta + 2)(5\cos\theta - 4) = 0$ **d.** $\tan^2\theta - 5\tan\theta - 4 = 0$

Solution

a. $5\sin(\theta + 1) = \tfrac{1}{2}$

$\underbrace{\sin(\theta + 1)}_{\uparrow} = .1$ *Divide both sides by 5.*

The angle is $(\theta + 1)$.

$\theta + 1 = \sin^{-1} .1$

$\theta = \sin^{-1} .1 - 1$ *This is solved for the unknown; notice that this is not the same as* $\sin^{-1}(.1 - 1)$; *to avoid this possible confusion, we usually write* $\theta = -1 + \sin^{-1} .1$.

b. $\sin 5(\theta + 1) = \frac{1}{2}$ *This is solved for the trigonometric function as given.*

The angle is $5(\theta + 1)$.

$$5(\theta + 1) = \sin^{-1}\tfrac{1}{2}$$
$$5\theta + 5 = \sin^{-1}\tfrac{1}{2}$$
$$5\theta = -5 + \sin^{-1}\tfrac{1}{2}$$
$$\theta = -1 + \tfrac{1}{5}\sin^{-1}\tfrac{1}{2} \quad \textit{This is solved for the unknown.}$$

c. $(3\cos\theta + 2)(5\cos\theta - 4) = 0$

This is solved by setting each factor equal to 0:

$$3\cos\theta + 2 = 0 \quad \text{and} \quad 5\cos\theta - 4 = 0$$
$$3\cos\theta = -2 \qquad\qquad 5\cos\theta = 4$$
$$\cos\theta = -\tfrac{2}{3} \qquad\qquad \cos\theta = \tfrac{4}{5} \quad \textit{This is solved for the function.}$$

The angle in both of these equations is the same as the unknown, namely θ. Finally, solve for θ:

$$\theta = \cos^{-1}\left(-\tfrac{2}{3}\right) \quad \text{and} \quad \cos^{-1}\tfrac{4}{5}$$

d. $\tan^2\theta - 5\tan\theta - 4 = 0$ cannot be solved by factoring, so use the quadratic formula*

$$\tan\theta = \frac{5 \pm \sqrt{25 - 4(1)(-4)}}{2}$$

$$= \frac{5 \pm \sqrt{41}}{2} \quad \textit{This is solved for the function.}$$

The angle is θ and is the same as the unknown:

$$\theta = \tan^{-1}\left(\frac{5 \pm \sqrt{41}}{2}\right) \qquad\qquad \square$$

Solving General Trigonometric Equations

We conclude this section by solving a variety of trigonometric equations. Notice that the procedure is the same as in Example 6 except for one added step—namely, the evaluation step. Since we know that multiple values are found by adding 2π to the angles found between 0 and 2π, we often ask for only those positive solutions less than one revolution.

---

* Remember the quadratic formula: If $ax^2 + bx + c = 0$, and $a \neq 0$, then $x = \dfrac{-b \pm \sqrt{b^2 - 4ac}}{2a}$.

Example 7

Solve $2\cos\theta\sin\theta = \sin\theta$ for $0 \le \theta < 2\pi$ (exact values).

Solution

This problem is solved by factoring.

$2\cos\theta\sin\theta - \sin\theta = 0$	*Obtain a 0 on one side.*
$\sin\theta(2\cos\theta - 1) = 0$	*Factor.*
$\sin\theta = 0$ and $2\cos\theta - 1 = 0$	*Set each factor equal to 0.*

WARNING: Do not divide both sides by $\sin\theta$. Notice that if $\theta = 0$ or π, then $\sin\theta = 0$ and you cannot divide by 0.

$$\cos\theta = \frac{1}{2}$$

$$\theta = 0, \pi \qquad\qquad \theta = \frac{\pi}{3}, \frac{5\pi}{3}$$

$$\left\{0, \pi, \frac{\pi}{3}, \frac{5\pi}{3}\right\}$$

Example 8

Solve $2\sin^2\theta = 1 + 2\sin\theta$ for $0 \le \theta < 2\pi$ (real-number solution correct to four decimal places).

Solution

This problem is solved by the quadratic formula, since you cannot factor the left-hand side.

$$2\sin^2\theta - 2\sin\theta - 1 = 0 \qquad \textit{First obtain a 0 on one side.}$$

$$\sin\theta = \frac{2 \pm \sqrt{4 - 4(2)(-1)}}{2(2)} \qquad \textit{Remember the quadratic formula.}$$

$$= \frac{1 \pm \sqrt{3}}{2}$$

$$\approx -.3660254038$$

$$1.366025404 \quad \longleftarrow \quad \textit{Reject, since } -1 \le \sin\theta \le 1$$

Solve $\theta' = \sin^{-1}(.3660254038)$ by using Appendix D Table or a calculator to find a reference angle of .3747. Since the sine is negative in Quadrants III and IV, the solutions are:

Quadrant III: Quadrant IV:

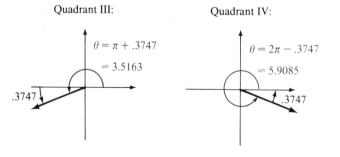

$\theta = \pi + .3747$ $\theta = 2\pi - .3747$

≈ 3.5163 ≈ 5.9085

The entire calculator process starting with $\theta = \sin^{-1}\left(\dfrac{1 - \sqrt{3}}{2}\right)$ is shown here:

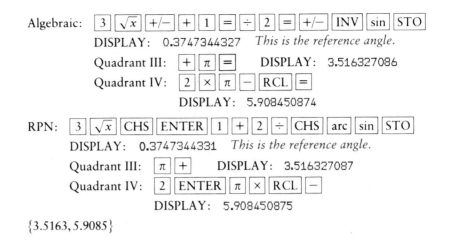

Algebraic: $\boxed{3}\;\boxed{\sqrt{x}}\;\boxed{+/-}\;\boxed{+}\;\boxed{1}\;\boxed{=}\;\boxed{\div}\;\boxed{2}\;\boxed{=}\;\boxed{+/-}\;\boxed{\text{INV}}\;\boxed{\sin}\;\boxed{\text{STO}}$

DISPLAY: 0.3747344327   *This is the reference angle.*

Quadrant III: $\boxed{+}\;\boxed{\pi}\;\boxed{=}$ DISPLAY: 3.516327086

Quadrant IV: $\boxed{2}\;\boxed{\times}\;\boxed{\pi}\;\boxed{-}\;\boxed{\text{RCL}}\;\boxed{=}$

DISPLAY: 5.908450874

RPN: $\boxed{3}\;\boxed{\sqrt{x}}\;\boxed{\text{CHS}}\;\boxed{\text{ENTER}}\;\boxed{1}\;\boxed{+}\;\boxed{2}\;\boxed{\div}\;\boxed{\text{CHS}}\;\boxed{\text{arc}}\;\boxed{\sin}\;\boxed{\text{STO}}$

DISPLAY: 0.3747344331   *This is the reference angle.*

Quadrant III: $\boxed{\pi}\;\boxed{+}$ DISPLAY: 3.516327087

Quadrant IV: $\boxed{2}\;\boxed{\text{ENTER}}\;\boxed{\pi}\;\boxed{\times}\;\boxed{\text{RCL}}\;\boxed{-}$

DISPLAY: 5.908450875

$\{3.5163, 5.9085\}$

In the process of solving for the function, you may need to use one or more identities, as illustrated by Example 9.

Example 9

Solve $\cos^2 x - \sin^2 x = \sin x$ for $0 \le x < 2\pi$ (exact values).

Solution

$$\cos^2 x - \sin^2 x = \sin x$$
$$(1 - \sin^2 x) - \sin^2 x = \sin x$$
$$1 - 2\sin^2 x = \sin x$$
$$2\sin^2 x + \sin x - 1 = 0$$
$$(2\sin x - 1)(\sin x + 1) = 0$$

$$\sin x = \tfrac{1}{2} \qquad \text{and} \quad \sin x = -1$$
$$x = \sin^{-1}(\tfrac{1}{2}) \qquad \qquad x = \sin^{-1}(-1)$$
$$= \tfrac{\pi}{6}, \tfrac{5\pi}{6} \qquad \qquad = \tfrac{3\pi}{2} \qquad \text{*Remember } 0 \le x < 2\pi.\text{*}$$

$\{\tfrac{\pi}{6}, \tfrac{5\pi}{6}, \tfrac{3\pi}{2}\}$

As illustrated by Example 9, if different functions appear in the equation, use the fundamental identities to express the equation in terms of a single function. If it is necessary to multiply both sides by a trigonometric function or to square both sides, be sure to check for extraneous roots.

Example 10

Solve $\tan x + \sqrt{3} = \sec x$ for $0 \le x < 2\pi$ (exact values).

Solution

A better procedure is to square both sides rather than to introduce radicals.

$$\tan^2 x + 2\sqrt{3}\tan x + 3 = \sec^2 x$$
$$\tan^2 x + 2\sqrt{3}\tan x + 3 = 1 + \tan^2 x \qquad \textit{Write the equations in terms of a}$$
$$2\sqrt{3}\tan x = -2 \qquad \qquad \textit{single function, if possible.}$$
$$\tan x = \frac{-1}{\sqrt{3}} = -\frac{1}{3}\sqrt{3}$$

Since the tangent is negative in Quadrants II and IV, use exact values to find $x' = \tan^{-1}(\frac{1}{3}\sqrt{3}) = \frac{\pi}{6}$, which means the reference angle is $\frac{\pi}{6}$. Place this reference angle in the appropriate quadrants:

$$x = \frac{5\pi}{6} \quad \textit{Quadrant II } (\pi - x' \text{ or } \pi - \tfrac{\pi}{6})$$

$$x = \frac{11\pi}{6} \quad \textit{Quadrant IV } (2\pi - x' \text{ or } 2\pi - \tfrac{\pi}{6})$$

Check for extraneous roots in the original equation by substituting possible solutions into the original equation.

$$x = \frac{5\pi}{6}: \quad \tan\frac{5\pi}{6} + \sqrt{3} = -\frac{\sqrt{3}}{3} + \sqrt{3} \quad \text{and} \quad \sec\frac{5\pi}{6} = \frac{-2\sqrt{3}}{3}$$

$$= \frac{2\sqrt{3}}{3}$$

These are not the same, so $\frac{5\pi}{6}$ is extraneous.

$$x = \frac{11\pi}{6}: \quad \tan\frac{11\pi}{6} + \sqrt{3} = -\frac{\sqrt{3}}{3} + \sqrt{3} \quad \text{and} \quad \sec\frac{11\pi}{6} = \frac{2\sqrt{3}}{3}$$

$$= \frac{2\sqrt{3}}{3}$$

These are the same, so $\frac{11\pi}{6}$ checks.

$$\left\{ \frac{11\pi}{6} \right\}$$

The third step in the procedure for solving trigonometric equations is used when the angle is not the same as the unknown. This process will be discussed in the next section.

Problem Set 4.1

A   *Evaluate each of the expressions in Problems 1–5 to the nearest degree (use exact values when possible).*

1. a. $\sin^{-1}\frac{1}{2}$ **b.** $\arccos\frac{1}{2}$ **2. a.** $\cos^{-1}(-\frac{\sqrt{3}}{2})$ **b.** $\arcsin(-\frac{\sqrt{2}}{2})$

3. a. $\arcsin .3907$ **b.** $\arccos .2924$ **4. a.** $\arctan\sqrt{3}$ **b.** $\text{arcsec}\sqrt{2}$

5. a. $\text{arccsc}\frac{2}{3}\sqrt{3}$ **b.** $\text{arccot } 1$

Evaluate each of the expressions in Problems 6–10 to the nearest hundredth of a radian (use exact values when possible).

6. a. $\arcsin(-\frac{1}{2})$ **b.** $\cos^{-1}(-\frac{1}{2})$ **7. a.** $\arccos\frac{\sqrt{3}}{2}$ **b.** $\sin^{-1}\frac{\sqrt{2}}{2}$

8.   **a.** $\arcsin(-\frac{\sqrt{3}}{2})$ **b.** $\cos^{-1}(-\frac{\sqrt{2}}{2})$ 9.   **a.** $\operatorname{arccot}(-1)$ **b.** $\operatorname{arccsc}(-\sqrt{2})$

10.   **a.** $\operatorname{arcsec}(-\frac{2}{3}\sqrt{3})$ **b.** $\arctan(-\frac{1}{3}\sqrt{3})$

Solve each of the equations in Problems 11–19. Use exact values.

11.   **a.** $\sin x = \frac{1}{2}$ $(0 \le x < \frac{\pi}{2})$ **b.** $\sin x = \frac{1}{2}$ $(0° \le x < 90°)$

12.   **a.** $\sin x = \frac{1}{2}$ (radians)   **b.** $\sin x = -\frac{1}{2}$ (radians)

13.   **a.** $\sin x = \frac{1}{2}$ $(0 \le x < \pi)$ **b.** $\sin x = -\frac{1}{2}$ $(0 \le x < \pi)$

14. $(\sin x)(\cos x) = 0$ $(0° \le x < 360°)$ 15. $(\sin x)(\tan x) = 0$ $(0° \le x < 360°)$

16. $(2\cos x + \sqrt{2})(2\cos x - 1) = 0$ $(0 \le x < 2\pi)$

17. $(3\tan x + \sqrt{3})(3\tan x - \sqrt{3}) = 0$ $(0 \le x < 2\pi)$

18. $(\sec x - 2)(2\sin x - 1) = 0$ $(0 \le x < 2\pi)$

19. $(\csc x - 2)(2\cos x - 1) = 0$ $(0 \le x < 2\pi)$

B

Solve each of the equations in Problems 20–29 for $0 \le x < 2\pi$. Use exact values where possible, but state approximate answers to four decimal places.

20. $\tan^2 x = \tan x$ 21. $\tan^2 x = \sqrt{3}\tan x$ 22. $\cos^2 x = \frac{1}{2}$

23. $\sin^2 x = \frac{1}{2}$ 24. $2\cos x \sin x = \sin x$ 25. $3\sin x \cos x = \sin x$

26. $\cos^2 x - 1 - \cos x = 0$ 27. $\sin^2 x - \sin x - 2 = 0$

28. $\tan^2 x - 3\tan x + 1 = 0$ 29. $4\tan^2 x - 8\tan x + 3 = 0$

Refraction. If a light ray passes from one medium to another of greater or lesser density, the ray will bend. A measure of this tendency is called the *index of refraction, R*, and is defined by

$$R = \frac{\sin \alpha}{\sin \beta}$$

For example, if the angle of incidence, α, of a light ray entering a diamond is 30.0° and the angle of refraction, β, is 11.9°, then the diamond's index of refraction is

$$R = \frac{\sin 30.0°}{\sin 11.9°} \approx 2.42$$

Since the index of refraction for any particular substance is a constant (see Table 4.1), we can compare our calculated number to the known index number to determine if this diamond is an imitation. The index of refraction of an imitation diamond is rarely greater than 2. If a stone is a

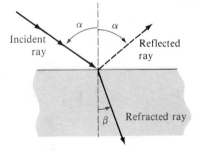

TABLE 4.1

Substance	Index of refraction
Diamond	2.42
Topaz	1.63
Garnet	1.78
Glass	1.50
Water	1.33
Ice	1.31

diamond and the angle of incidence is 45.0°, what should be its refracted angle?

$$R = \frac{\sin \alpha}{\sin \beta} \quad \text{where } R \approx 2.42, \alpha = 45°$$

Solve for β and substitute known values to find the refracted angle β:

$$\sin \beta = \frac{\sin \alpha}{R}$$

$$\beta = \sin^{-1}\left(\frac{\sin \alpha}{R}\right) \approx 16.95° \quad (16.99° \text{ if you work with the rounded value for } R)$$

To the nearest tenth, the refracted angle is 17.0°.

30. If a stone is a topaz and the angle of incidence is 20.0°, what should be its refracted angle (correct to the nearest tenth of a degree)?

31. If a stone is a garnet and the angle of incidence is 35.0°, what should be its refracted angle (correct to the nearest tenth of a degree)?

C   *Solve each of the equations in Problems 32–39, for $0 \leq x < 2\pi$. Use exact values where possible, but state approximate answers to four decimal places.*

32. $\sin^2 x - \sin x - 1 = 0$ 33. $\cos^2 x - \cos x - 1 = 0$ 34. $\sin x + 1 = \sqrt{3}$

35. $\cot x + \sqrt{3} = \csc x$ 36. $\cos^2 x - 3 \sin x + 3 = 0$ 37. $\sec^2 x - 1 = \sqrt{3} \tan x$

38. $2 \sin^2 x - 2 \cos^2 x = 1$ 39. $\sin^2 x + \cos x = 0$

*40. For $0 \leq \theta < \frac{\pi}{2}$, solve

$$\left(\frac{16}{81}\right)^{\sin^2\theta} + \left(\frac{16}{18}\right)^{1-\sin^2\theta} = \frac{26}{27}$$

41. How many solutions are there for the equation

$$\frac{x}{100} = \sin x$$

---

†4.2   TRIGONOMETRIC EQUATIONS WITH MULTIPLE ANGLES

There are two main procedures for solving trigonometric equations with multiple angles. The first is by substitution, and the second is by using identities. Neither method will be adequate for all equations, so you need to study both procedures.

If the unknown and the angle are not the same, you can let θ equal the angle and then solve the resulting equation as shown in the last section. In doing this, you will need to pay special attention to the domain, as shown in Example 1.

---

* Problem 40 is from *The Mathematics Student Journal* of the National Council of Teachers of Mathematics, April 1973, 20(4). Reprinted by permission.
† Optional section.

Example 1

Solve for $0 \leq x < 2\pi$, giving both exact values and calculator approximations correct to two decimal places.

a. $\sin x = \frac{1}{2}$ **b.** $\sin 2x = \frac{1}{2}$

c. $\sin 3x = -\frac{1}{2}$ **d.** $\sin(2x + 1) = \frac{1}{2}$

Solution

a. $\sin x = \frac{1}{2}$ *The reference angle (from the table of exact values) is $\frac{\pi}{6}$. This reference angle should be placed in Quadrants I and II since sine is positive.*

$$x = \tfrac{\pi}{6}, \tfrac{5\pi}{6} \longleftarrow \text{Quadrant II}$$
$$\phantom{x = \tfrac{\pi}{6}, \tfrac{5\pi}{6}} \longleftarrow \text{Quadrant I}$$

These are the only values of x such that $0 \leq x < 2\pi$. Giving the calculator approximations, the solution set is

$$\{.52, 2.62\} \quad \textit{All of these answers should be between 0 and 6.28}$$
$$\textit{(approximate values correct to two decimal places).}$$

b. The equation in part a was the type of equation solved in the last section. This example, $\sin 2x = \frac{1}{2}$, has an angle $(2x)$ different from the unknown (which is x). Let $\theta = 2x$. Solve $\sin \theta = \frac{1}{2}$. From part a, the reference angle is $\frac{\pi}{6}$. **But you must now consider the domain:** if $\theta = 2x$, then $x = \frac{\theta}{2}$. That is, if

$$0 \leq x < 2\pi$$

then, in terms of θ, this is

$$0 \leq \tfrac{\theta}{2} < 2\pi$$
$$0 \leq \theta < 4\pi \quad \textit{Multiply all three parts of the inequality by 2.}$$

You need to find values of θ over a larger domain than that of x:

2π added to the second-quadrant solution

$$\theta = \tfrac{\pi}{6}, \tfrac{5\pi}{6}, \tfrac{13\pi}{6}, \tfrac{17\pi}{6} \quad \textit{Additional multiples of 2π are greater than 4π.}$$

2π added to the first-quadrant solution

Now, substitute values for x:

$$2x = \tfrac{\pi}{6} \qquad 2x = \tfrac{5\pi}{6} \qquad 2x = \tfrac{13\pi}{6} \qquad 2x = \tfrac{17\pi}{6}$$
$$x = \tfrac{\pi}{12} \qquad x = \tfrac{5\pi}{12} \qquad x = \tfrac{13\pi}{12} \qquad x = \tfrac{17\pi}{12}$$

$$\{.26, 1.31, 3.40, 4.45\} \quad \textit{Notice that all of these roots are between 0 and 6.28.}$$

c. $\sin 3x = -\frac{1}{2}$; let $\theta = 3x$. Then $x = \frac{\theta}{3}$ and

$$0 \leq x < 2\pi$$
$$0 \leq \tfrac{\theta}{3} < 2\pi$$
$$0 \leq \theta < 6\pi \quad \textit{Multiply by 3.}$$

From part a, the reference angle is $\frac{\pi}{6}$, but this time it needs to be placed in Quadrants III and IV (because of the negative value):

Quadrant III, *reference angle* $\frac{\pi}{6}$

Quadrant IV, *reference angle* $\frac{\pi}{6}$

$\theta = \frac{7\pi}{6}, \frac{11\pi}{6}, \frac{19\pi}{6}, \frac{23\pi}{6}, \frac{31\pi}{6}, \frac{35\pi}{6}$ *Additional multiples of 2π are greater than 6π.*

$+2\pi$ $+2\pi$

$+2\pi$ $+2\pi$

Since $\theta = 3x$, replace θ by $3x$ and solve for x in each of the six equations to obtain $\left\{ \frac{7\pi}{18}, \frac{11\pi}{18}, \frac{19\pi}{18}, \frac{23\pi}{18}, \frac{31\pi}{18}, \frac{35\pi}{18} \right\}$. The solution set using approximate values can now be found by using a calculator: $\{1.22, 1.92, 3.32, 4.01, 5.41, 6.11\}$.

d. $\sin(2x + 1) = \frac{1}{2}$; let $\theta = 2x + 1$. Then $\frac{\theta - 1}{2} = x$ and $0 \le x < 2\pi$.

$0 \le \dfrac{\theta - 1}{2} < 2\pi$

$0 \le \theta - 1 < 4\pi$ *Multiply all parts by 2.*

$1 \le \theta < 4\pi + 1$ *Add 1 to all parts.*

This says that you will include only values of θ that are between 1 and about 13.57.

From part a   *Multiples of 2π added to answers form part a*

$\theta = \dfrac{\pi}{6}, \dfrac{5\pi}{6}, \dfrac{13\pi}{6}, \dfrac{17\pi}{6}, \dfrac{25\pi}{6}, \dfrac{29\pi}{6}, \ldots$

Include

Reject $\frac{29\pi}{6} \approx 15.18$, *which is greater than 13.57. Also reject additional larger multiples of 2π.*

Reject (less than 1)

Now each of the included values for θ is used in an equation to find x:

$\theta = \dfrac{5\pi}{6}$ $\theta = \dfrac{13\pi}{6}$ $\theta = \dfrac{17\pi}{6}$ $\theta = \dfrac{25\pi}{6}$

$2x + 1 = \dfrac{5\pi}{6}$ $2x + 1 = \dfrac{13\pi}{6}$ $2x + 1 = \dfrac{17\pi}{6}$ $2x + 1 = \dfrac{25\pi}{6}$

$2x = \dfrac{5\pi}{6} - 1$ $2x = \dfrac{13\pi - 6}{6}$ $x = \dfrac{17\pi - 6}{12}$ $x = \dfrac{25\pi - 6}{12}$

$x = \dfrac{5\pi - 6}{12}$ $x = \dfrac{13\pi - 6}{12}$

The solution set (using approximate values) is

$$\{.81, 2.90, 3.95, 6.04\} \quad \textit{Notice that all of the roots are between } 0 \textit{ and } 2\pi.$$ □

The second method for handling multiple angles is to use one or more trigonometric identities, as illustrated by Example 2.

Example 2

Solve $\sin 2x = \cos x$ for $0 \le x < 2\pi$, giving both exact values and calculator approximations correct to two decimal places.

Solutions

$$\sin 2x = \cos x \quad \textit{Given}$$

$$2 \sin x \cos x = \cos x \quad \textit{Use the double-angle identity } \sin 2x = 2 \sin x \cos x.$$

$$2 \sin x \cos x - \cos x = 0$$

$$\cos x (2 \sin x - 1) = 0 \quad \textit{Complete the solution as shown in Section 4.1.}$$

$$\cos x = 0 \qquad \sin x = \tfrac{1}{2}$$

$$x = \tfrac{\pi}{2}, \tfrac{3\pi}{2} \qquad x = \tfrac{\pi}{6}, \tfrac{5\pi}{6}$$

$$\{\tfrac{\pi}{2}, \tfrac{3\pi}{2}, \tfrac{\pi}{6}, \tfrac{5\pi}{6}\} \quad \text{or} \quad \{.52, 1.57, 2.62, 4.71\}$$ □

Problem Set 4.2

A   *Solve each of the equations in Problems 1–9 for exact values such that $0 \le x < 2\pi$.*

1. $\cos 2x = \tfrac{1}{2}$ 2. $\cos 3x = \tfrac{1}{2}$ 3. $\cos 2x = -\tfrac{1}{2}$

4. $\sin 2x = \tfrac{\sqrt{2}}{2}$ 5. $\sin 2x = -\tfrac{\sqrt{3}}{2}$ 6. $\sin 3x = \tfrac{\sqrt{2}}{2}$

7. $\tan 3x = 1$ 8. $\tan 3x = -1$ 9. $\sec 2x = -\tfrac{2\sqrt{3}}{3}$

B   *Solve each of the equations in Problems 10–30 for $0 \le x < 2\pi$ correct to two decimal places.*

10. $2 \cos 2x \sin 2x = \sin 2x$ 11. $\sin 2x + 2 \cos x \sin 2x = 0$

12. $\cos 3x + 2 \sin 2x \cos 3x = 0$ 13. $\sin 2x = \cos 2x$

14. $\tan 2x = \cot 2x$ 15. $\sin \tfrac{x}{2} = \cos \tfrac{x}{2}$

16. $\sin 2x + 1 = \sqrt{3}$ 17. $\cos 3x - 1 = \sqrt{2}$

18. $\tan 2x + 1 = \sqrt{3}$ 19. $\sin(2x + 1) = \tfrac{1}{2}$

20. $\cos(3x - 1) = \tfrac{1}{2}$ 21. $\tan(2x + 1) = \sqrt{3}$

22. $\cos 2x = \cos x$ 23. $\cos 2x = \sin x$

24. $\cos 2x + \cos x = 0$ 25. $1 - \sin x = \cos 2x$

26. $\cos x = \sin 2x$ 27. $\sin^2 3x + \sin 3x + 1 = 1 - \sin^2 3x$

28. $\sin^2 3x + \sin 3x = \cos^2 3x - 1$

29. $3 \sin 4x - \sin 3x + \sqrt{3} = 2 \sin 2x - \sin 3x + 3 \sin 4x$

30. $3 \cos x - \cos 3x + 2 \cos 5x = 1 + 3 \cos x - \cos 3x$

C   31.   *Sound waves.* A tuning fork vibrating at 264 Hz ($f = 264$) with an amplitude of .0050 cm produces C on the musical scale and can be described by an equation of the form

$$y = .0050 \sin 528\pi x$$

Find the smallest positive value of x (correct to four decimal places) for which $y = .0020$.

32.   *Electrical.* In a certain electric circuit, the electromotive force V (in volts) and the time t (in seconds) are related by an equation of the form

$$V = \cos 2\pi t$$

Find the smallest positive value for t (correct to three decimal places) for which $V = .400$.

33.   *Space science.* The orbit of a certain satellite alternates above and below the equator according to the equation

$$y = 4,000 \sin\left(\tfrac{\pi}{45}t + \tfrac{5\pi}{18}\right)$$

where t is the time (in minutes) and y is the distance (in kilometers) from the equator. Find the times the satellite crosses the equator during the first hour and a half (that is, for $0 \le t \le 90$).

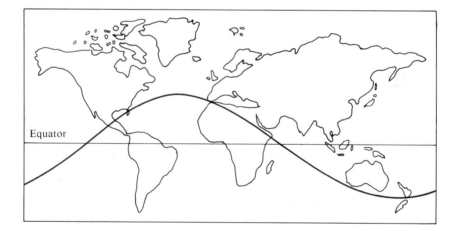

Equator

4.3 INVERSE TRIGONOMETRIC FUNCTIONS

Principal Values

One of the troublesome aspects about solving trigonometric equations is the multiple solutions. For example, in Section 4.1, we found the solution of $\cos\theta = .5$ is $\theta = \cos^{-1}.5 = \frac{\pi}{3}$ (Quadrant I), $\frac{5\pi}{3}$ (Quadrant IV), $\frac{7\pi}{3}$ (Quadrant I, first value $+ 2\pi$), $\frac{11\pi}{3}$ (Quadrant IV, second value $+ 2\pi$), ... One of the simplifying agreements we used was to limit our answers to positive angles less than one revolution. However, even with that restriction, the inverse relation has more than one solution: $\theta = \cos^{-1}.5 = \frac{\pi}{3}, \frac{5\pi}{3}$. Since there is more than one value, the inverse cosine is not a function.*

Our goal now is to limit the domain of the inverse cosine so that it *is* a function. We begin by defining a **principal value** for each of the inverse relations. The designation that we want principal values rather than all values is denoted by using a capital letter, as shown in Table 4.2. When the capital letter is used it tells us to give a single value, and as such can properly be called an **inverse trigonometric function** to distinguish it from an inverse trigonometric relation. With this definition we see $\text{Cos}^{-1}.5 = \frac{\pi}{3}$ (.5 is positive, so choose the single value in Quadrant I).

TABLE 4.2   Definition of Principal Values

| | | Quadrant of Principal Value, y | | |
| | | Value of x can be | | |
Inverse Relation	**Inverse Function**	POSITIVE	NEGATIVE	0
$y = \cos^{-1} x$ or $y = \arccos x$	$y = \text{Cos}^{-1} x$ or $y = \text{Arccos}\, x$	I	II	$\frac{\pi}{2}$
$y = \sin^{-1} x$ or $y = \arcsin x$	$y = \text{Sin}^{-1} x$ or $y = \text{Arcsin}\, x$	I	IV	0
$y = \tan^{-1} x$ or $y = \arctan x$	$y = \text{Tan}^{-1} x$ or $y = \text{Arctan}\, x$	I	IV	0
$y = \sec^{-1} x$ or $y = \text{arcsec}\, x$	$y = \text{Sec}^{-1} x$ or $y = \text{Arcsec}\, x$	I	II	undefined
$y = \csc^{-1} x$ or $y = \text{arccsc}\, x$	$y = \text{Csc}^{-1} x$ or $y = \text{Arccsc}\, x$	I	IV	undefined
$y = \cot^{-1} x$ or $y = \text{arccot}\, x$	$y = \text{Cot}^{-1} x$ or $y = \text{Arccot}\, x$	I	II	undefined

Answers in Quadrant I should be between 0 and $\frac{\pi}{2}$ (that is, between 0 and about 1.57 or, in degrees, between 0° and 90°); answers in Quadrant II should be between $\frac{\pi}{2}$ and $\frac{3\pi}{2}$ (that is, between about 1.57 and 3.14 or, in degrees, between 90° and 180°); answers in Quadrant IV should be negative numbers between $-\frac{\pi}{2}$ and 0 (that is, between about -1.57 and 0 or, in degrees, between $-90°$ and 0°).

The definition of principal value means that instead of giving multiple values for the inverse relations you will give only one value chosen from the appropriate quadrant as shown in Table 4.2.

Example 1

Evaluate the given inverse trigonometric functions.

a. $\text{Arccos}\, \frac{\sqrt{2}}{2}$ **b.** $\text{Arccos}\left(-\frac{\sqrt{2}}{2}\right)$
c. $\text{Cos}^{-1}(-.4695)$ **d.** $\text{Cos}^{-1} 2$

Solution

a. $\text{Arccos}\, \frac{\sqrt{2}}{2}$ Notice that x is positive so $\text{Arccos}\, \frac{\sqrt{2}}{2}$ is the first-quadrant angle whose cosine is $\frac{\sqrt{2}}{2}$; that is $\frac{\pi}{4} \approx .7854$.

b. $\text{Arccos}\left(-\frac{\sqrt{2}}{2}\right)$ Notice that x is negative so $\text{Arccos}\left(-\frac{\sqrt{2}}{2}\right)$ is the second-quadrant angle whose reference angle is $\frac{\sqrt{2}}{2}$; this is $\frac{3\pi}{4} \approx 2.356$.

___

* Remember the definition of a function. For each member of the domain there can be only one value in the range. In everyday language, this means there can be only one answer if it is a function.

c. $\text{Cos}^{-1}(-.4695)$ Notice that x is negative so $\text{Cos}^{-1}(-.4695)$ is the second-quadrant angle whose value is $-.4695$. Your calculator is preprogrammed to give you the correct principal values, so you can input negative values when using your calculator:

$\boxed{.4695}$ $\boxed{+/-}$ $\boxed{\text{INV}}$ $\boxed{\cos}$ DISPLAY: 2.0595207

d. $\text{Cos}^{-1} 2$ Notice that x is positive and also that this represents the angle whose cosine is 2; there is no such angle (remember, cosine is between -1 and 1). If you press $\boxed{2}$ $\boxed{\text{INV}}$ $\boxed{\cos}$, your calculator will indicate that an error has been made.

You can check you work by making sure you have appropriate answers for the inverse cosine function. All answers for the inverse cosine function should be between 0 and 3.14; more formally, if $y = \text{Arccos}\, x$, then $0 \le y \le \pi$ and $\cos y = x$.   □

WARNING! Remember, with the inverse *functions* we are looking for *one* value; that value is either positive or negative.

From Table 4.2 you can see that if the x value is positive the inverse trigonometric functions are in the first quadrant. For negative values we need to specify either the second or the fourth quadrant, depending on the function. In Example 1 we saw that for negative x values the inverse cosine is a second-quadrant angle. In Example 2 we can see that for negative x values the inverse sine and inverse tangent must be in Quadrant IV rather than Quadrant II. If we tried, for example, to put the inverse sine in Quadrants I and II we see that it would no longer be a function.

Example 2   Evaluate the given inverse trigonometric functions.

a. $\text{Sin}^{-1}\frac{\sqrt{3}}{2}$ **b.** $\text{Sin}^{-1}(-\frac{\sqrt{3}}{2})$ **c.** $\text{Arcsin}(-.4695)$

d. $\text{Arctan}\, 3.5$ **e.** $\text{Tan}^{-1}(-2.3)$

Solution

a. $\text{Sin}^{-1}\frac{\sqrt{3}}{2}$ Notice that x is positive so $\text{Sin}^{-1}\frac{\sqrt{3}}{2}$ is the first-quadrant angle whose sine is $\frac{\sqrt{3}}{2}$; this is $\frac{\pi}{3} \approx 1.0472$.

b. $\text{Sin}^{-1}(-\frac{\sqrt{3}}{2})$ Notice that x is negative so $\text{Sin}^{-1}(-\frac{\sqrt{3}}{2})$ is the fourth-quadrant angle whose reference angle is $\frac{\pi}{3}$; this is $-\frac{\pi}{3}$. Fourth-quadrant angles for the inverse sine function are negative numbers between $-\frac{\pi}{2}$ and 0.

c. $\text{Arcsin}(-.4695)$ Notice that x is negative so $\text{Arcsin}(-.4695)$ is the fourth-quadrant angle whose value is $-.4695$. Your calculator is preprogrammed to give you the correct principal values, so you can input negative values when using your calculator:

$\boxed{.4695}$ $\boxed{+/-}$ $\boxed{\text{INV}}$ $\boxed{\sin}$ DISPLAY: −0.4887244

d. $\text{Arctan}\, 3.5$ PRESS: $\boxed{3.5}$ $\boxed{\text{INV}}$ $\boxed{\tan}$ DISPLAY: 1.2924967

e. $\text{Tan}^{-1}(-2.3)$ PRESS: $\boxed{2.3}$ $\boxed{+/-}$ $\boxed{\text{INV}}$ $\boxed{\tan}$ DISPLAY:
−1.160668986   □

To evaluate arcsecant, arccosecant, or arccotangent relations or functions when their exact values are not known, you need to use the reciprocal identities. First find the reciprocal, then find the inverse of the reciprocal. That is, find the inverse of the reciprocal function. This procedure is summarized by the following *inverse identities*.

INVERSE
TRIGONOMETRIC
IDENTITIES

$$35. \ \text{Arccot} \ x = \begin{cases} \text{Arctan} \dfrac{1}{x} & \text{if } x \text{ is positive} \\[2ex] \pi + \text{Arctan} \dfrac{1}{x} & \text{if } x \text{ is negative} \\[2ex] \dfrac{\pi}{2} & \text{if } x = 0 \end{cases}$$

$$36. \ \text{Arcsec} \ x = \text{Arccos} \dfrac{1}{x} \quad \text{if } x \geq 1 \text{ or } x \leq -1$$

$$37. \ \text{Arccsc} \ x = \text{Arcsin} \dfrac{1}{x} \quad \text{if } x \geq 1 \text{ or } x \leq -1$$

Proof of Identity 35:   Let $\theta = \text{Arccot} \ x$, where x is positive. Then

$$\cot \theta = x, \quad 0 < \theta < \frac{\pi}{2} \qquad \textit{Definition of } \text{Arccot} \ x$$

$$\frac{1}{\tan \theta} = x, \quad 0 < \theta < \frac{\pi}{2} \qquad \textit{Fundamental identity}$$

$$\tan \theta = \frac{1}{x}, \quad 0 < \theta < \frac{\pi}{2} \qquad \textit{Solve for } \tan \theta$$

$$\theta = \text{Tan}^{-1} \frac{1}{x}, \quad 0 < \theta < \frac{\pi}{2} \qquad \textit{Definition of } \text{Arctan} \ x$$

Thus, $\text{Arccot} \ x = \text{Arctan} \frac{1}{x}$ where x is positive. The proof of the second part of Identity 35 is left as an exercise. It is needed because the range of the inverse tangent function does not coincide with the range of the inverse cotangent function. The third part of the identity is obvious since $\cot \frac{\pi}{2} = 0$.

Proofs of Identities 36 and 37 are left as exercises.

Example 3

Find θ using the inverse identities and a calculator.

a. $\text{Arcsec}(-3) = \theta$

b. $\text{Arccsc} \ 7.5 = \theta$

c. $\text{Arccot} \ 2.4747$

d. $\text{Arccot}(-4.852)$

Solution

a. $\text{Arcsec}(-3) = \theta$. This uses Identity 36.

PRESS: $\boxed{3}$ $\boxed{+/-}$ $\boxed{1/x}$ $\boxed{\text{INV}}$ $\boxed{\cos}$ DISPLAY: 1.910633236

b. Arccsc $7.5 = \theta$. This uses Identity 37.

PRESS: ⬚7.5⬚ ⬚1/x⬚ ⬚INV⬚ ⬚sin⬚   DISPLAY: 0.1337315894

c. Arccot $2.4747 = \theta$. This uses the first part of Identity 35 since 2.4747 is positive.

PRESS: ⬚2.4747⬚ ⬚1/x⬚ ⬚INV⬚ ⬚tan⬚   DISPLAY: 0.3840267299

d. Arccot(-4.852). This uses the second part of Identity 35 since -4.852 is negative.

PRESS: ⬚4.852⬚ ⬚+/−⬚ ⬚1/x⬚ ⬚INV⬚ ⬚tan⬚ ⬚+⬚ ⬚π⬚ ⬚=⬚

DISPLAY: 2.938338095   ⬚

We can now summarize the calculator steps for the inverse trigonometric functions.

Inverse Function	Calculator
	ENTER THE VALUE OF x, THEN PRESS:
$y = \text{Arccos}\, x$	⬚x⬚ ⬚inv⬚ ⬚cos⬚
$y = \text{Arcsin}\, x$	⬚x⬚ ⬚inv⬚ ⬚sin⬚
$y = \text{Arctan}\, x$	⬚x⬚ ⬚inv⬚ ⬚tan⬚
$y = \text{Arcsec}\, x$	⬚x⬚ ⬚1/x⬚ ⬚inv⬚ ⬚cos⬚
$y = \text{Arccsc}\, x$	⬚x⬚ ⬚1/x⬚ ⬚inv⬚ ⬚sin⬚
$y = \text{Arccot}\, x$	If $x > 0$: ⬚x⬚ ⬚1/x⬚ ⬚inv⬚ ⬚tan⬚
	If $x < 0$: ⬚x⬚ ⬚1/x⬚ ⬚inv⬚ ⬚tan⬚ ⬚+⬚ ⬚π⬚ ⬚=⬚*

* Use 180 instead of π if working in degrees.

Graphs of the Inverse Trigonometric Functions

The graphs of $y = \cos x$ and $y = \cos^{-1} x$ are shown in Figure 4.1.

FIGURE 4.1   Graphs of $y = \cos x$ and $y = \cos^{-1} x$ (shown in color)

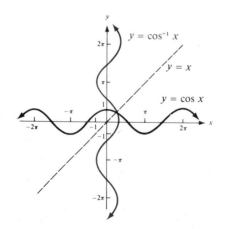

You can see from Figure 4.1 that $y = \cos x$ is a function but $y = \cos^{-1} x$ is not. However, if a suitable restriction is placed on the domain of $y = \cos x$, we will have a function whose *inverse is also a function*. This function with the restricted domain is denoted by $y = \text{Cos } x$ (note the capital letter) and is defined by $y = \cos x$ with the domain restricted to $0 \leq x \leq \pi$. The graph of $y = \text{Cos } x$ is shown in Figure 4.2 in color. Ordered pairs that satisfy $y = \text{Cos } x$ are called the *principal values* of the cosine function.

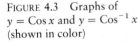

FIGURE 4.2   Graphs of $y = \cos x$ and $y = \text{Cos } x$ (shown in color)

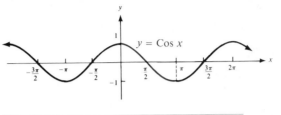

Function	Domain	Range
$y = \cos x$	$-\infty < x < \infty$	$-1 \leq y \leq 1$
$y = \text{Cos } x$	$0 \leq x \leq \pi \longleftarrow$	$-1 \leq y \leq 1$

Restricted domain ⏌

The inverse of $y = \text{Cos } x$ is $y = \text{Cos}^{-1} x$ with domain $-1 \leq x \leq 1$ and range $0 \leq y \leq \pi$. The graphs of $y = \text{Cos } x$ and $y = \text{Cos}^{-1} x$ are shown in Figure 4.3.

FIGURE 4.3   Graphs of $y = \text{Cos } x$ and $y = \text{Cos}^{-1} x$ (shown in color)

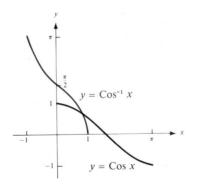

Function	Domain	Range
$y = \text{Cos } x$	$0 \leq x \leq \pi$	$-1 \leq y \leq 1$
$y = \text{Cos}^{-1} x$	$-1 \leq x \leq 1$	$0 \leq y \leq \pi$

The inverse of the sine function, like the inverse cosine, is not a function, so a function $y = \text{Sin } x$ (note the capital letter) is defined as $y = \sin x$ with the domain restricted so that $-\frac{\pi}{2} \leq x \leq \frac{\pi}{2}$. The graph of $y = \sin x$ and the graph of $y = \text{Sin } x$ are compared in Figure 4.4. Ordered pairs that satisfy $y = \text{Sin } x$ are called the *principal values* of the sine function.

FIGURE 4.4   Graphs of
$y = \sin x$ and $y = \operatorname{Sin} x$ (shown
in color)

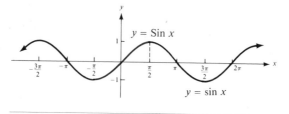

Function	Domain	Range
$y = \sin x$	$-\infty \leq x \leq \infty$	$-1 \leq y \leq 1$
$y = \operatorname{Sin} x$	$-\frac{\pi}{2} \leq x \leq \frac{\pi}{2}$	$-1 \leq y \leq 1$

The inverse of $y = \operatorname{Sin} x$ is $y = \operatorname{Sin}^{-1} x$ with domain $-1 \leq x \leq 1$ and range $-\frac{\pi}{2} \leq y \leq \frac{\pi}{2}$. These graphs are shown in Figure 4.5.

FIGURE 4.5   Graphs of
$y = \operatorname{Sin} x$ and $y = \operatorname{Sin}^{-1} x$

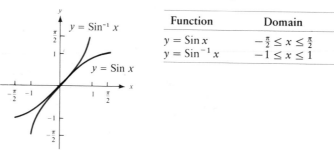

Function	Domain	Range
$y = \operatorname{Sin} x$	$-\frac{\pi}{2} \leq x \leq \frac{\pi}{2}$	$-1 \leq y \leq 1$
$y = \operatorname{Sin}^{-1} x$	$-1 \leq x \leq 1$	$-\frac{\pi}{2} \leq y \leq \frac{\pi}{2}$

The graphs of the other inverse relations and inverse functions are compared in Figure 4.6. Take special note of the range.*

FIGURE 4.6   Graphs of inverse
trigonometric relations and
inverse trigonometric functions

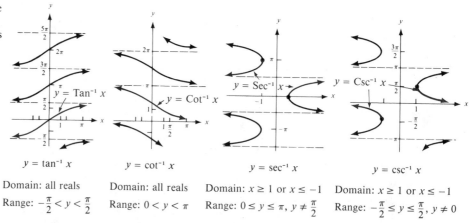

$y = \tan^{-1} x$	$y = \cot^{-1} x$	$y = \sec^{-1} x$	$y = \csc^{-1} x$
Domain: all reals	Domain: all reals	Domain: $x \geq 1$ or $x \leq -1$	Domain: $x \geq 1$ or $x \leq -1$
Range: $-\frac{\pi}{2} < y < \frac{\pi}{2}$	Range: $0 < y < \pi$	Range: $0 \leq y \leq \pi$, $y \neq \frac{\pi}{2}$	Range: $-\frac{\pi}{2} \leq y \leq \frac{\pi}{2}$, $y \neq 0$

* There is no universal agreement about the range of the inverse secant and inverse cosecant functions. In this book, we used those that correspond to the preprogrammed values of your calculator. However, be sure to check these ranges carefully when you take calculus.

Evaluating Functions of Inverse Trigonometric Functions

Remember that the inverse trigonometric functions designate single angles, and as such can be used in any context that calls for angles.

Example 4

Find:

a. Arccos(cos 2)

b. cos(Arccos .5)

c. Arccos(cos 4)

d. sin(Arcsin 2.463)

Solution

When finding these values you must pay attention to the range of the inverse functions.

a. Since Arccosine and cosine are inverse functions and since the angle, 2, is between 0 and π, Arccos(cos 2) = 2.

b. Let θ = Arccos .5. By definition of Arccosine, $\cos\theta$ = .5. Then, by substitution, cos(Arccos .5) = $\cos\theta$ = .5.

c. The angle 4 radians is in the third quadrant, and, therefore, cos 4 is negative. The Arccosine of a negative angle will be a second-quadrant angle. This is the second-quadrant angle having the same reference angle as 4, as shown in Figure 4.7. The reference angle is $4 - \pi \approx .8584073464$. In the second quadrant: π − reference angle ≈ 2.283185307. Therefore, Arccos(cos 4) ≈ 2.283185307. By calculator:

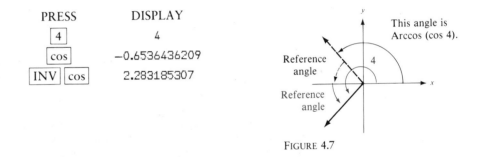

PRESS	DISPLAY
4	4
cos	−0.6536436209
INV cos	2.283185307

FIGURE 4.7

d. sin(Arcsin 2.463) is not defined since 2.463 is not between −1 and +1.　□

Example 5

Evaluate $\cot[\operatorname{Csc}^{-1}(-3)]$ without tables or a calculator.

Solution

Let $\theta = \operatorname{Csc}^{-1}(-3)$.　　　　　　　θ *is in Quadrant* IV.

Then $\csc\theta = -3$.

$$\text{Find } \cot[\operatorname{Csc}^{-1}(-3)] = \cot\theta \qquad \textit{Substitution } \theta = \operatorname{Csc}^{-1}(-3)$$

$$= \pm\sqrt{\csc^2\theta - 1} \qquad \textit{Fundamental identity}$$

$$= -\sqrt{(-3)^2 - 1} \qquad \textit{Substitute } \csc\theta = -3; \textit{ choose the}$$
$$\textit{negative value for the radical because } \theta$$
$$= -\sqrt{8} \qquad\qquad \textit{is in Quadrant } IV \textit{ and cotangent is}$$
$$= -2\sqrt{2} \qquad\qquad \textit{negative in Quadrant } IV. \qquad \square$$

Example 6

Evaluate $\cos(\mathrm{Sin}^{-1}\frac{4}{5} + \mathrm{Cos}^{-1}\frac{3}{5})$ without tables or a calculator.

Solution

Let $x = \mathrm{Sin}^{-1}\frac{4}{5}$ and $y = \mathrm{Cos}^{-1}\frac{3}{5}$; then

$$\sin x = \frac{4}{5} \quad \text{and} \quad \cos y = \frac{3}{5}$$

where x and y are in Quadrant I. Find $\cos(x + y) = \cos x \cos y - \sin x \sin y$.

We know $\sin x$, so we need to find $\cos x$:

$$\cos x = \sqrt{1 - \sin^2 x} \qquad \cos x \text{ is positive}$$
$$= \sqrt{1 - 16/25} \qquad \text{in Quadrant I}$$
$$= \frac{3}{5}$$

We also know $\cos y$, so we need to find $\sin y$:

$$\sin y = \sqrt{1 - \cos^2 y} \qquad \sin y \text{ is positive}$$
$$= \sqrt{1 - 9/25} \qquad \text{in Quadrant I}$$
$$= \frac{4}{5}$$

Therefore,

$$\cos(x + y) = \left(\frac{3}{5}\right)\left(\frac{3}{5}\right) - \left(\frac{4}{5}\right)\left(\frac{4}{5}\right)$$

$$= \frac{9}{25} - \frac{16}{25}$$

$$= -\frac{7}{25}$$

Example 7

Let x be positive. Prove $\sin(\mathrm{Arccos}\,x) = \sqrt{1 - x^2}$.

Solution

Let $\theta = \mathrm{Arccos}\,x$; then $\cos\theta = x$ and θ is in Quadrant I. Find

$$\sin(\mathrm{Arccos}\,x) = \sin\theta \qquad \text{Substitution}$$
$$= \pm\sqrt{1 - \cos^2\theta} \quad \text{Fundamental identity}$$
$$= \sqrt{1 - x^2} \qquad \text{Substitute } \cos\theta = x \text{ and choose the positive value for the radical because } \theta \text{ is in Quadrant I.}$$

Problem Set 4.3

A

In each of Problems 1–6 obtain the given angle (in radians) from memory. Use exact values.

1. **a.** $\mathrm{Arcsin}\,0$ **b.** $\mathrm{Cos}^{-1}\frac{\sqrt{3}}{2}$ **c.** $\mathrm{Sin}^{-1}1$ **d.** $\mathrm{Arccos}\,1$

2. **a.** $\mathrm{Cos}^{-1}\frac{\sqrt{2}}{2}$ **b.** $\mathrm{Arcsin}\frac{1}{2}$ **c.** $\mathrm{Arcsin}\frac{1}{2}\sqrt{2}$ **d.** $\mathrm{Sin}^{-1}\frac{\sqrt{3}}{2}$

3. **a.** $\mathrm{Arcsin}(-1)$ **b.** $\mathrm{Arccos}(-\frac{\sqrt{2}}{2})$ **c.** $\mathrm{Sin}^{-1}(-\frac{1}{2}\sqrt{2})$ **d.** $\mathrm{Cos}^{-1}(-\frac{1}{2})$

4. a. $\mathrm{Tan}^{-1}\frac{\sqrt{3}}{3}$ b. $\mathrm{Arccot}\sqrt{3}$ c. $\mathrm{Arctan}\sqrt{3}$ d. $\mathrm{Tan}^{-1}1$

5. a. $\mathrm{Arccot}\frac{\sqrt{3}}{3}$ b. $\mathrm{Arccot}\,1$ c. $\mathrm{Arccsc}\sqrt{2}$ d. $\mathrm{Cot}^{-1}(-1)$

6. a. $\mathrm{Arcsec}\,2$ b. $\mathrm{Arccsc}(-2)$ c. $\mathrm{Arcsec}(-\sqrt{2})$ d. $\mathrm{Arccot}\frac{\sqrt{3}}{3}$

Use Appendix D Table or a calculator to find the values in Problems 7–12 to the nearest degree.

7. a. $\mathrm{Sin}^{-1}.2588$ b. $\mathrm{Arccos}.3907$ c. $\mathrm{Cos}^{-1}.9397$

8. a. $\mathrm{Sin}^{-1}(-.4695)$ b. $\mathrm{Cos}^{-1}.3584$ c. $\mathrm{Arccos}(-.2672)$

9. a. $\mathrm{Arccos}(-.9598)$ b. $\mathrm{Arcsin}(-.6858)$ c. $\mathrm{Cos}^{-1}(-.3584)$

10. a. $\mathrm{Tan}^{-1}.123$ b. $\mathrm{Cot}^{-1}.341$ c. $\mathrm{Sec}^{-1}3.415$

11. a. $\mathrm{Csc}^{-1}2.816$ b. $\mathrm{Arccsc}\,3.945$ c. $\mathrm{Arccot}(-1)$

12. a. $\mathrm{Arccot}(-2)$ b. $\mathrm{Arctan}(-3)$ c. $\mathrm{Arcsec}(-6)$

Use Appendix D Table or a calculator to find the values (in radians) given in each of Problems 13–18. Round your answers to the nearest hundredth.

13. a. $\mathrm{Arcsin}.20846$ b. $\mathrm{Cos}^{-1}.83646$ c. $\mathrm{Arccos}.94604$

14. a. $\mathrm{Sin}^{-1}(-.32404)$ b. $\mathrm{Arcsin}(-.35227)$ c. $\mathrm{Cos}^{-1}(-.7074)$

15. a. $\mathrm{Cos}^{-1}(-.36236)$ b. $\mathrm{Arccos}(-.80210)$ c. $\mathrm{Arcsin}(-.38942)$

16. a. $\mathrm{Tan}^{-1}1.489$ b. $\mathrm{Cot}^{-1}3.451$ c. $\mathrm{Sec}^{-1}4.315$

17. a. $\mathrm{Csc}^{-1}5.791$ b. $\mathrm{Arccsc}\,2.985$ c. $\mathrm{Arccot}(-3)$

18. a. $\mathrm{Arccot}(-4)$ b. $\mathrm{Arctan}(-2)$ c. $\mathrm{Arcsec}(-5)$

B   *Evaluate the expressions in Problems 19–30 without using tables or a calculator.*

19. a. $\mathrm{Arccos}(\cos\frac{\pi}{6})$ b. $\sin(\mathrm{Arcsin}\frac{1}{3})$ c. $\mathrm{Arcsin}(\sin\frac{2\pi}{15})$

20. a. $\cos(\mathrm{Arccos}\frac{2}{3})$ b. $\sin(\mathrm{Arcsin}.5)$ c. $\cos(\mathrm{Arccos}.4)$

21. $\tan[\mathrm{Sec}^{-1}(-3)]$ 22. $\tan[\mathrm{Sec}^{-1}(-4)]$

23. $\cos[\mathrm{Tan}^{-1}(-\frac{3}{4})]$ 24. $\sin[\mathrm{Tan}^{-1}(-\frac{3}{4})]$

25. $\sin[\mathrm{Cot}^{-1}(-\frac{3}{4})]$ 26. $\cos[\mathrm{Cot}^{-1}(-\frac{3}{4})]$

27. $\sin(\mathrm{Sin}^{-1}\frac{4}{5}-\mathrm{Cos}^{-1}\frac{3}{5})$ 28. $\cos(\mathrm{Sin}^{-1}\frac{3}{5}+\mathrm{Cos}^{-1}\frac{4}{5})$

29. $\cos(\mathrm{Cos}^{-1}\frac{1}{4}-\mathrm{Sin}^{-1}\frac{1}{3})$ 30. $\sin(\mathrm{Cos}^{-1}\frac{2}{3}+\mathrm{Sin}^{-1}\frac{1}{2})$

C   31. Prove Identity 36.   32. Prove Identity 37.

33. Prove $\sec(\mathrm{Arctan}\,x)=\sqrt{x^2+1}$ for $x\geq 0$.

34. Prove $\tan(\mathrm{Arcsec}\,x)=\sqrt{x^2-1}$ for $x\geq 0$.

35. Prove $\cos(\mathrm{Arctan}\,x)=1/\sqrt{1+x^2}$ for $x\geq 0$.

36. Prove $\mathrm{Arccot}\,x=\pi+\mathrm{Arctan}\frac{1}{x}$ if x is negative.

*4.4   A REDUCTION IDENTITY

One application of inverse functions is useful in graphing, in calculus, and in certain engineering applications. It involves an identity for changing the form of a trigonometric expression.

REDUCTION IDENTITY

> **38.** $a \sin\theta + b \cos\theta = \sqrt{a^2 + b^2} \sin(\theta + \alpha)$, where α is chosen so that
>
> $$\cos\alpha = \frac{a}{\sqrt{a^2 + b^2}} \quad \text{and} \quad \sin\alpha = \frac{b}{\sqrt{a^2 + b^2}}$$

The proof of this identity is straightforward after you multiply by 1, written in the form

$$\frac{\sqrt{a^2 + b^2}}{\sqrt{a^2 + b^2}}$$

$$a \sin\theta + b \cos\theta = \frac{\sqrt{a^2 + b^2}}{\sqrt{a^2 + b^2}} (a \sin\theta + b \cos\theta)$$

$$= \sqrt{a^2 + b^2} \left(\frac{a}{\sqrt{a^2 + b^2}} \sin\theta + \frac{b}{\sqrt{a^2 + b^2}} \cos\theta \right)$$

$$= \sqrt{a^2 + b^2} \left(\sin\theta \frac{a}{\sqrt{a^2 + b^2}} + \cos\theta \frac{b}{\sqrt{a^2 + b^2}} \right)$$

Now, let $\alpha = \cos^{-1}\left(\dfrac{a}{\sqrt{a^2 + b^2}} \right)$ so that $\cos\alpha = \dfrac{a}{\sqrt{a^2 + b^2}}$. Then,

$$\sin^2\alpha = 1 - \cos^2\alpha$$

$$= 1 - \left(\frac{a}{\sqrt{a^2 + b^2}} \right)^2$$

$$= 1 - \frac{a^2}{a^2 + b^2}$$

$$= \frac{a^2 + b^2 - a^2}{a^2 + b^2}$$

$$= \frac{b^2}{a^2 + b^2}$$

---

* Optional section.

and

$$\sin \alpha = \frac{b}{\sqrt{a^2 + b^2}}$$ *If α is in Quadrant I or II, b is positive; if α is in Quadrant III or IV, b is negative; so $\pm$ signs are not necessary.*

Therefore, by substitution,

$$a \sin \theta + b \cos \theta = \sqrt{a^2 + b^2} \, (\sin \theta \cos \alpha + \cos \theta \sin \alpha)$$
$$= \sqrt{a^2 + b^2} \sin (\theta + \alpha)$$

Example 1

Change $\dfrac{\sqrt{3}}{2} \sin \theta + \dfrac{1}{2} \cos \theta$ using the reduction identity.

Solution

By inspection, $a = \frac{\sqrt{3}}{2}$ and $b = \frac{1}{2}$, so $\sqrt{a^2 + b^2} = \sqrt{\frac{3}{4} + \frac{1}{4}} = 1$. Find α so that $\cos \alpha = \frac{\sqrt{3}}{2}$ and $\sin \alpha = \frac{1}{2}$. The reference angle α' is $\frac{\pi}{6}$, and since cosine and sine are both positive, α is in Quadrant I: $\alpha = \frac{\pi}{6}$. Therefore,

$$\frac{\sqrt{3}}{2} \sin \theta + \frac{1}{2} \cos \theta = \sin \left(\theta + \frac{\pi}{6} \right)$$

☐

Example 2

Change $-\sin \dfrac{\theta}{2} - \cos \dfrac{\theta}{2}$ by using the reduction identity.

Solution

$a = -1$ and $b = -1$ so that $\sqrt{a^2 + b^2} = \sqrt{2}$. Find α so that $\cos \alpha = \frac{-1}{\sqrt{2}}$ and $\sin \alpha = \frac{-1}{\sqrt{2}}$. The reference angle α' is $\frac{\pi}{4}$, and since cosine and sine are both negative, α is in Quadrant III: $\alpha = \frac{5\pi}{4}$. Therefore,

$$-\sin \frac{\theta}{2} - \cos \frac{\theta}{2} = \sqrt{2} \sin \left(\frac{\theta}{2} + \frac{5\pi}{4} \right)$$

☐

Graphing Sums of Sines and Cosines

You can use the reduction identity when graphing sums of sines and cosines, as shown in Example 3.

Example 3

Graph $y = \sqrt{3} \cos \theta - \sin \theta$ by first using the reduction identity.

Solution

Compare $y = \sqrt{3} \cos \theta - \sin \theta$ to $a \sin \theta + b \cos \theta$ to see that $a = -1$ and $b = \sqrt{3}$. Thus, $\sqrt{a^2 + b^2} = \sqrt{1 + 3} = 2$. Find α so that

$$\cos \alpha = \frac{-1}{2} \quad \text{and} \quad \sin \alpha = \frac{\sqrt{3}}{2}$$

The reference angle for α is $\frac{\pi}{3}$, and sine is positive while cosine is negative, so α must be in Quadrant II: $\alpha = \frac{2\pi}{3}$. Thus,

$$y = \sqrt{3}\cos\theta - \sin\theta = 2\sin\left(\theta + \frac{2\pi}{3}\right)$$

The graph is now easy to frame and complete, as shown in Figure 4.8.

FIGURE 4.8   Graph of
$y = \sqrt{3}\cos\theta - \sin\theta$

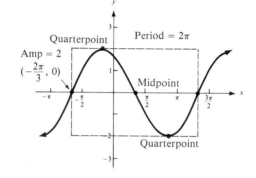

Problem Set 4.4

A   *Change each expression in Problems 1–12 by using the reduction identity to write it as a sine function.*

1. $\sin\theta + \cos\theta$

2. $\sin\theta - \cos\theta$

3. $-\sin\frac{\theta}{2} + \cos\frac{\theta}{2}$

4. $\sin\frac{\theta}{2} - \cos\frac{\theta}{2}$

5. $\frac{1}{2}\sin\theta + \frac{\sqrt{3}}{2}\cos\theta$

6. $\frac{1}{2}\sin\theta - \frac{\sqrt{3}}{2}\cos\theta$

7. $-\frac{1}{2}\sin\theta + \frac{\sqrt{3}}{2}\cos\theta$

8. $\cos\theta - \sin\theta$

9. $-\cos\theta - \sin\theta$

10. $\cos 2\theta - \sqrt{3}\sin 2\theta$

11. $\sqrt{3}\cos\pi\theta - \sin\pi\theta$

12. $\frac{\sqrt{3}}{2}\cos\pi\theta - \frac{1}{2}\sin\pi\theta$

B   *Change each expression in Problems 13–21 by using the reduction identity. Write α as a real number rounded to the nearest tenth.*

13. $2\sin\theta + 3\cos\theta$

14. $3\sin\theta - 2\cos\theta$

15. $-3\sin\theta + 2\cos\theta$

16. $3\cos\theta - 4\sin\theta$

17. $4\cos\theta - 3\sin\theta$

18. $8\sin\theta + 15\cos\theta$

19. $-8\sin\theta - 15\cos\theta$

20. $15\cos\theta - 8\sin\theta$

21. $5\cos\theta - 2\sin\theta$

Use the reduction identity when graphing each equation in Problems 22–31.

22. $y = \sin\theta + \cos\theta$

23. $y = \sin\theta - \cos\theta$

24. $y = \frac{1}{2}\sin\theta + \frac{\sqrt{3}}{2}\cos\theta$

25. $y = \frac{\sqrt{3}}{2}\sin\theta - \frac{1}{2}\cos\theta$

26. $y = 2\sin\theta - 3\cos\theta$

27. $y = 3\sin\theta + 2\cos\theta$

28. $y = \cos\theta - \sin\theta$

29. $y = -\sin\theta - \cos\theta$

30. $y = 8\sin\theta + 15\cos\theta$

31. $y = -8\sin\theta - 15\cos\theta$

4.5 SUMMARY AND REVIEW

	OBJECTIVES	PAGES/EXAMPLES
4.1 INTRODUCTION TO TRIGONOMETRIC EQUATIONS	**1.** Solve trigonometric equations in linear form.	p. 149; Examples 1–6, Problems 11–13.
	2. Solve trigonometric equations by factoring.	p. 152; Examples 6–7; Problems 14–21; 29.
	3. Solve trigonometric equations by using the quadratic formula.	p. 152; Examples 6 and 8; Problems 26; 28.
	4. Solve trigonometric equations by using trigonometric identities.	p. 155; Examples 9 and 10; Problems 35–39.
*4.2 TRIGONOMETRIC EQUATIONS WITH MULTIPLE ANGLES	**5.** Solve trigonometric equations with multiple angles.	p. 158; Example 1; Problems 1–12, 16–21.
	6. Solve trigonometric equations with multiple angles by using trigonometric identities.	p. 161; Example 2; Problems 13–15; 22–30.
4.3 INVERSE TRIGONOMETRIC FUNCTIONS	**7.** Evaluate inverse trigonometric functions by knowing the quadrants for principal values.	p. 162; Examples 1–3; Problems 1–18.
	8. Define the inverse trigonometric relations and the inverse trigonometric functions, including the domain and the range. Draw a quick sketch of each.	p. 163 and pp. 167–168.
	9. Evaluate functions of inverse trigonometric functions.	p. 169; Examples 4 and 5; Problems 19–30.
*4.4 A REDUCTION IDENTITY	**10.** Use the reduction identity to simplify certain trigonometric equations.	p. 172; Examples 1 and 2; Problems 1–21.
	11. Use the reduction identity to graph certain trigonometric equations.	p. 173; Example 3; Problems 22–31.

Terms

arccosecant [4.1]	Arccotangent [4.3]	arctangent [4.1]
Arccosecant [4.3]	arcsecant [4.1]	Arctangent [4.3]
arccosine [4.1]	Arcsecant [4.3]	Inverse function [4.3]
Arccosine [4.3]	arcsine [4.1]	Reduction identity [4.4]
arccotangent [4.1]	Arcsine [4.3]	

---

* Optional section.

Chapter 4 Review of Objectives

The problem numbers correspond to the objectives listed in Section 4.5.

1. Find all real values of θ so that $0 \leq \theta < 2\pi$ and $\tan\theta = 1.557407725$.
2. Solve $4\cos^2\theta = 1$ for $0 \leq \theta < 2\pi$.
3. Solve $\tan^2\theta - 2\tan\theta - 3 = 0$ for $0 \leq \theta < 2\pi$ using radians correct to two decimal places.
4. Solve $\cos^2\theta = 3\sin\theta$ for $0 \leq \theta < 2\pi$ using real numbers correct to two decimal places.
5. Solve $3\cos^2 2\theta = 1$ for $0 \leq \theta < 6.28$, correct to two decimal places.
6. Solve $2\sin^2 2\theta - 2\cos^2 2\theta = 1$ for $0 \leq \theta < 2\pi$ (exact values).
7. Simplify the expressions; in parts a–d use exact values in degrees. In parts e–h use exact values in radians. In parts i and j use real numbers correct to two decimal places.
   - **a.** $\operatorname{Arccos}\frac{\sqrt{3}}{2}$
   - **b.** $\operatorname{Arctan} 1$
   - **c.** $\operatorname{Arccot}\frac{\sqrt{3}}{3}$
   - **d.** $\operatorname{Cos}^{-1}\frac{\sqrt{2}}{2}$
   - **e.** $\operatorname{Cos}^{-1}(-\frac{\sqrt{2}}{2})$
   - **f.** $\operatorname{Cot}^{-1}(-\sqrt{3})$
   - **g.** $\operatorname{Arcsin}\frac{1}{2}$
   - **h.** $\operatorname{Arctan}\frac{\sqrt{3}}{3}$
   - **i.** $\operatorname{Tan}^{-1} 2.310$
   - **j.** $\operatorname{Arcsec} 3.485$
8. Fill in the blanks indicated by the lowercase letters.

Function	Sign of x	Quadrant of θ	Approximate value of θ
$\theta = \operatorname{Arccos} x$	Positive	I	θ between 0 and 1.57
$\theta = \operatorname{Arccos} x$	Negative	II	θ between 1.57 and 3.14
$\theta = \operatorname{Arcsin} x$	Positive	I	θ between 0 and 1.57
$\theta = \operatorname{Arcsin} x$	Negative	IV	θ between -1.57 and 0 (a negative number or angle)
$\theta = \operatorname{Arctan} x$	Positive	a. _____	θ between 0 and 1.57
$\theta = \operatorname{Arctan} x$	Negative	IV	b. _____
$\theta = \operatorname{Arcsec} x$	Positive	c. _____	d. _____
$\theta = \operatorname{Arcsec} x$	Negative	e. _____	θ between 1.57 and 3.14
$\theta = \operatorname{Arccsc} x$	Positive	I	f. _____
$\theta = \operatorname{Arccsc} x$	Negative	g. _____	h. _____
$\theta = \operatorname{Arccot} x$	Positive	I	θ between 0 and 1.57
$\theta = \operatorname{Arccot} x$	Negative	i. _____	j. _____

9. Evaluate $\tan(\operatorname{Sec}^{-1} 3)$ without using tables or a calculator.
*10. In the reduction identity $a\sin\theta + b\cos\theta = \sqrt{a^2 + b^2}\sin(\theta + \alpha)$, what is the proper quadrant for α, given the conditions in parts a–d?
   - **a.** a and b are both positive.
   - **b.** a and b are both negative.
   - **c.** a is positive, b is negative.
   - **d.** a is negative, b is positive.
   - **e.** Use the reduction identity to write $12\sin\theta - 5\cos\theta$ as a function of sine; approximate α to the nearest tenth.
*11. Use the results of Problem 10 to graph $y = 12\sin\theta - 5\cos\theta$.

---

* Optional section.

Miscellaneous Problems

Fill in the domain and range for each of the given inverse trigonometric functions in Problems 1–6.

		DOMAIN	RANGE
1.	$y = \text{Arccos}\, x$	_____	_____
2.	$y = \text{Arcsin}\, x$	_____	_____
3.	$y = \text{Arctan}\, x$	_____	_____
4.	$y = \text{Arcsec}\, x$	_____	_____
5.	$y = \text{Arccsc}\, x$	_____	_____
6.	$y = \text{Arccot}\, x$	_____	_____

Evaluate the expressions in Problems 7–30. Use exact values (radians) where possible; otherwise give your answer as a real number correct to two decimal places.

7. $\sin^{-1}\left(-\frac{\sqrt{3}}{2}\right)$ 8. $\arccos\frac{1}{2}$ 9. $\arcsin .813$

10. $\arccos(-4.521)$ 11. $\cos^{-1}(-.3592)$ 12. $\arctan(-1)$

13. $\text{Arccot}(-1)$ 14. $\text{Cot}^{-1}\pi$ 15. $\text{Cot}^{-1}(-\pi)$

16. $\text{Tan}^{-1}(-\pi)$ 17. $\text{Tan}^{-1}\pi$ 18. $\text{Arcsec}(-\sqrt{2})$

19. $\text{Arccsc}(-\sqrt{2})$ 20. $\text{Arccsc}(-3.4589)$ 21. $\text{Sec}^{-1}(-4.3298)$

22. $\text{Sec}^{-1}\frac{1}{2}$ 23. $\text{Arccot}(-\frac{3}{5})$ 24. $\text{Arctan}(-\frac{3}{5})$

25. $\text{Arcsin}(\sin 4)$ 26. $\text{Arccos}(\cos 5)$ 27. $\text{Arcsin}(\sin 6)$

28. $\sin(\text{Arcsin} .7568)$ 29. $\cos(\text{Arccos} .2836)$ 30. $\cos(\text{Arccos} .4567)$

Solve the equations in Problems 31–50 for $0 \leq \theta < 2\pi$. Use exact values (radians) where possible. Otherwise give your answer as a real number correct to two decimal places.

31. $\sin \theta = \frac{1}{2}$ 32. $\frac{1}{2}\sin \theta = 0$ *33. $\sin\frac{1}{2}\theta = 0$

34. $\cos \theta = \frac{1}{5}$ 35. $4\cos \theta = 1$ *36. $\cos 4\theta = 1$

37. $(3\cos \theta - 1)(2\cos \theta + 1) = 0$ 38. $(4\sin \theta + 1)(3\sin \theta - 1) = 0$

39. $\sin^2\theta - 1 = 0$ 40. $2\cos^2\theta = \cos \theta$

41. $6\cos^2\theta - 7\cos \theta + 2 = 0$ 42. $12\sin^2\theta - \sin \theta - 6 = 0$

43. $6\sec^2\theta - 7\sec \theta + 2 = 0$ 44. $12\tan^2\theta - \tan \theta - 6 = 0$

45. $3\sin^2\theta - 8\sin \theta + 4 = 0$ 46. $5\cos^2\theta + 14\cos \theta - 4 = 0$

47. $3\cot^2\theta - 8\cot \theta + 4 = 0$ 48. $5\csc^2\theta + 14\csc \theta - 4 = 0$

49. **a.** $2\cos \theta - 1 = 0$ **b.** $2\cos(\theta - 1) = 0$

50. **a.** $5\sin \theta + 2 = 0$ **b.** $5\sin(\theta + 2) = 0$

---

* Optional section.

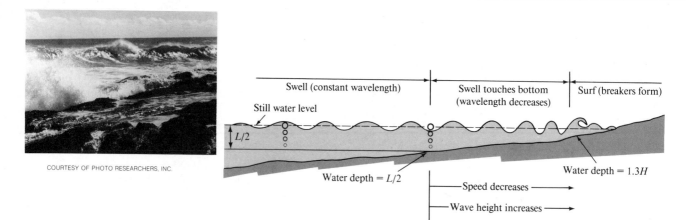

COURTESY OF PHOTO RESEARCHERS, INC.

Swell (constant wavelength)

Swell touches bottom
(wavelength decreases)

Surf (breakers form)

Still water level

$L/2$

Water depth = $L/2$

Water depth = $1.3H$

Speed decreases

Wave height increases

Water Waves

Wave motion is one of the most common examples of periodic motion. There are water waves, sound waves, and radio waves, as well as many other electromagnetic waves. Sometimes waves are classified by considering how the motions of the particles of matter are related to the direction of the waves themselves. This classification system yields waves called **transverse** waves and others called **longitudinal** waves. In a transverse wave, the particles vibrate at right angles to the direction in which the wave itself is propagated. In a longitudinal wave, the particles vibrate in the same direction as that in which the wave is propagated. Both types of waves are shown in Figure 4.9.

FIGURE 4.9   Transverse and longitudinal waves

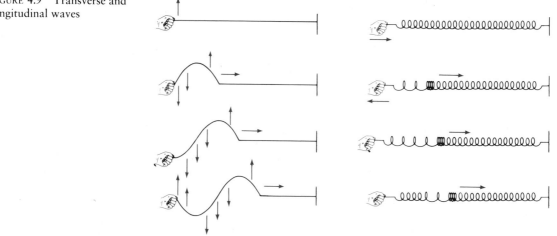

Transverse waves

Longitudinal waves

If we look at water waves, we find the wave action to be a combination of both transverse and longitudinal waves. The water molecules move back and forth and, at the same time, up and down. The path of the molecules makes a circle or an ellipse. (See Figure 4.10.) In their simplest form, water waves can be shown to be sine curves represented by an equation of the form

$$y = a \sin \frac{2\pi}{\lambda} (x - \lambda n t)$$

where

a = amplitude of the wave;

n = frequency (defined as the number of waves passing any given point per second; it is customary to express frequency in vibrations per second or in cycles per second);

λ = wavelength (representing the distance between two adjacent points in the wave that have the same phase);

t = time.

FIGURE 4.10 Water waves

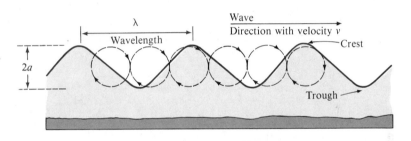

If T is the period of the wave, then

$$T = \frac{1}{n}$$

and it can be shown that the phase velocity (the speed of the wave) is

$$v = \sqrt{\frac{g\lambda}{2\pi}} \quad \text{or} \quad v = \frac{\lambda}{T}$$

in feet per second, where $g = 32$ ft/sec^2. Thus,

$$v \approx \sqrt{5.09\lambda} \quad \text{or} \quad v = \lambda n$$

Example 1

The wave equation can also be written as

$$y = a \sin \frac{2\pi}{\lambda}(x - vt)$$

Solve for v.

Solution

$$\frac{y}{a} = \sin \frac{2\pi}{\lambda}(x - vt)$$

$$\sin^{-1}\left(\frac{y}{a}\right) = \frac{2\pi}{\lambda}(x - vt)$$

$$\lambda \sin^{-1}\left(\frac{y}{a}\right) = 2\pi(x - vt)$$

$$\lambda \sin^{-1}\left(\frac{y}{a}\right) = 2\pi x - 2\pi vt$$

$$2\pi vt = 2\pi x - \lambda \sin^{-1}\left(\frac{y}{a}\right)$$

$$v = \frac{2\pi x - \lambda \sin^{-1}(\frac{y}{a})}{2\pi t}$$

▯

Example 2

Find λ in terms of T.

Solution

$v \approx \sqrt{5.09\lambda}$ and $v = \frac{\lambda}{T}$, so

$$\frac{\lambda}{T} \approx \sqrt{5.09\lambda}$$

$$\frac{\lambda^2}{T^2} \approx 5.09\lambda$$

$$\lambda^2 - 5.09\lambda T^2 \approx 0$$

$$\lambda(\lambda - 5.09T^2) \approx 0$$

$$\lambda \approx 0, \lambda \approx 5.09T^2$$

↑ *Trivial solution*   ▯

Example 3

Suppose you are watching waves pass a pier piling, and you count 20 waves per minute, each with an amplitude of 2 ft. Find its wavelength and phase velocity (in mph), and write the equation for one of these waves.

Solution 20 waves per minute is $\frac{1}{3}$ wave per second, so $n = \frac{1}{3}$ and $T = 3$. Then, from Example 2,

$$\lambda = 5.09(3^2)$$
$$\approx 45.81 \qquad \textit{This is the wavelength (in ft)}.$$

and

$$v = \frac{\lambda}{T}$$
$$= \frac{45.81}{3}$$
$$= 15.27 \text{ ft/sec} \quad \textit{This is the phase velocity}.$$

To change to miles per hour, write

$$\frac{15.27 \text{ ft}}{\text{sec}} = \frac{15.27 \text{ ft}}{\text{sec}} \cdot \frac{60 \text{ sec}}{1 \text{ min}} \cdot \frac{60 \text{ min}}{1 \text{ hr}} \cdot \frac{1 \text{ mi}}{5{,}280 \text{ ft}}$$

These fractions are all one, so multiplication does not change the original value.

$$= \frac{15.27 \cdot 60 \cdot 60}{5{,}280} \cdot \frac{\text{sec}}{\text{sec}} \cdot \frac{\text{min}}{\text{min}} \cdot \frac{\text{ft}}{\text{ft}} \cdot \frac{\text{mi}}{\text{hr}}$$

Use a calculator. | *These are all one. They are from a rearrangement of the factors from the line above.* | *This is abbreviated mph.*

$$\approx 10.41 \text{ mph} \qquad \textit{This is the phase velocity (in mph)}.$$

The equation is

$$y = 2 \sin \frac{2\pi}{45.81} (x - 15.27t)$$
$$y = 2 \sin .14(x - 15.27t)$$

Harmonic Waves

In Section 2.4, we spoke of modulation and beats. Harmonic wave motion and resonance are related to these ideas. The displacement of a particle that can be expressed in terms of an equation of the form

$$y = a \cos(\omega t + h) \quad \text{or} \quad y = a \sin(\omega t + h)$$

where a, ω, and h are constants, is said to be in **simple harmonic motion**. However, many moving bodies do not move back and forth between precisely fixed limits

It might be interesting to look at a comparison of the sun and the moon as tide-generating bodies. Even though the gravitational attraction between the earth and the sun is much greater than between the earth and moon (over 177 times greater), the moon dominates the tides, as shown by the calculation below, which uses the formula

$$\text{tide-generating force} \approx \frac{m_1 m_2}{r^3}$$

where m_1 and m_2 are the masses of the bodies and r is the distance between the bodies.

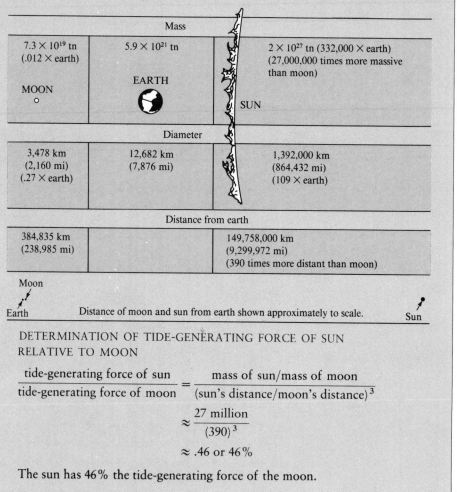

Mass		
7.3×10^{19} tn (.012 × earth) MOON ○	5.9×10^{21} tn EARTH	2×10^{27} tn (332,000 × earth) (27,000,000 times more massive than moon) SUN
Diameter		
3,478 km (2,160 mi) (.27 × earth)	12,682 km (7,876 mi)	1,392,000 km (864,432 mi) (109 × earth)
Distance from earth		
384,835 km (238,985 mi)		149,758,000 km (9,299,972 mi) (390 times more distant than moon)

Moon

Earth   Distance of moon and sun from earth shown approximately to scale.   Sun

DETERMINATION OF TIDE-GENERATING FORCE OF SUN RELATIVE TO MOON

$$\frac{\text{tide-generating force of sun}}{\text{tide-generating force of moon}} = \frac{\text{mass of sun/mass of moon}}{(\text{sun's distance/moon's distance})^3}$$

$$\approx \frac{27 \text{ million}}{(390)^3}$$

$$\approx .46 \text{ or } 46\%$$

The sun has 46% the tide-generating force of the moon.

* Historical Note is from *Introductory Oceanography*, 5th ed., by H. V. Thurman, p. 248. Reprinted with permission of Merrill, an imprint of Macmillan Publishing Company. Copyright ©1988 Merrill Publishing Company, Columbus, Ohio.

because friction slows down their motion. Such motions are called **damped harmonic motions**. The period T of a harmonic motion is the time required to complete one round trip of the motion, and the frequency n is the number of oscillations or cycles per unit of time so that

$$n = \frac{1}{T} = \frac{\omega}{2\pi}$$

The effect of damping can be seen by looking at the graphs shown in Figure 4.11.

FIGURE 4.11 (a) Harmonic motion (amplitude a constant) and (b) damped harmonic motion (amplitude a decreasing variable)

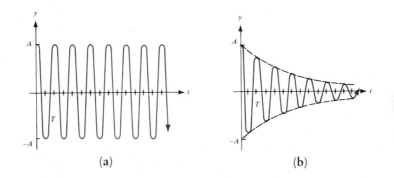

(a)   (b)

Example 4

If a particle is moving in simple harmonic motion according to the equation

$$y = 2 \cos 18\pi t$$

what are the period, frequency, and maximum distance from the origin when $t = 0$?

Solution

By inspection, $\omega = 18\pi$, so

$$n = \frac{\omega}{2\pi} = \frac{18\pi}{2\pi} = 9 \quad \text{and} \quad T = \frac{1}{n} = \frac{1}{9}$$

Therefore the period is $\frac{1}{9}$, and the frequency is 9 cycles per unit of time. If $t = 0$, then $y = 2$, so the maximum distance from the origin when $t = 0$ is 2 units. ☐

Resonance

Notice in Figure 4.11 that, for damped harmonic motion, the amplitude decreases toward zero. The reverse situation, when the amplitude increases to a maximum, is called **resonance** and has a graph as shown in Figure 4.12.

In physics it is shown that the amplitude of the vibrations is large compared to the amount of static displacement. For example, it would be possible for soldiers marching in step across a suspension bridge to make it vibrate with a very large destructive amplitude, provided that the frequency of their steps happened to be the same as the natural frequency of the bridge. This is why marching soldiers break step when crossing a bridge.

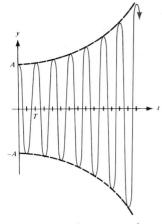

FIGURE 4.12 Resonance: Amplitude an increasing variable

183

A striking example of the destructive force of resonance is the collapse of the Tacoma Narrows Bridge, which was opened on July 1, 1940. Early on the morning of November 7, 1940, the wind velocity was 40 to 45 mi/hr—perhaps a larger velocity than any previously encountered by the bridge. Traffic was shut down. By 9:30 A.M. the span was vibrating in eight or nine segments, with frequency 36 vib/min and amplitude about 3 ft. While measurements were underway, at about 10:00 A.M. the main span abruptly began to vibrate torsionally in two segments with frequency 14 vib/min. The amplitude of torsional vibrations quickly built up to 35° each direction from horizontal. The main span broke up shortly after 11:00 A.M. During most of the catastrophic torsional vibration there was a transverse nodal line at midspan and a longitudinal nodal line down the center of the roadway.* (See Figure 4.13.)

FIGURE 4.13 The Tacoma Narrows Bridge at Puget Sound, Washington

COURTESY OF WIDE WORLD PHOTOS, INC.

Problems for Further Study: Waves

1. *Physics.* If a water wave has an amplitude of 3 ft and you count 10 waves per minute, what are the wavelength, phase velocity (in mph), and equation of the wave?

2. *Physics.* If a water wave has an amplitude of 35 ft with a period of 10 seconds, what are its wavelength, its phase velocity (in mph), and its equation?

3. *Physics.* If a water wave has the equation

   $$y = 12 \sin 1.32(x - 4.75t)$$

   how high is the wave from trough to crest? What are the wavelength and phase velocity (in mph) of the wave?

4. *Physics.* If a water wave has the equation

   $$y = 18 \sin .62(x - 4.50t)$$

---

* A filmstrip *The Tacoma Narrows Bridge Collapse* (#80-218) is available from The Ealing Corporation, Cambridge, MA 02140, which provided the information about the bridge.

how high is the wave from trough to crest? What are the wavelength and phase velocity (in mph)?

5. Solve the equation given in Problem 3 for t.
6. Solve the equation given in Problem 4 for t.
7. *Physics.* If a particle is moving in simple harmonic motion according to the equation

$$y = \sqrt{3} \sin\left(10\pi t - \frac{\pi}{2}\right)$$

what are the period and frequency, and what is the maximum distance from the origin when $t = 0$?

8. *Physics.* If a particle is moving in simple harmonic motion with period 4π and constant $h = 0$, write an equation describing the distance y of the particle from the origin if the maximum distance from the origin is 5 units.

9. *Physics.* Graph the given equations for $t \geq 0$, and classify each motion as harmonic motion, damped harmonic motion, or resonance.
   a. $y = 3 \cos 4\pi t$ b. $y = 3t \cos 4\pi t$ c. $y = \frac{3}{t} \cos 4\pi t$

10. In calculus it is shown that if a particle is moving in simple harmonic motion according to the equation

$$y = a \cos(\omega t + h)$$

then the velocity of the particle at any time t is given by

$$v = a\omega \sin(\omega t + h)$$

Graph this velocity curve, where the frequency is $\frac{1}{4}$ cycle per unit of time, $h = \frac{\pi}{6}$, and $a = \frac{120}{\pi}$.

11. Use Problem 10 to find a maximum value and a minimum value for the velocity.

CUMULATIVE REVIEW FOR CHAPTERS 3 AND 4

SAMPLE TEST I: OBJECTIVE FORM

1. *State* the eight fundamental identities.

2. *Prove* $\cot^2\theta + 1 = \csc^2\theta$.

3. *Given* a unit circle, as shown in the figure below.

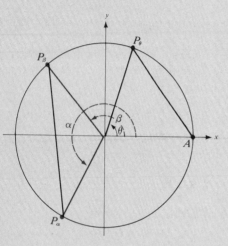

   a. What are the coordinates of the points A and P_θ?

   b. Use the distance formula to find the length of chord AP_θ.

4. *Prove* $(\cos\beta - \cos\alpha)^2 + (\sin\beta - \sin\alpha)^2$
$$= 2 - 2(\cos\alpha\cos\beta + \sin\alpha\sin\beta).$$

5. *Prove*
$$\frac{\sin 3\theta + \cos 3\theta + 1}{\cos 3\theta} = \tan 3\theta + \sec 3\theta + 1.$$

6. *Prove*
$$\sec A - \tan^2 A = \sec^2 A(\cos^2 A + \cos A - 1).$$

7. *Prove* or *disprove*
$$\tan 2\theta \cos 2\theta = 2\sin\theta\cos\theta.$$

8. *Solve* $\cos^2\theta = 3\sin\theta$ for $0 \le \theta < 2\pi$ (in radians correct to three places).

9. *Graph* $y - 3 = 2\sin(\frac{\pi}{3} - x)$.

10. *Find* the other circular functions so that $\cos\theta = \frac{4}{5}$ when $\sin\theta < 0$.

SAMPLE TEST II: FILL IN AND MULTIPLE CHOICE

Fill in Problems 1–5 with the word or words to make the statements complete and correct.

1. The three reciprocal identities are

   _____, _____, and

   _____.

2. The two ratio identities are _____ and

   _____.

3. The three Pythagorean identities are

   _____, _____, and

   _____.

4. The first step in the proof of $\cos^2\theta + \sin^2\theta = 1$ is

   _____.

5. To multiply $\dfrac{\cos\theta}{1 - \sin\theta}$ by $\dfrac{1 + \sin\theta}{1 + \sin\theta}$ is called

   _____.

*Choose the **best** answer from the choices given.*

6. If $\sin\theta = -\sqrt{1 - \cos^2\theta}$ then θ could be in Quadrants
   A. I and II   B. III and IV
   C. I and IV   D. II and III
   E. none of these

7. $\cot^2 173° - \csc^2 173°$ is
   A. 16.35   B. 0
   C. 1   D. 2
   E. none of these

8. If $\csc\theta = -\sqrt{\frac{10}{3}}$ and $\cos\theta > 0$ then $\tan\theta$ is
   A. $-\frac{1}{3}$ B. $\frac{3}{10}\sqrt{10}$
   C. $-\frac{\sqrt{21}}{3}$ D. $-\frac{\sqrt{21}}{7}$
   E. none of these

9. If $\dfrac{\cos^2\theta(\tan^2\theta + 1)}{\cot^2\theta}$ is changed to sines and co-
    sines and then completely simplified, the result is
    A. $\sin^4\theta + \sin^2\theta$ B. 1
    C. $\dfrac{\cos^4\theta \cdot}{\sin^4\theta}$ D. $\dfrac{\sin^2\theta}{\cos^2\theta}$
    E. none of these

10. $\dfrac{1}{1 + \sin 2\theta} + \dfrac{1}{1 - \sin 2\theta}$ is identical to
    A. $2\sec^2 2\theta$ B. $2\csc^2 2\theta$
    C. 1   D. $\dfrac{2}{\cos^4\theta}$
    E. none of these

11. $\sec^2 A(\cos^2 A + \cos A - 1)$ is identical to
    A. $\sec A - \tan^2 A$
    B. $\sec^2 A - \sec A - 1$
    C. $1 + \sec A - 2\sec^2 A$
    D. $\sec A - \tan^2 A - 1$
    E. none of these

12. $\dfrac{\cot\theta + \tan\theta}{\sec\theta\csc\theta}$ is identical to
    A. 1   B. $\cos 2\theta$
    C. $\sin\theta - \cos\theta$ D. $\sin\theta\cos\theta$
    E. none of these

13. If $\cos^2\theta + \sin\theta = 0$, then a solution is
    A. $\dfrac{1 \pm \sqrt{5}}{2}$ B. 3.808
    C. .666   D. 38.2
    E. none of these

14. Which of the following statements are false?
    A. $\cos^2\theta + \sin^2\theta = 1$
    B. $\tan^2\theta = \sec^2\theta - 1$
    C. $\cos\theta\sec\theta = 1$
    D. $\cot^2\theta - 1 = \csc^2\theta$
    E. none of these

15. If $\cos(\alpha + \beta) = \cos\alpha\cos\beta - \sin\alpha\sin\beta$, then an
    exact value for $\cos 165°$ is
    A. $\dfrac{\sqrt{2} - \sqrt{6}}{4}$ B. $\dfrac{\sqrt{6} - \sqrt{2}}{4}$
    C. $\dfrac{\sqrt{2} + \sqrt{6}}{4}$ D. $-.9659$
    E. none of these

16. If $\alpha = \frac{\pi}{2}$ and $\beta = -\theta$ and these values are sub-
    stituted into the identity given in Problem 15, then
    the result is
    A. $\cos(\frac{\pi}{2} - \theta) = \sin(-\theta)$
    B. $\cos(\frac{\pi}{2} - \theta) = -\cos\theta$
    C. $\cos(\frac{\pi}{2} - \theta) = -\sin(-\theta)$
    D. $\cos(\frac{\pi}{2} - \theta) = -\sin\theta$
    E. none of these

17. If you use the identity given in Problem 15 to
    simplify $\cos 3\theta$ the result is
    A. $3\cos\theta$
    B. $\cos 2\theta\cos\theta - \sin 2\theta\sin\theta$
    C. $2\cos^3\theta - 2\sin^2\theta - 1$
    D. $2\cos^3\theta - \sin 2\theta\sin\theta - 1$
    E. none of these

18. The leftmost point (yet still within the frame) of
    the graph of $y = -\cos\theta$ is
    A. $(0, \frac{\pi}{2})$ B. $(0, -\frac{\pi}{2})$
    C. $(0, 1)$ D. $(0, -1)$
    E. none of these

19. The (h, k) point for the graph of
    $y + 2 = \sin(\frac{\pi}{3} - \theta)$ is
    A. $(\theta, -2)$ B. $(\frac{\pi}{3}, -2)$
    C. $(-\frac{\pi}{3}, -2)$ D. $(\frac{\pi}{3}, 2)$
    E. none of these

20. The amplitude for the graph of $y = \sin\theta + \cos\theta$ is
    A. 1   B. $\sqrt{2}$
    C. $\frac{\pi}{4}$ D. $-\frac{\pi}{4}$
    E. none of these

SOLVING TRIANGLES

CHAPTER OVERVIEW

In the first four chapters we were concerned with one use of trigonometry—using identities to treat the circular functions as important mathematical relationships. In this chapter, we are concerned with another use of trigonometry—solving triangles to find unknown parts. This procedure of *solving a triangle* is useful in surveying, in measuring distances indirectly, in navigation and aviation, as well as in finding areas and volumes. We begin by solving right triangles using the Pythagorean theorem. We then turn to the solution of *oblique* triangles (triangles that are not right triangles). We will derive two trigonometric laws, the *Law of Cosines* and the *Law of Sines*, that allow us to solve certain triangles using a calculator. There is a third trigonometric law that lends itself to logarithmic calculation—the *Law of Tangents*, which is useful when no calculator is at hand. This law is discussed in Problems 33–38 of Problem Set 5.3. The triangle shown in Figure 5.1 illustrates the way we will designate the parts of a triangle, with points A, B, and C; sides a, b, and c; and angles α, β, and γ.

5.1 RIGHT TRIANGLES

Solving Triangles

One of the most important uses of trigonometry is for the solution of triangles. Recall from geometry that every triangle has three sides and three angles, which are called the six *parts* of the triangle. We say that a **triangle is solved** when all six parts are known and listed. Typically, three parts will be given, or known, and it will be our task to find the other three parts.

A triangle is usually labeled as shown in Figure 5.1.

FIGURE 5.1 Correctly labeled triangle

Vertices	Sides	Angles
A	a	α
B	b	β
C	c	γ

The vertices are labeled A, B, and C, with the sides opposite those vertices labeled a, b, and c, respectively. The angles are labeled α, β, and γ, respectively. In this section, we assume that γ denotes the right angle and c the hypotenuse.

Right Triangle Definition

According to the definition of the trigonometric functions, the angle under consideration must be in standard position. This requirement is sometimes inconvenient,

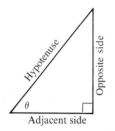

Adjacent side

so we use that definition to give us the following special case, which applies to any acute angle θ of a right triangle. Notice from Figure 5.1 that θ might be α or β but would not be γ, since γ is not an acute angle. Also notice that the **hypotenuse** is one of the sides of both acute angles. The other side making up the angle is called the **adjacent side**. Thus side a is adjacent to β and side b is adjacent to α. The third side of the triangle (the one that is not part of the angle) is called the **opposite side**. Thus side a is opposite α and side b is opposite β.

RIGHT TRIANGLE
DEFINITION OF THE
TRIGONOMETRIC RATIOS

If θ is an acute angle in a right triangle,

$$\cos \theta = \frac{\text{adjacent side}}{\text{hypotenuse}} \qquad \sec \theta = \frac{\text{hypotenuse}}{\text{adjacent side}}$$

$$\sin \theta = \frac{\text{opposite side}}{\text{hypotenuse}} \qquad \csc \theta = \frac{\text{hypotenuse}}{\text{opposite side}}$$

$$\tan \theta = \frac{\text{opposite side}}{\text{adjacent side}} \qquad \cot \theta = \frac{\text{adjacent side}}{\text{opposite side}}$$

When solving triangles, you will not need to use the secant, cosecant, or cotangent ratios, so it is necessary to remember only the cosine, sine, and tangent ratios.

Example 1

Let $\theta = \beta$ in Figure 5.1. Write the trigonometric ratios in terms of a, b, and c.

Solution

$$\cos \beta = \frac{\text{adjacent side}}{\text{hypotenuse}} = \frac{a}{c} \qquad \sec \beta = \frac{\text{hypotenuse}}{\text{adjacent side}} = \frac{c}{a}$$

$$\sin \beta = \frac{\text{opposide side}}{\text{hypotenuse}} = \frac{b}{c} \qquad \csc \beta = \frac{\text{hypotenuse}}{\text{opposite side}} = \frac{c}{b}$$

$$\tan \beta = \frac{\text{opposite side}}{\text{adjacent side}} = \frac{b}{a} \qquad \cot \beta = \frac{\text{adjacent side}}{\text{opposite side}} = \frac{a}{b} \qquad \Box$$

Example 2

Given the triangle with sides 3, 4, and 5, as shown in Figure 5.2, find the six trigonometric ratios of the angle β.

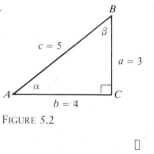

FIGURE 5.2

Solution

$$\cos \beta = \frac{3}{5} = .6 \qquad \sec \beta = \frac{5}{3} = 1.\overline{6}$$

$$\sin \beta = \frac{4}{5} = .8 \qquad \csc \beta = \frac{5}{4} = 1.25$$

$$\tan \beta = \frac{4}{3} = 1.\overline{3} \qquad \cot \beta = \frac{3}{4} = .75 \qquad \Box$$

When solving triangles, you are dealing with measurements that are never exact. In "real life," measurements are made with a certain number of digits of precision, and results are not claimed to have more digits of precision than the least accurate number in the input data. There are rules for working with significant digits (see Appendix B), but it is important for you to focus first on the trigonometry. Just remember that your answers should not be more accurate than any of the given measurements. However, be careful not to round your answers when doing your calculations; you should round only when stating your answers.

WARNING! Note this agreement on rounding.

Example 3

Solve the right triangle with $a = 50.0$ and $\alpha = 35°$.

Solution

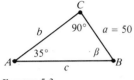

FIGURE 5.3

Begin by drawing a triangle (not necessarily to scale) in order to see the given relationships. When solving triangles, you should state all six parts, including those parts given in the problem. These parts should be stated to the correct number of significant digits. These parts are shown in color below.

$\alpha = 35°$ Given.

$\beta = 55°$ Since $\alpha + \beta = 90°$ for any right triangle with right angle at C

$\gamma = 90°$ Given.

$a = 50$ Given. Notice that we write the answer to the nearest unit because the given angle is measured to the nearest degree and we use the *least accurate* of the given measurements to determine the accuracy of the answer (see Appendix B).

b: $\tan 35° = \dfrac{50}{b}$ *See Figure 5.3; tangent is opp/adj. Do not forget that c is the hypotenuse, which forces b to be the adjacent side.*

$$b = \dfrac{50}{\tan 35°}$$

i. By table: $b = \dfrac{50}{\tan 35°} \approx \dfrac{50}{.7002}$ or $b = 50 \cot 35°$

$$\approx 71.41 \qquad\qquad \approx 50(1.4281)$$

$$= 71.405$$

From Appendix D Table, $\tan 35° \approx .7002$; for the division you can use a four-function calculator or long division. By using the cotangent (shown at the right), you can avoid division, which makes this method easier if you do not have a calculator.

ii. By calculator with algebraic logic:

$\boxed{50}\ \boxed{\div}\ \boxed{35}\ \boxed{\tan}\ \boxed{=}$

DISPLAY: 71.40740034

iii. By calculator with RPN logic:

$\boxed{50}\ \boxed{\text{ENTER}}\ \boxed{35}\ \boxed{\tan}\ \boxed{\div}$

DISPLAY: 71.40740034

The correct answer should be stated to two significant digits: $b = 71$.

$$\sin 35° = \frac{50}{c} \qquad \textit{See Figure 5.3; sine is opp/hyp.}$$

$$c = \frac{50}{\sin 35°} \qquad \begin{array}{l} \textit{The intermediate steps of finding c by table} \\ \textit{or by calculator are left for you.} \end{array}$$

$$c \approx 87 \qquad \textit{To two significant digits}$$

Example 4

Find the distance PA across the river shown in Figure 5.4.

FIGURE 5.4   Right-triangle trigonometry for finding the distance across a river

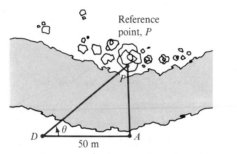

Solution

You can find the distance across a river by finding some reference point P on the other side, as shown in Figure 5.4. Next, fix a point A directly across from P and measure out a fixed distance (say 50 m) to a point D so that $\angle PAD$ is a right angle. Now measure the angle θ (say 38°) and use the definition of tangent to write

$$\tan \theta = \frac{PA}{DA}$$

$$\tan 38° = \frac{PA}{50}$$

$$PA = 50 \tan 38°$$

i. By table:

$$PA = 50(.7813)$$
$$= 39.065$$

ii. By calculator with algebraic logic:

[50] [×] [38] [tan] [=]   DISPLAY: 39.06428133

iii. By calculator with RPN logic:

[50] [ENTER] [38] [tan] [×]   DISPLAY: 39.06428132

To two significant digits, the distance across the river is 39 m.

Angles of Depression and Elevation

A common application of right triangles involves an observer looking at an object. The **angle of depression** is the acute angle measured from a horizontal line down to the line of sight, as shown in Figure 5.5. On the other hand, if we measure from a horizontal line up to the line of sight, we call that angle the **angle of elevation**, also shown in Figure 5.5.

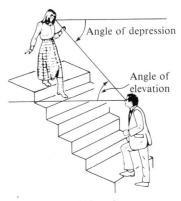

FIGURE 5.5   Angles of depression and elevation

Example 5

The angle of elevation of a tree from a point on the ground 42 m from its base is 33°. Find the height of the tree (see Figure 5.6).

Solution

Let θ = angle of elevation; h = height of tree.

$$\tan \theta = \frac{h}{42}$$
$$h = 42 \tan 33°$$
$$\approx 27.28$$

The tree is 27 m tall.

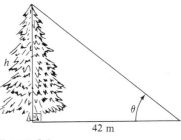

FIGURE 5.6

Bearing

Another application of the solution of right triangles involves the **bearing** of a line, which is defined as an acute angle made with a north–south line. When giving the bearing of a line, first write N or S to determine whether to measure the angle from the north or the south side of a point on the line. Then give the measure of the angle followed by E or W, denoting which side of the north–south line you are measuring. Some examples are shown in Figure 5.7.

FIGURE 5.7

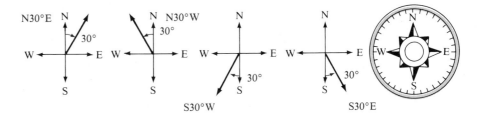

Historical Note

Courtesy of Teledyne Gurley

You may have seen surveyors using an instrument called a transit, or perhaps you have wondered how one could decide on the angle of elevation for a problem like Example 5. A transit is a telescope that allows for accurate line-of-sight measurements. After fixing the line of sight, you can read directions from both horizontal and vertical circles. Look at the transit pictured at the left and find these horizontal and vertical circles. These circles are graduated every 10′, 20′, or 30′ (depending on the instrument). A transit also contains a device called a *vernier*, which is used to measure fractional parts of the scale shown on the horizontal and vertical circles. Finally, the transit contains a level bubble to ensure that the transit is level when readings are taken.

Example 6

To find the width $|AB|$ of a canyon, a surveyor measures 100 m from A in the direction of N42.6°W to locate point C. The surveyor then determines that the bearing of CB is N73.5°E. Find the width of the canyon if the point B is situated so that $\angle BAC = 90.0°$. (See Figure 5.8.)

FIGURE 5.8

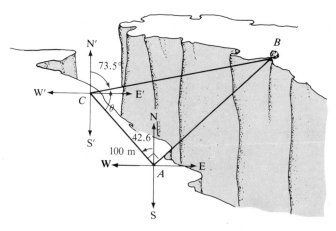

Solution

Let $\theta = \angle BCA$ in Figure 5.8. Then $\tan \theta = \dfrac{|AB|}{|AC|}$. So

$$|AB| = |AC| \tan \theta = 100 \tan \theta$$

Now find θ:

$$\angle BCE' = 16.5° \quad \textit{Complementary angles}$$
$$\angle ACS' = 42.6° \quad \textit{Alternate interior angles}$$
$$\angle E'CA = 47.4° \quad \textit{Complementary angles}$$
$$\theta = \angle BCA = \angle BCE' + \angle E'CA$$
$$= 16.5° + 47.4°$$
$$= 63.9°$$

Thus

$$|AB| = 100 \tan 63.9°$$
$$\approx 204.125.$$

The canyon is 204 m across.   ☐

Problem Set 5.1

Solve each of the right triangles (γ is a right angle) in Problems 1–18.

A
1. $a = 80; \beta = 60°$
2. $b = 37; \alpha = 69°$
3. $a = 29; \alpha = 76°$
4. $b = 90; \beta = 13°$
5. $a = 49; \alpha = 45°$
6. $b = 82; \alpha = 50°$
7. $c = 28.3; \alpha = 69.2°$
8. $\beta = 57.4°; a = 70.0$
9. $\alpha = 56.00°; b = 2,350$
10. $\beta = 16.4°; b = 2,580$
11. $\alpha = 42°; b = 350$
12. $c = 75.4; \alpha = 62.5°$
13. $c = 28.7; \alpha = 67.8°$
14. $c = 343.6; \beta = 32.17°$
15. $c = 418.7; \beta = 61.05°$
16. $b = .8024; \beta = 24.16°$
17. $b = .316; \alpha = 48.32°$
18. $b = .3596; \beta = 76.06°$

19. *Surveying.* Find the distance PA across the river shown in Figure 5.4 if $\theta = 42°$.

20. *Surveying.* Find the distance PA across the river shown in Figure 5.4 if $\theta = 16°$.

21. *Navigation.* The most powerful lighthouse is on the coast of Brittany, France, and is 50 m tall. Suppose you are in a boat just off the coast, as shown in Figure 5.9. Determine your distance from the base of the lighthouse if $\theta = 12°$.

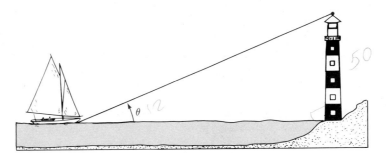

FIGURE 5.9   Distance from a boat to the *Créac'h d' Ouessant* on the coast of Brittany, France

22. *Navigation.* Use Problem 21 to determine your distance from the lighthouse if $\theta = 32°$.

23. *Surveying.* The angle of elevation of a building from a point on the ground 30 m from its base is 38°. Find the height of the building.

24. *Surveying.* The angle of elevation of the top of the Great Pyramid of Khufu (or Cheops) from a point on the ground 351 ft from a point directly below the top is 52.0°. Find the height of the pyramid.

Courtesy of Photo Researchers, Inc.

B

25. *Surveying.* To find the northern boundary of a piece of land, a surveyor must divert his path from point C on the boundary by proceeding due south for 300 ft to a point A. Point B, which is due east of point C, is now found to be in the direction of N49°E from point A. What is the distance CB?

26. *Surveying.* To find the distance across a river that runs east–west, a surveyor locates points P and Q on a north–south line on opposite sides of the river. She then paces out 150 ft from Q due east to a point R. Next she determines that the bearing of RP is N58°W. How far is it across the river (distance PQ)?

27. *Surveying.* To find the boundary of a piece of land, a surveyor must divert his path from a point A on the boundary for 500 ft in the direction S50°E. He then determines that the bearing of a point B located directly south of A is S40°W. Find the distance AB.

28. *Surveying.* To find the distance across a river, a surveyor locates points P and Q on either side of the river. Next she measures 100 m from point Q in the direction of S35.0°E to point R. Then she determines that point P is now in the direction of N25.0°E from point R and that angle PQR is a right angle. Find the distance across the river.

29. *Surveying.* On the top of the Empire State Building is a TV tower. From a point 1,000 ft from a point on the ground directly below the top of the tower the angle of elevation to the bottom of the tower is 51.34° and to the top of the tower is 55.81°. What is the length of the TV tower?

Historical Note   The angle of elevation for a pyramid is the angle between the edge of the base and the slant height, the line from the apex of the pyramid to the midpoint of any side of the base. It is the maximum possible angle of ascent for anyone trying to climb the pyramid to the top. In an article "Angles of Elevation of the Pyramids of Egypt," in *The Mathematics Teacher* (February 1982, pp. 124–127), author Arthur F. Smith notes that the angle of elevation for these pyramids is either 43.5° or 52°. Why did the Egyptians build pyramids using these angles of elevation? See Problem 34, Problem Set 5.5, page 223.

Smith states that (according to Kurt Mendelssohn, *The Riddle of the Pyramids*; New York: Praeger Publications, 1974) Egyptians might have measured long horizontal distances by means of a circular drum with some convenient diameter such as one cubit. The circumference would then have been pi cubits. In order to design a pyramid of convenient and attractive proportions, the Egyptians used a 1:4 ratio for the rise relative to one revolution of the drum. Smith then shows that the angle of elevation of the slant height is

$$\text{Arctan}\,\frac{4}{\pi} \approx 52°$$

If a smaller angle of elevation was desired (as in the case of the Red Pyramid), a 1:3 ratio might have been used. In that case, the angle of elevation of the slant height is

$$\text{Arctan}\,\frac{3}{\pi} \approx 43.5°$$

30. *Surveying.* The distance across a canyon can be determined from an airplane. Using the information given in Figure 5.10, what is the distance across the canyon?

FIGURE 5.10   Determining the distance across a canyon from the air

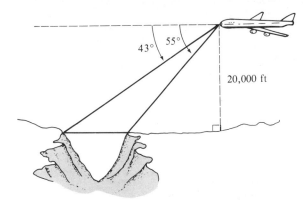

31. In the movie *Close Encounters of the Third Kind*, Devil's Tower in Wyoming figured prominently. There was a scene in which the star, Richard Dreyfuss, was approaching the Tower. He could have determined his distance from Devil's Tower by stopping at a point *P* and estimating the angle *P*, as shown in Figure 5.11.

FIGURE 5.11   Procedure for determining the height of Devil's Tower

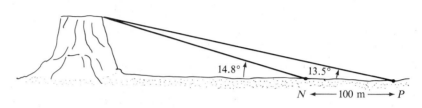

After moving 100 m toward Devil's Tower, he could estimate the angle *N*, as shown in Figure 5.11. How far away from Devil's Tower is point *N*?

32. Using Problem 31, determine the height of Devil's Tower.

C

33. *Astronomy*. If the distance from the earth to the sun is 92.9 million miles and the angle formed between Venus, the earth, and the sun (as shown in Figure 5.12) is 47.0°, find the distance from the sun to Venus.

34. *Astronomy*. Use the information in Problem 33 to find the distance from the earth to Venus.

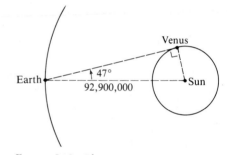

FIGURE 5.12   Planetary positions

35. The largest ground area covered by any office building is that of the Pentagon in Arlington, Virginia. If the radius of the circumscribed circle is 783.5 ft, find the length of one side of the Pentagon.

36. Use the information in Problem 35 to find the radius of the circle inscribed in the Pentagon.

37. In algebra, the slope of a nonvertical line *L* passing through points $P(x_1, y_1)$ and $Q(x_2, y_2)$ is defined by

$$m = \frac{y_2 - y_1}{x_2 - x_1}$$

as shown in Figure 5.13. If we let θ be the angle that the line *L* makes with the *x*-axis such that $0 \le \theta < 180°$, then θ is called the **angle of inclination** of *L*. It follows that

$$m = \tan \theta$$

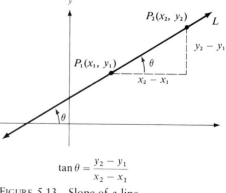

$$\tan \theta = \frac{y_2 - y_1}{x_2 - x_1}$$

FIGURE 5.13   Slope of a line

a. Find the slope of a line whose angle of inclination is 35°.

b. If the equation of a line passing through (h, k) is $y - k = m(x - h)$, find the equation of a line through $(2, -3)$ having an angle of inclination of $60°$.

c. If the equation of a line passing through (h, k) is $y - k = m(x - h)$, find the equation of a line through $(4, -1)$ having an angle of inclination of $30°$.

38. Show that, in every right triangle, the value of c lies between $\dfrac{a + b}{\sqrt{2}}$ and $a + b$.

39. Show that the area of a right triangle is given by $\frac{1}{2}bc \sin A$.

40. A given triangle ABC is *not* a right triangle. Draw BD perpendicular to AC forming right triangles ABD and BDC (with right angle at D). Show that

$$\frac{\sin A}{a} = \frac{\sin C}{c}$$

5.2   LAW OF COSINES

Oblique Triangles

In the last section we solved right triangles. We now extend that study to triangles with no right angles. Such triangles are called **oblique triangles**. Consider a triangle labeled as in Figure 5.14, and notice that γ is not now restricted to $90°$.

FIGURE 5.14   Correctly labeled triangle

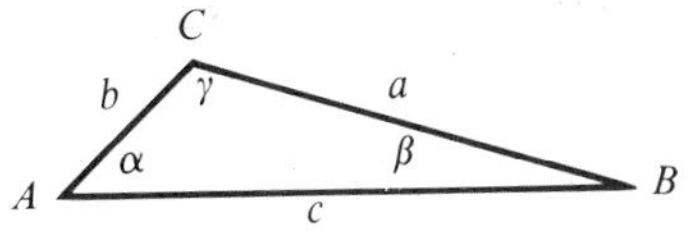

In general, you will be given three parts of a triangle and be asked to find the remaining three parts. But can you do so given *any* three parts? Consider the possibilities.

1. SSS—By SSS we mean that you are given three sides and want to find the three angles.
2. SAS—You are given two sides and an included angle.
3. AAA—You are given three angles.
4. ASA or AAS—You are given two angles and a side.
5. SSA—You are given two sides and the angle opposite one of them.

We will consider these possibilities one at a time.

SSS

To solve a triangle given SSS, it is necessary for the sum of the lengths of the two smaller sides to be greater than the length of the largest side. In this case, we use a

Historical Note

Francois Viète
1540–1603

The transition from Renaissance mathematics to modern mathematics was aided, in large part, by the Frenchman François Viète (1540–1603). He was a lawyer who practiced mathematics as a hobby. In his book *Canon mathematicus*, he solved oblique triangles by breaking them into right triangles. He was probably the first person to develop and use the Law of Tangents. He also was the first to prove (in 1579) the Law of Cosines. When he proved this he did not use the form used in this book. Instead:

$$\frac{2ab}{a^2 + b^2 - c^2} = \frac{1}{\sin(90° - C)}$$

However, the Law of Cosines was first stated by Euclid in Book II of his *Elements*.

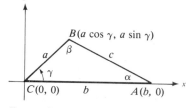

generalization of the Pythagorean theorem called the **Law of Cosines**:

LAW OF COSINES

> In triangle *ABC*,
>
> $$c^2 = a^2 + b^2 - 2ab\cos\gamma$$

Proof: Let γ be an angle in standard position with A on the positive x-axis, as shown in Figure 5.15. The coordinates of the vertices are as follows.

$C(0,0)$ *Since C is in standard position*

$A(b,0)$ *Since A is on the x-axis a distance of b units from the origin*

$B(a\cos\gamma, a\sin\gamma)$ *Let $B(x,y)$; then by definition of the trigonometric functions, $\cos\gamma = \frac{x}{a}$ and $\sin y = \frac{y}{a}$. Thus, $x = a\cos\gamma$ and $y = a\sin\gamma$.*

FIGURE 5.15

Use the distance formula for the distance between $A(b,0)$ and $B(a\cos\gamma, a\sin\gamma)$:

$$c^2 = (a\cos\gamma - b)^2 + (a\sin\gamma - 0)^2$$
$$= a^2\cos^2\gamma - 2ab\cos\gamma + b^2 + a^2\sin^2\gamma$$
$$= a^2(\cos^2\gamma + \sin^2\gamma) + b^2 - 2ab\cos\gamma$$
$$= a^2 + b^2 - 2ab\cos\gamma$$

Notice that for a right triangle, $\gamma = 90°$. This means that

$$c^2 = a^2 + b^2 - 2ab \cos 90° \quad \text{or} \quad c^2 = a^2 + b^2$$

since $\cos 90° = 0$. But this last equation is simply the Pythagorean theorem.

By letting A and B, respectively, be in standard position, it can also be shown that

$$a^2 = b^2 + c^2 - 2bc \cos \alpha$$

and

$$b^2 = a^2 + c^2 - 2ac \cos \beta$$

To find the angles when you are given three sides, solve for α, β, or γ.

**LAW OF COSINES
ALTERNATE FORMS**

$$a^2 = b^2 + c^2 - 2bc \cos \alpha \qquad \cos \alpha = \frac{b^2 + c^2 - a^2}{2bc}$$

$$b^2 = a^2 + c^2 - 2ac \cos \beta \qquad \cos \beta = \frac{a^2 + c^2 - b^2}{2ac}$$

$$c^2 = a^2 + b^2 - 2ab \cos \gamma \qquad \cos \gamma = \frac{a^2 + b^2 - c^2}{2ab}$$

Example 1

What is the smallest angle of a triangular patio whose sides measure 25, 18, and 21 feet?

Solution

If γ represents the smallest angle, then c (the side opposite γ) must be the smallest side, so $c = 18$. Then

$$\cos \gamma = \frac{a^2 + b^2 - c^2}{2ab}$$

$$= \frac{25^2 + 21^2 - 18^2}{2(25)(21)} \quad \textit{Use this number and trigonometric tables if you have only a four-function calculator.}$$

$$\gamma = \cos^{-1}\left(\frac{25^2 + 21^2 - 18^2}{2(25)(21)}\right)$$

1. By calculator with algebraic logic:

$\boxed{25}\ \boxed{x^2}\ \boxed{+}\ \boxed{21}\ \boxed{x^2}\ \boxed{-}\ \boxed{18}\ \boxed{x^2}\ \boxed{=}\ \boxed{\div}\ \boxed{2}\ \boxed{\div}\ \boxed{25}\ \boxed{\div}\ \boxed{21}\ \boxed{=}$
$\boxed{\text{INV}}\ \boxed{\cos}$ DISPLAY: 45.03565072

2. By calculator with RPN logic:

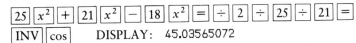
$\boxed{25}\ \boxed{\text{ENTER}}\ \boxed{\times}\ \boxed{21}\ \boxed{\text{ENTER}}\ \boxed{\times}\ \boxed{+}\ \boxed{18}\ \boxed{\text{ENTER}}\ \boxed{\times}\ \boxed{-}\ \boxed{2}$
$\boxed{\div}\ \boxed{25}\ \boxed{\div}\ \boxed{21}\ \boxed{\div}\ \boxed{\text{arc}}\ \boxed{\cos}$ DISPLAY: 45.03565071

To two significant digits, the answer is $45°$.

SAS

The second possibility listed for solving oblique triangles is that of being given two sides and an included angle. It is necessary that the given angle be less than 180°. Again use the Law of Cosines for this possibility, as shown by Example 2.

Example 2

Find c where $a = 52.0$, $b = 28.3$, and $\gamma = 28.5°$.

Solution

By the Law of Cosines:

$$c^2 = a^2 + b^2 - 2ab \cos \gamma$$
$$= (52.0)^2 + (28.3)^2 - 2(52.0)(28.3) \cos 28.5°$$

By calculator:

$$c^2 \approx 918.355474$$
$$c \approx 30.30438044$$

To three significant digits, $c = 30.3$.   ☐

AAA

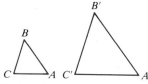

FIGURE 5.16   Similar triangles

The third case supposes that three angles are given. However, from what you know of similar triangles (see Figure 5.16)—that they have the same shape but not necessarily the same size, and thus their corresponding angles have equal measure—you can conclude that the triangle cannot be solved without knowing the length of at least one side.

We will discuss ASA and AAS in the next section.

Problem Set 5.2

A

Solve triangle ABC for the parts requested in Problems 1–12.

1. $a = 7.0$; $b = 8.0$; $c = 2.0$. Find α.

2. $a = 7.0$; $b = 5.0$; $c = 4.0$. Find β.

3. $a = 10$; $b = 4.0$; $c = 8.0$. Find γ.

4. $a = 4.0$; $b = 5.0$; $c = 6.0$. Find α.

5. $a = 11$; $b = 9.0$; $c = 8.0$.
   Find the largest angle.

6. $a = 12$; $b = 6.0$; $c = 15$.
   Find the smallest angle.

7. $a = 18$; $b = 25$; $\gamma = 30°$. Find c.

8. $a = 18$; $c = 11$; $\beta = 63°$. Find b.

9. $a = 15$; $b = 8.0$; $\gamma = 38°$. Find c.

10. $b = 21$; $c = 35$; $\alpha = 125°$. Find a.

11. $b = 14$; $c = 12$; $\alpha = 82°$. Find a.

12. $a = 31$; $b = 24$; $\gamma = 120°$. Find c.

Solve triangle ABC for the parts requested in Problems 13–18. If the triangle cannot be solved, tell why.

13. $a = 38$; $b = 41$; $c = 25$.
    Find the largest angle.

14. $a = 45$; $b = 92$; $c = 41$.
    Find the smallest angle.

15. $a = 241; b = 187; c = 100$. Find β.   **16.** $a = 38.2; b = 14.8; \gamma = 48.2°$. Find c.

17. $b = 123; c = 485; \alpha = 163.0°$. Find a.   **18.** $a = 48.3; c = 35.1; \beta = 215.0°$. Find b.

Solve triangle ABC in Problems 19–24. If the triangle cannot be solved, tell why.

19. $b = 5.2; c = 3.4; \alpha = 54.6°$ **20.** $a = 81; c = 53; \beta = 85.2°$

21. $a = 214; b = 320; \gamma = 14.8°$ **22.** $a = 18; b = 12; c = 23.3$

23. $a = 140; b = 85.0; c = 105$ **24.** $\alpha = 83°; \beta = 52°; \gamma = 45°$

25. *Aviation.* New York City is approximately 210 mi N9°E of Washington, D.C., and Buffalo, New York, is N49°W of Washington, D.C. How far is Buffalo from New York City if the distance from Buffalo to Washington, D.C., is approximately 285 mi?

26. *Aviation.* New Orleans, Louisiana, is approximately 1080 mi S56°E of Denver, Colorado, and Chicago, Illinois, is N76°E of Denver. How far is Chicago from New Orleans if it is approximately 950 mi from Denver to Chicago?

27. *Surveying.* A vertical tower is located on a hill whose inclination is 6°. From a point P 100 ft down the hill from the base of the tower, the angle of elevation to the top of the tower is 28°. What is the height of the tower? [*Hint:* Use right triangles.]

Problems 28–30 refer to the baseball stadium shown in Figure 5.17. Find the requested distances to the nearest foot.

Figure 5.17   Baseball stadium

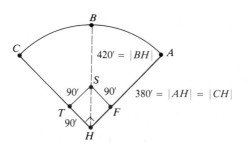

28. Right field (point A) to left field (point C)

29. Right field (point A) to center field (point B)

30. First base (point F) to center field (point B)

31. Prove that $a^2 = b^2 + c^2 - 2bc \cos \alpha$.

32. Prove that $b^2 = a^2 + c^2 - 2ac \cos \beta$.

33. *Geometry.* A dime, a penny, and a quarter are placed on a table so that they just touch each other, as shown. Let D, P, and Q be the respective centers. Solve $\triangle DPQ$. [*Hint:* If you measure the coins, the diameters are 1.75 cm, 2.00 cm, and 2.50 cm.]

34. *Geometry.* An equilateral triangle is inscribed in a circle of radius 5.0 in. Find the perimeter of the triangle.

35. *Geometry.* A square is inscribed in a circle of radius 5.0 in. Find the area of the square.

C

36. In $\triangle ABC$, use the law of cosines to prove $a = b \cos \gamma + c \cos \beta$.

37. In $\triangle ABC$, use the law of cosines to prove $b = a \cos \gamma + c \cos \alpha$.

38. In $\triangle ABC$, use the law of cosines to prove $c = a \cos \beta + b \cos \alpha$.

39. From the equation given in Problem 31 show that

$$\cos \alpha + 1 = \frac{(b + c - a)(a + b + c)}{2bc}$$

40. From the equation given in Problem 32 show that

$$\cos \beta + 1 = \frac{(a + c - b)(a + b + c)}{2ac}$$

41. *Historical Question* In *The Mathematics Teacher*, November 1958 (pp. 544–546), Howard Eves published an article, "Pappus's Extension of the Pythagorean Theorem." In this article, he points out that if you combine Propositions 12 and 13 of Book II and Proposition 47 of Book I of Euclid's *Elements*, you can derive the Law of Cosines from the Pythagorean theorem. Do some research and write a paper showing how this can be done.

5.3   LAW OF SINES

In the last section, we listed five possibilities for given information in solving oblique triangles: SSS, SAS, AAA, ASA or AAS, and SSA. We discussed three of those cases: SSS, SAS, and AAA. In this section we will consider the fourth case, ASA or AAS. Notice that ASA and AAS both give us the same information about a triangle because knowing any two angles is the same as knowing all three angles (the sum of the angles of any triangle is always 180°).

ASA or AAS

Case 4 supposes that two angles and a side are given. For a triangle to be formed, the sum of the two given angles must be less than 180°, and the given side must be greater than 0. If you know two angles, you can easily find the third angle, since the sum of the three angles is 180°. The Law of Cosines is not sufficient in this case because at least two sides are needed for the application of that law. We state and prove a result called the **Law of Sines**.

LAW OF SINES

> In any $\triangle ABC$,
>
> $$\frac{\sin \alpha}{a} = \frac{\sin \beta}{b} = \frac{\sin \gamma}{c}$$

The equation in the Law of Sines means that you can use any of the following equations to solve a triangle:

$$\frac{\sin \alpha}{a} = \frac{\sin \beta}{b} \qquad \frac{\sin \alpha}{a} = \frac{\sin \gamma}{c} \qquad \frac{\sin \beta}{b} = \frac{\sin \gamma}{c}$$

Proof: Consider any oblique triangle, as shown in Figure 5.18.

FIGURE 5.18   Oblique triangles for Law of Sines

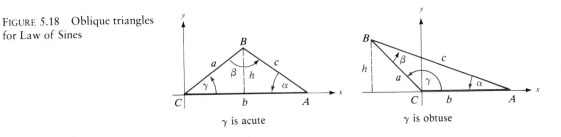

γ is acute γ is obtuse

Let h = height of the triangle with base CA. Then

$$\sin \alpha = \frac{h}{c} \quad \text{and} \quad \sin \gamma = \frac{h}{a}$$

Solving for h,

$$h = c \sin \alpha \quad \text{and} \quad h = a \sin \gamma$$

Thus,

$$c \sin \alpha = a \sin \gamma$$

Dividing by ac,

$$\frac{\sin \alpha}{a} = \frac{\sin \gamma}{c}$$

Repeat these steps for the height of the same triangle with base AB:

$$\frac{\sin \alpha}{a} = \frac{\sin \beta}{b}$$

Example 1   Solve the triangle in which $a = 20$, $\alpha = 38°$, and $\beta = 121°$.

Solution

$\alpha = 38°$ Given.

$\beta = 121°$ Given.

$\gamma = 21°$ Since $\alpha + \beta + \gamma = 180°$, $\gamma = 180° - 38° - 121° = 21°$.

$a = 20$ Given.

$b = 28$ Use the Law of Sines: $\dfrac{\sin 38°}{20} = \dfrac{\sin 121°}{b}$; then

$$b = \frac{20 \sin 121°}{\sin 38°} \quad \textit{Use tables or a calculator.}$$

$$\approx \frac{20(.8572)}{.6157} \quad \textit{Use logarithms or a calculator.}$$

$$\approx 27.8454097 \quad \textit{Give answer to two significant digits.}$$

$c = 12$ Use the Law of Sines: $\dfrac{\sin 38°}{20} = \dfrac{\sin 21°}{c}$; then

$$c = \frac{20 \sin 21°}{\sin 38°}$$

$$\approx 11.64172078$$

Notice that the answers are given to two significant digits.   □

Example 2   A boat traveling at a constant rate due west passes a buoy that is 1.0 kilometer from a lighthouse. The lighthouse is N30°W of the buoy. After the boat has traveled for one-half hour, its bearing to the lighthouse is N74°E. How fast is the boat traveling?

Solution   The angle at the lighthouse (see Figure 5.19) is $180° - 60° - 16° = 104°$.

FIGURE 5.19   Finding the rate of travel

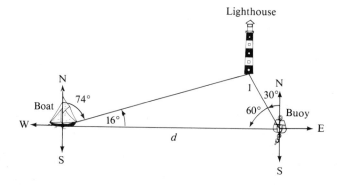

Therefore, by the Law of Sines,

$$\frac{\sin 104°}{d} = \frac{\sin 16°}{1.0}$$

$$d = \frac{\sin 104°}{\sin 16°} \approx 3.520189502 \qquad \text{\textit{Do not round answers in your work;}}$$
$$\text{\textit{round only when stating the answer.}}$$

After one hour the distance is $2d$, so the rate of the boat is about 7.0 kph.   ▢

Problem Set 5.3

A   *Solve triangle ABC with the parts given in Problems 1–12. If the triangle cannot be solved, tell why.*

1. $a = 10;\ \alpha = 48°;\ \beta = 62°$
2. $a = 18;\ \alpha = 65°;\ \gamma = 115°$
3. $a = 30;\ \beta = 50°;\ \gamma = 100°$
4. $b = 23;\ \alpha = 25°;\ \beta = 110°$
5. $b = 40;\ \alpha = 50°;\ \gamma = 60°$
6. $b = 90;\ \beta = 85°;\ \gamma = 25°$
7. $c = 53;\ \alpha = 82°;\ \beta = 19°$
8. $c = 43;\ \alpha = 120°;\ \gamma = 7°$
9. $c = 115;\ \beta = 81.0°;\ \gamma = 64.0°$
10. $\alpha = 18.3°;\ \beta = 54.0°;\ a = 107$
11. $\beta = 85°;\ \gamma = 24°;\ b = 223$
12. $a = 85;\ \alpha = 48.5°;\ \gamma = 72.4°$

B   *Solve △ABC with the parts given in Problems 13–24. If the triangle cannot be solved, tell why.*

13. $a = 41.0;\ \alpha = 45.2°;\ \beta = 21.5°$
14. $b = 55.0;\ c = 92.0;\ \alpha = 98.0°$
15. $b = 58.3;\ \alpha = 120°;\ \gamma = 68.0°$
16. $c = 123;\ \alpha = 85.2°;\ \beta = 38.7°$
17. $a = 26;\ b = 71;\ c = 88$
18. $\alpha = 48°;\ \beta = 105°;\ \gamma = 27°$
19. $a = 25.0;\ \beta = 81.0°;\ \gamma = 25.0°$
20. $a = 25.0;\ b = 45.0;\ c = 102$
21. $a = 80.6;\ b = 23.2;\ \gamma = 89.2°$
22. $b = 1,234;\ \alpha = 85.26°;\ \beta = 24.45°$
23. $c = 28.36;\ \beta = 42.10°;\ \gamma = 102.30°$
24. $a = 481;\ \beta = 28.6°;\ \gamma = 103.0°$

25. *Surveying.* In San Francisco, a certain hill makes an angle of 20° with the horizontal and has a tall building at the top. At a point 100 ft down the hill from the base of the building, the angle of elevation to the top of the building is 72°. What is the height of the building?

26. In the movie *Star Wars*, the hero, Luke, must hit a small target on the Death Star by flying a horizontal distance to reach the target. When the target is sighted, the onboard computer calculates the angle of depression to be 28.0°. If after 150 km the target has an angle of depression of 42.0°, how far is the target from Luke's spacecraft at that instant?

27. The Galactic Empire's computers on the Death Star are monitoring the positions of the invading forces (see Problems 26). At a particular instant, two observation points 2,500 m apart make a fix on Luke's spacecraft, which is between the observation points and in the same vertical plane. If the angle of elevation from the first observation point is 3.00° and the angle of elevation from the second is 1.90°, find the distance from Luke's spacecraft to each of the observation points.

28. Solve Problem 27 if both observation points are on the same side of the spacecraft and all the other information is unchanged.

29. Mr. T, who is 6 ft tall, is standing on a hill with an inclination of 5° and needs to throw a rope to the top of a nearby building. If the angle of elevation to the top of the building is 20° and the angle of depression to the base of the building is 11°, how tall is the building and how far is it from Mr. T to the top of the building?*

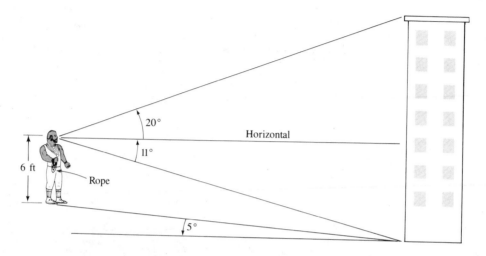

30. Johnny Carson is standing on the deck of his Malibu home and is looking down a hillside with an inclination of 35° to a small boat on the beach. If the angles of depression to the bow (point *B*) and stern (point *S*) are $\beta = 32°$ and $\theta = 25°$, how long is the boat if it is known that distance *JF* is 50 ft and *F*, *B*, and *S* are on level ground?

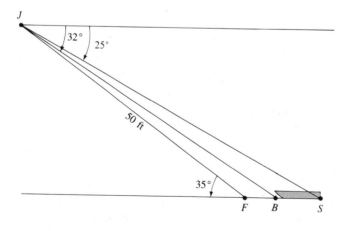

31. Given $\triangle ABC$, show that

$$\frac{\sin \alpha}{a} = \frac{\sin \beta}{b}$$

32. Given $\triangle ABC$, show that

$$\frac{\sin \beta}{b} = \frac{\sin \gamma}{c}$$

* Thanks to Jean Lane for the idea for this problem.

33. Given $\triangle ABC$, show that

$$\frac{\sin \alpha}{\sin \beta} = \frac{a}{b}$$

C   **34.** Using Problem 33, show that

$$\frac{\sin \alpha - \sin \beta}{\sin \alpha + \sin \beta} = \frac{a - b}{a + b}$$

35. Using Problem 34 and the formulas for the sum and difference of sines, show that

$$\frac{2 \cos \frac{1}{2}(\alpha + \beta) \sin \frac{1}{2}(\alpha - \beta)}{2 \sin \frac{1}{2}(\alpha + \beta) \cos \frac{1}{2}(\alpha - \beta)} = \frac{a - b}{a + b}$$

36. Using Problem 35, show that

$$\frac{\tan \frac{1}{2}(\alpha - \beta)}{\tan \frac{1}{2}(\alpha + \beta)} = \frac{a - b}{a + b}$$

This result is known as the **Law of Tangents**.

37. Another form of the Law of Tangents (see Problem 36) is

$$\frac{\tan \frac{1}{2}(\beta - \gamma)}{\tan \frac{1}{2}(\beta + \gamma)} = \frac{b - c}{b + c}$$

Derive this formula.

38. A third form of the Law of Tangents (see Problems 36 and 37) is

$$\frac{\tan \frac{1}{2}(\alpha - \gamma)}{\tan \frac{1}{2}(\alpha + \gamma)} = \frac{a - c}{a + c}$$

Derive this formula.

39. Newton's formula involves all six parts of a triangle. It is not useful in solving a triangle, but it is helpful in checking your results after you have done so. Show that

$$\frac{a + b}{c} = \frac{\cos \frac{1}{2}(\alpha - \beta)}{\sin \frac{1}{2}\gamma}$$

40. Mollweide's formula involves all six parts of a triangle. It is not useful in solving a triangle, but it is helpful in checking your results after you have done so. Show that

$$\frac{a - b}{c} = \frac{\sin \frac{1}{2}(\alpha - \beta)}{\cos \frac{1}{2}\gamma}$$

41. The following letter to the editor appeared in the February 1977 issue of *Popular Science.** See if you can solve this puzzle. As a hint, you might guess from its placement in this book that it uses the Law of Cosines or the Law of Sines.

---

* Reprinted with permission from *Popular Science*. ©1977.

Thirty-seven-year-old puzzle

For more than 37 years I have tried, on and off, to solve a problem that appeared in *Popular Science*.

All right, I give up. What's the answer? In the September, 1939 issue (page 14) the following letter appeared:

"Have enjoyed monkeying with the problems you print, for the last couple of years. Here's one that may make the boys scratch their heads a little: A man always drives at the same speed (his car probably has a governor on it). He makes it from *A* direct to *C* in 30 minutes; from *A* through *B* to *C*

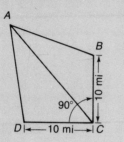

in 35 minutes; and from *A* through *D* to *C* in 40 minutes. How fast does he drive?

D.R.C., Sacramento, Calif."

It certainly looked easy, and I started to work on it. By 3 A.M. I had filled a lot of sheets of paper, both sides, with notations. In subsequent issues of *Popular Science*, all I could find on the subject was the following, in the November, 1939 issue:

"In regard to the problem submitted by D.R.C. of Sacramento, Calif. in the September issue. According to my calculations, the speed of the car must be 38.843 miles an hour. W.L.B., Chicago, Ill."

But the reader gave no hint of how he had arrived at that figure. It may or may not be correct. Over the years, I probably have shown the problem to more than 500 people. The usual reaction was to nod the head knowingly and start trying to find out what the hypotenuse of the right angle is. But those few who remembered the formula then found there is little you can do with it when you know the hypotenuse distance.

Now I'd like to challenge *Popular Science*. If your readers of today cannot solve this little problem (and prove the answer) how about your finding the answer, if there is one, editors?

R. F. Davis, Sun City, Ariz.

Darrell Huff, master of the pocket calculator and frequent PS contributor [Feb. '76, June '75, Dec. '74] has come up with a couple of solutions. We'd like to see what you readers can do before we publish his answers.

5.4   THE AMBIGUOUS CASE: SSA

The remaining case of solving oblique triangles as given in Section 5.1 is Case 5, in which we are given two sides and an angle that is not an included angle. Call the given angle θ (which may be α, β, or γ, depending on the problem). Since we are given SSA, one of the given sides must not be one of the sides of θ. Call this side OPP. The given side that is one of the sides of θ is called ADJ. You might find it helpful to refer to Table 5.1 on page 214 as you read this section.

1. Suppose $\theta > 90°$. There are two possibilities.
  i. OPP $\leq$ ADJ.
   No triangle is formed (see Figure 5.20).
  ii. OPP > ADJ.
   One triangle is formed (see Figure 5.21).
   Use the Law of Sines as shown in Example 1.

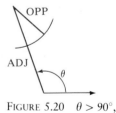

FIGURE 5.20 $\theta > 90°$, OPP $\leq$ ADJ

FIGURE 5.21 $\theta > 90°$, OPP > ADJ

Example 1

Let $a = 3.0$, $b = 2.0$, and $\alpha = 110°$. Solve the triangle.

Solution

$\alpha = 110°$ Given.

$\beta = 39°$ Work shown at right.

$\gamma = 31°$ $\gamma = 180° - 110° - \beta$

$\approx 31.2104436°$

$a = 3.0$ Given.

$b = 2.0$ Given.

$c = 1.7$ Use the Law of Sines:

$$\frac{\sin 110°}{3} = \frac{\sin \gamma}{c}$$

$$c = \frac{3 \sin \gamma}{\sin 110°}$$

$$\approx 1.654$$

$$\frac{\sin \alpha}{a} = \frac{\sin \beta}{b}$$

$$\frac{\sin 110°}{3} = \frac{\sin \beta}{2}$$

$$\sin \beta = \frac{2}{3} \sin 110°$$

$$\beta \approx 38.78955642°$$

☐

Do not work with rounded results. Notice, when finding c in Example 1, if you work with

$$\gamma \approx 31.2104436°$$

WARNING! Once again, a remainder about rounding.

you obtain $c \approx 1.654$, or 1.7, to two significant digits. However, if you work with $\gamma \approx 31°$, you obtain $c \approx 1.644$, or 1.6, to two significant digits. The proper procedure is to round only when stating answers.

2. **Suppose $\theta < 90°$. There are four possibilities (see i–iv below).** Let h be the altitude of the triangle drawn from the vertex connecting the OPP and ADJ sides. Now, to find h use the right-triangle definition of sine:

$$h = (ADJ) \sin \theta$$

   i. OPP < h < ADJ.
     No triangle is formed (see Figure 5.22).

   ii. OPP = h < ADJ.
     A right triangle is formed (see Figure 5.23).
     Use the methods of Section 5.1 to solve the
     triangle.

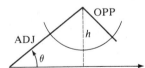

FIGURE 5.22 $\theta < 90°$, OPP < h < ADJ

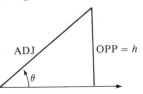

FIGURE 5.23 $\theta < 90°$, OPP = h < ADJ

iii. $h <$ OPP $<$ ADJ.

This situation is called the **ambiguous case** and is really the only special case you must watch for. All the other cases can be determined from the calculations with out any special consideration. Notice from Figure 5.24 that two *different* triangles are formed with the given information. The process for finding both solutions is shown in Example 2.

FIGURE 5.24 The ambiguous case

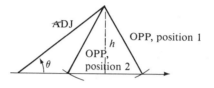

Example 2

Solve the triangle with $a = 1.50$, $b = 2.00$, and $\alpha = 40.0°$.

Solution

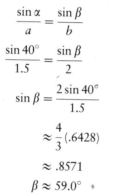

$$\frac{\sin \alpha}{a} = \frac{\sin \beta}{b}$$

$$\frac{\sin 40°}{1.5} = \frac{\sin \beta}{2}$$

$$\sin \beta = \frac{2 \sin 40°}{1.5}$$

$$\approx \frac{4}{3}(.6428)$$

$$\approx .8571$$

$$\beta \approx 59.0°$$

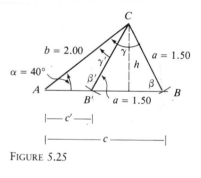

FIGURE 5.25

But from Figure 5.25, you can see that this is only the acute-angle solution. For the obtuse-angle solution—call it β'—find

$$\beta' = 180° - \beta$$
$$\approx 121°$$

Finish the problem by working two calculations, which are presented side by side.

SOLUTION 1		SOLUTION 2	
$\alpha = 40.0°$	Given.	$\alpha = 40.0°$	Given.
$\beta = 59.0°$	See above.	$\beta' = 121.0°$	See above.
$\gamma = 81.0°$	$\gamma = 180° - \alpha - \beta$	$\gamma' = 19.0°$	$\gamma' = 180° - \alpha - \beta'$
$a = 1.50$	Given.	$a = 1.50$	Given.
$b = 2.00$	Given.	$b = 2.00$	Given.

SOLUTION 1   SOLUTION 2

$c = 2.30$ $\dfrac{\sin \alpha}{a} = \dfrac{\sin \gamma}{c}$ $c' = .759$ $\dfrac{\sin \alpha}{a} = \dfrac{\sin \gamma'}{c'}$

$c = \dfrac{1.5 \sin \gamma}{\sin 40°}$ $c' = \dfrac{1.5 \sin \gamma'}{\sin 40°}$

≈ 2.3049 $\approx .7592$ ▯

iv. OPP ≥ ADJ.
One triangle is formed as shown in Figure 5.26.

Example 3   Solve the triangle given by $a = 3.0$, $b = 2.0$, and $\alpha = 40°$.

Solution

$$\dfrac{\sin \alpha}{a} = \dfrac{\sin \beta}{b}$$

$$\dfrac{\sin 40°}{3} = \dfrac{\sin \beta}{2}$$

$$\sin \beta = \dfrac{2}{3} \sin 40°$$

$$\approx \dfrac{2}{3}(.6428)$$

$$\approx .4285$$

$$\beta \approx 25.374°$$

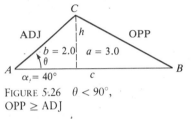

FIGURE 5.26 $\theta < 90°$,
OPP ≥ ADJ

$\alpha = 40°$ Given.

$\beta = 25°$ See work shown above.

$\gamma = 115°$ $\gamma = 180° - \alpha - \beta$

$a = 3.0$ Given.

$b = 2.0$ Given.

$c = 4.2$ Since $\dfrac{\sin 40°}{3} = \dfrac{\sin \gamma}{c}$

$$c = \dfrac{3 \sin \gamma}{\sin 40°}$$

$$\approx 4.2427$$ ▯

Solving Triangles: Summary

The most important skill to be learned from this section is the ability to select the proper trigonometric law when given a particular problem. In the rest of this section, you will encounter applications of right triangles, the Law of Cosines, and the Law of Sines. A review of various types of problems may be helpful; see Table 5.1.

TABLE 5.1   Summary for Solving Triangles

Given	Conditions on given information	Law to use for solution
1. SSS	a. The sum of the lengths of the two smaller sides is less than or equal to the length of the larger side.	No solution
	b. The sum of the lengths of the two smaller sides is greater than the length of the larger side.	Law of Cosines
2. SAS	a. The given angle is greater than or equal to 180°.	No solution
	b. The given angle is less than 180°.	Law of Cosines
3. ASA or AAS	a. The sum of the given angles is greater than or equal to 180°.	No solution
	b. The sum of the given angles is less than 180°.	Law of Sines
4. SSA	Let θ be the given angle with adjacent (ADJ) and opposite (OPP) sides given.	
	a. $\theta > 90°$	
	i. OPP $\leq$ ADJ	No solution
	ii. OPP $>$ ADJ	Law of Sines
	b. $\theta = 90°$	Right-triangle solution
	c. $\theta < 90°$	
	i. OPP $<$ ADJ	
	Find the height, h, by	
	$h = $ (ADJ) $\sin \theta$	
	$h < $ OPP $<$ ADJ	**Ambiguous case:** Use the Law of Sines to find two solutions.
	OPP $< h <$ ADJ	No solution
	OPP $= h <$ ADJ	Right-triangle solution
	ii. OPP $\geq$ ADJ	Law of Sines
5. AAA		No solution

Remember, *when given two sides and an angle that is not an included angle, check to see whether the length of one side is between the length of the altitude from the vertex determined by the two sides and the length of the other side. If so, there will be two solutions.*

Example 4

An airplane is 100 km N40°E of a Loran station, traveling due west of 240 kph. How long will it be (to the nearest minute) before the plane is 90 km from the Loran station? (See Figure 5.27.)

Solution

You are given SSA; check the other conditions for the ambiguous case.

Let $\theta = 50°$; 100 km is the side adjacent to θ; 90 km is the side opposite to θ;

$$h = (ADJ) \sin \theta$$
$$= 100 \sin 50°$$
$$\approx 76.6$$
$$h < OPP < ADJ$$
$$76.6 < \ \ 90 \ \ < 100$$

which is the ambiguous case.

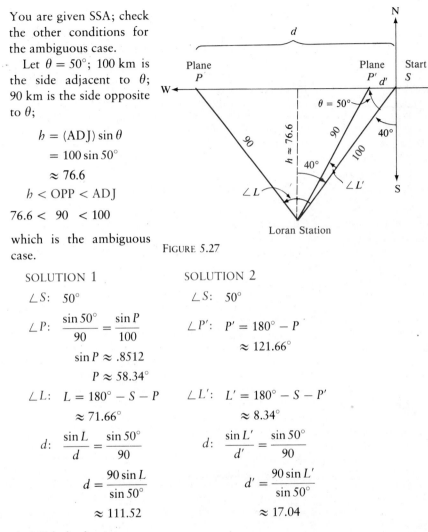

FIGURE 5.27

SOLUTION 1

$\angle S$: 50°

$\angle P$: $\dfrac{\sin 50°}{90} = \dfrac{\sin P}{100}$

$\qquad \sin P \approx .8512$

$\qquad P \approx 58.34°$

$\angle L$: $L = 180° - S - P$

$\qquad \approx 71.66°$

d: $\dfrac{\sin L}{d} = \dfrac{\sin 50°}{90}$

$\qquad d = \dfrac{90 \sin L}{\sin 50°}$

$\qquad \approx 111.52$

SOLUTION 2

$\angle S$: 50°

$\angle P'$: $P' = 180° - P$

$\qquad \approx 121.66°$

$\angle L'$: $L' = 180° - S - P'$

$\qquad \approx 8.34°$

d: $\dfrac{\sin L'}{d'} = \dfrac{\sin 50°}{90}$

$\qquad d' = \dfrac{90 \sin L'}{\sin 50°}$

$\qquad \approx 17.04$

At 240 kph, the times are

$$\frac{d}{240} \approx .4647 \quad \text{and} \quad \frac{d'}{240} \approx .0710$$

To convert these to minutes, multiply by 60 and round to the nearest minute. The times are 28 min and 4 min.

COMPUTER APPLICATION

There are many computer packages that will help you solve triangles. For example, TRIGPAK by John Mowbray will give you very worthwhile practice solving triangles. When you load TRIGPAK you will first see a *main menu* that gives you several choices; select O (oh) to see the next appropriate menu, called the *functions menu*. At this menu select T (solving triangles) and press the "enter" key. You will see a page of explanations and then the program will ask if you want computer-selected values or if you want to input the values. The computer values will give you practice solving triangles. On the other hand, if you want to practice by solving the problems in this text you can input the values. You might be thinking "this is great, I can let the computer do my work for me!" Not so; the computer will help YOU solve the triangle. After you input the values you must make one of the following choices:

1-SSS

2-SAS

3-SSA

4-AAS or ASA

5-AAA

Notice that these choices are based on Table 5.1. Then the program will tell if you have made the correct choice. If not, it will tell you where you made your mistake; if correct, the computer will then offer you the following choices:

1-no solution

2-infinitely many solutions

3-one solution using the Law of Cosines

4-one solution using the Law of Sines

5-two solutions using the Law of Sines

If you select the incorrect method the program will help you find where you made your mistake. If you select the correct method of solution, the program will show you an analysis of the problem and then, finally, will show you the answer.

The logic used by this computer application is summarized in Table 5.1. As an exercise, you might try writing a flowchart that could be used to write a computer program to solve triangles. Even if you do not know anything about computers or programming, this exercise might help you understand all of the information in Table 5.1.

Problem Set 5.4

A   *Solve triangle ABC in Problems 1–12. If two triangles are possible, give both solutions. If the triangle does not have a solution, tell why.*

1. $a = 3.0; b = 4.0; \alpha = 125°$

2. $a = 3.0; b = 4.0; \alpha = 80°$

3. $a = 5.0; b = 7.0; \alpha = 75°$

4. $a = 5.0; b = 7.0; \alpha = 135°$

5. $a = 5.0; b = 4.0; \alpha = 125°$

6. $a = 5.0; b = 4.0; \alpha = 80°$

7. $a = 9.0; b = 7.0; \alpha = 75°$

8. $a = 9.0; b = 7.0; \alpha = 135°$

9. $a = 7.0; b = 9.0; \alpha = 52°$

10. $a = 12.0; b = 9.00; \alpha = 52.0°$

11. $a = 8.629973679; b = 11.8; \alpha = 47.0°$ 12. $a = 10.2; b = 11.8; \alpha = 47.0°$

Solve triangle ABC with the parts given in Problems 13–24. If the triangle cannot be solved, tell why.

13. $a = 14.2; b = 16.3; \beta = 115.0°$ 14. $a = 14.2; c = 28.2; \gamma = 135.0°$

15. $\beta = 15.0°; \gamma = 18.0°; b = 23.5$ 16. $b = 45.7; \alpha = 82.3°; \beta = 61.5°$

17. $b = 82.5; c = 52.2; \gamma = 32.1°$ 18. $a = 151; b = 234; c = 416$

19. $a = 68.2; \alpha = 145°; \beta = 52.4°$ 20. $\alpha = 82.5°; \beta = 16.9°; \gamma = 80.6°$

21. $a = 123; b = 225; c = 351$ 22. $a = 27.2; c = 35.7; \alpha = 43.7°$

23. $c = 196; \alpha = 54.5°; \gamma = 63.0°$ 24. $b = 428; c = 395; \gamma = 28.4°$

B

25. *Ballistics.* An artillery-gun observer must determine the distance to a target at point T. He knows that the target is 5.20 miles from point I on a nearby island. He also knows that he (at point H) is 4.30 miles from point I. If $\angle HIT$ is 68.4°, how far is he from the target? (See Figure 5.28.)

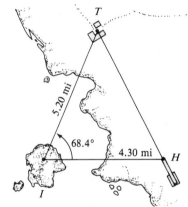

FIGURE 5.28 Determining the distance to a target

26. *Consumer.* A buyer is interested in purchasing a triangular lot with vertices LOT, but unfortunately, the marker at point L has been lost. The deed indicates that TO is 453 ft and LO is 112 ft, and that the angle at L is 82.6°. What is the distance from L to T?

27. *Navigation.* A UFO is sighted by people in two cities 2.300 miles apart. The UFO is between and in the same vertical plane as the two cities. The angle of elevation of the UFO from the first city is 10.48° and from the second is 40.79°. At what altitude is the UFO flying? What is the actual distance of the UFO from each city?

28. *Engineering.* At 500 ft in the direction that the Tower of Pisa is leaning, the angle of elevation is 20.24°. If the tower leans at an angle of 5.45° from the vertical, what is the length of the tower?

29. *Engineering.* What is the angle of elevation of the leaning Tower of Pisa (described in Problem 28) if you measure from a point 500 ft in the direction exactly opposite from the way it is leaning?

30. *Surveying.* The world's longest deepwater jetty is at Le Havre, France. Since access to the jetty is restricted, it was necessary for me to calculate its length by noting that it forms an angle of 85.0° with the shoreline. After pacing out 1,000 ft along the line making an 85.0° angle with the jetty, I calculated the angle to the end of the jetty to be 83.6°. What is the length of the jetty?

31. *Navigation.* From a blimp, the angle of depression to the top of the Eiffel Tower is 23.2° and to the bottom is 64.6°. After flying over the tower at the same height and at a distance of 1,000 ft from the first location, you determine that the angle of depression to the top of the tower is now 31.4°. What is the height of the Eiffel Tower given that these measurements are in the same vertical plane? (See Figure 5.29.)

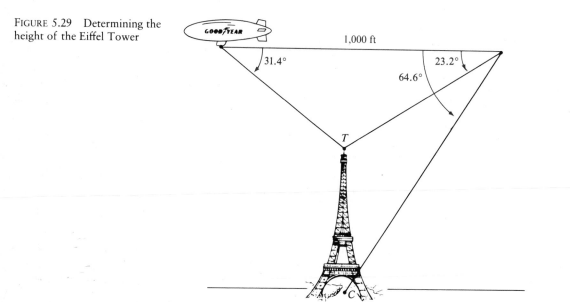

FIGURE 5.29   Determining the height of the Eiffel Tower

32. *Historical Question*   In January 1978, Martin Cohen, Terry Goodman, and John Benard published an article, "SSA and the Law of Cosines," in *The Mathematics Teacher*. In this article, they used the quadratic formula to solve the Law of Cosines for c:

$$c = b \cos \alpha \pm \sqrt{a^2 - b^2 \sin^2 \alpha}$$

Show this derivation.

33. Suppose $\alpha < 90°$. If $d = a^2 - b^2 \sin^2 \alpha$ in Problem 32, show that if $d < 0$, there is no solution and this equation corresponds to no triangle. If $d = 0$, there is one solution, which corresponds to one triangle; if $d > 0$, there are two solutions and two triangles are formed.

5.5 AREAS AND VOLUMES

Areas

In elementary school you learned that the area, K, of a triangle is $K = \frac{1}{2}bh$.* Now you sometimes need to find the area of a triangle given the measurements for the sides and angles but not for the height, and, therefore, you need trigonometry to find the area.

Using the orientation shown in Figure 5.30,

$$K = \tfrac{1}{2}bh \quad \text{and} \quad \sin \alpha = \tfrac{h}{c} \quad \text{or} \quad h = c \sin \alpha$$

---

* In elementary school you no doubt used A for area. In trigonometry, we use K for area because we have already used A to represent a vertex of a triangle.

FIGURE 5.30   Oblique triangles

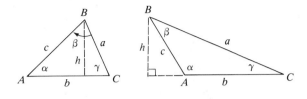

Thus,

$$K = \tfrac{1}{2}bc \sin \alpha$$

We could use any other pair of sides to derive the following area formulas.

AREA OF A TRIANGLE
(two sides and an
included angle are
known)

$$K = \frac{1}{2}bc \sin \alpha \qquad K = \frac{1}{2}ac \sin \beta \qquad K = \frac{1}{2}ab \sin \gamma$$

Example 1

Find the area of $\triangle ABC$ where $\alpha = 18.4°$, $b = 154$ ft, and $c = 211$ ft.

Solution

$$K = \tfrac{1}{2}(154)(211) \sin 18.4°$$
$$= 5,128.349903$$

To three significant figures, the area is 5,130 sq ft. ▯

The area formulas given above are useful for finding the area of a triangle when two sides and an included angle are known. Suppose, however, that you know the angles but only one side—say, side a. Then you can use either formula involving a and replace the other variable by using the Law of Sines:

$$\frac{\sin \alpha}{a} = \frac{\sin \beta}{b}$$

so

$$b = \frac{a \sin \beta}{\sin \alpha}$$

Therefore,

$$K = \frac{1}{2}ab \sin \gamma = \frac{1}{2}a \frac{a \sin \beta}{\sin \alpha} \sin \gamma = \frac{a^2 \sin \beta \sin \gamma}{2 \sin \alpha}$$

The other area formulas shown in the following box can be similarly derived.

AREA OF A TRIANGLE
(two angles and an
included side are known)

$$K = \frac{a^2 \sin \beta \sin \gamma}{2 \sin \alpha} \qquad K = \frac{b^2 \sin \alpha \sin \gamma}{2 \sin \beta} \qquad K = \frac{c^2 \sin \alpha \sin \beta}{2 \sin \gamma}$$

Example 2

Find the area of a triangle with angles 20°, 50°, and 110° if the side opposite the 50° angle is 24 in. long.

Solution

$$K = \frac{24^2 \sin 20° \sin 110°}{2 \sin 50°}$$

$$\approx 120.8303469$$

By calculator:

To two significant figures, the area is 120 sq in.

24 x^2 × 20 sin × 110 sin = ÷
2 ÷ 50 sin =

If three sides (but none of the angles) are known, you will need another formula to find the area of a triangle. The formula is derived from the Law of Cosines; this derivation is left as an exercise but is summarized in the following box. The result is known as **Heron's formula**.

AREA OF A TRIANGLE (three sides known)

$$K = \sqrt{s(s-a)(s-b)(s-c)} \qquad \text{where } s = \tfrac{1}{2}(a+b+c)$$

Example 3

Find the area of a triangle having sides of 43 ft, 89 ft, and 120 ft.

Solution

Let $a = 43$, $b = 89$, and $c = 120$. Then

$$s = \frac{1}{2}(43 + 89 + 120) = 126.$$

Thus,

$$K = \sqrt{126(126 - 43)(126 - 89)(126 - 120)}$$

$$\approx 1{,}523.704696$$

To two significant digits, the area is 1,500 sq ft.

Area of a Sector

A **sector of a circle** is the portion of the interior of a circle cut by a central angle, θ, as shown in Figure 5.31. Since the area of a circle is $A = \pi r^2$, the area of a sector is the fraction $\frac{\theta}{2\pi}$ of the entire circle.

In general,

area of sector = (fractional part of circle)(area of circle)

$$= \frac{\theta}{2\pi} \cdot \pi r^2$$

FIGURE 5.31   Sector of a circle (shaded portion)

AREA OF A SECTOR

Area of a sector $= \tfrac{1}{2}\theta r^2$ where θ is measured in radians

Example 4

What is the area of the sector of a circle of radius 12 in. whose central angle is 2?

Solution

$$\text{area of sector} = \tfrac{1}{2} \cdot 2 \cdot (12 \text{ in.})^2$$
$$= 144 \text{ in.}^2 \qquad \square$$

Example 5

What is the area of the sector of a circle of radius 420 m whose central angle is 2°?

Solution

First change 2° to radians:

$$2° = 2\left(\frac{\pi}{180}\right) = \frac{\pi}{90} \text{ radians}$$

$$\text{area of sector} = \frac{1}{2} \cdot \left(\frac{\pi}{90}\right)(420 \text{ m})^2$$

$$= 980\,\pi \text{ m}^2$$

$$\approx 3{,}079 \text{ m}^2 \qquad \square$$

Volume of a Cone

A third application in this section is finding the volume of a cone when you do not know its height. From geometry,

$$V = \frac{\pi r^2 h}{3}$$

Now suppose you know the radius of the base but not the height. If you know the angle of elevation α (see Figure 5.32), then

$$\tan \alpha = \frac{h}{r} \quad \text{or} \quad h = r \tan \alpha$$

Thus, the volume is

$$V = \frac{\pi r^3 \tan \alpha}{3} = \tfrac{1}{3}\pi r^3 \tan \alpha$$

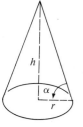

FIGURE 5.32   A cone with volume $V = \tfrac{1}{3}\pi r^2 h$

Example 6

If sand is dropped from the end of a conveyor belt, the sand will fall in conical heap such that the angle of elevation is about 33°. Find the volume of sand when the radius is exactly 10 ft.

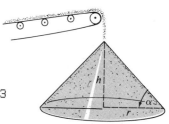

Solution

$$V = \tfrac{1}{3}\pi(10)^3 \tan 33° \qquad \text{DISPLAY: } 680.0580413$$
$$\approx 680 \qquad\qquad\qquad\qquad\qquad\qquad\qquad \square$$

Problem Set 5.5

A　　*Find the area of each triangle in Problems 1–12.*

1. $a = 15; b = 8.0; \gamma = 38°$
2. $a = 18; c = 11; \beta = 63°$
3. $b = 14; c = 12; \alpha = 82°$
4. $b = 21; c = 35; \alpha = 125°$
5. $a = 30; \beta = 50°; \gamma = 100°$
6. $b = 23; \alpha = 25°; \beta = 110°$
7. $b = 40; \alpha = 50°; \gamma = 60°$
8. $b = 90; \beta = 85°; \gamma = 25°$
9. $a = 7.0; b = 8.0; c = 2.0$
10. $a = 10; b = 4.0; c = 8.0$
11. $a = 11; b = 9.0; c = 8.0$
12. $a = 12; b = 6.0; c = 15$

B　　*Find the area of each triangle in Problems 13–24. If two triangles are formed, give the area of each, and if no triangle is formed, so state.*

13. $a = 12.0; b = 9.00; \alpha = 52.0°$
14. $a = 7.0; b = 9.0; \alpha = 52°$
15. $a = 10.2; b = 11.8; \alpha = 47.0°$
16. $a = 8.629973679; b = 11.8; \alpha = 47.0°$
17. $b = 82.5; c = 52.2; \gamma = 32.1°$
18. $a = 352; b = 230; c = 418$
19. $\beta = 15.0°; \gamma = 18.0°; b = 23.5$
20. $b = 45.7; \alpha = 82.3°; \beta = 61.5°$
21. $a = 68.2; \alpha = 145°; \beta = 52.4°$
22. $a = 151; b = 234; c = 416$
23. $a = 124; b = 325; c = 351$
24. $a = 27.2; c = 35.7; \alpha = 43.7°$

25. If the central angle subtended by the arc of a segment of a circle is 1.78 and the area is 54.4 sq cm, what is the radius of the circle?

26. If the area of a sector of a circle is 162.5 sq cm and the angle of the sector is .52, what is the radius of the circle?

27. A field is in the shape of a sector of a circle with a central angle of 20° and a radius of 320 m. What is the area of the field?

28. *Consumer.* A level lot has the dimensions shown in Figure 5.33. What is the total cost of treating the area for poison oak if the fee is $45 per acre (1 acre = 43,560 sq ft)?

29. If vulcanite is dropped from a conveyor belt, it will fall in a conical heap such that the angle of elevation is about 36°. Find the volume of vulcanite when the radius is 30 ft.

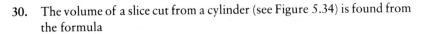

FIGURE 5.33　Area of a lot

30. The volume of a slice cut from a cylinder (see Figure 5.34) is found from the formula

$$V = hr^2 \left(\frac{\theta}{2} - \sin\frac{\theta}{2}\cos\frac{\theta}{2} \right)$$

for θ measured in radians such that $0 \le \theta \le \pi$. Find the volume of a slice cut from a log with a 6.0-in. radius that is 3.0 ft long when $\theta = \frac{\pi}{3}$.

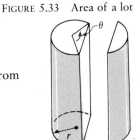

FIGURE 5.34　Slice of a cylinder

C

31. *Construction.* A 50-ft culvert carries water under a road. If there is 2 ft of water in a culvert with a 3-ft radius, use Problem 30 to find the volume of water in the culvert.

32. Derive the formula for finding the area when you know three sides of a triangle.

33. Derive the formula given in Problem 30.

34. *Historical Question* If we define the angle of elevation of a pyramid as the angle between the edge of the base and the slant height, we find that the major Egyptian pyramids are $43\frac{1}{2}°$ or $52°$. Write a paper explaining why the Egyptians built pyramids using angles of elevation of $43\frac{1}{2}°$ or $52°$.

References: Billard, Jules B., ed. *Ancient Egypt.* Washington, D.C.: National Geographic Society, 1978.

Edwards, I. E. S. *The Pyramids of Egypt.* New York: Penguin Books, 1961.

Fakhry, Ahmed. *The Pyramids* (2nd ed.). Chicago: University of Chicago Press, 1974.

Mendelssohn, Kurt. *The Riddle of the Pyramids.* New York: Praeger Publications, 1974.

Smith, Arthur F. "Angles of Elevation of the Pyramids of Egypt," *The Mathematics Teacher*, February 1982, pp. 124–127.

*5.6 VECTOR TRIANGLES

Many applications of mathematics involve quantities that have *both* magnitude and direction, such as forces, velocities, accelerations, and displacements.

Definition of a Vector

Example 1

State the magnitude and direction, and draw a diagram for each.

a. An airplane travels with an airspeed of 790 mph and a direction of 230°. [*Note:* In aviation, direction is measured clockwise from the north. This is called the **heading**. The **true course** is the direction of the path over the ground, which is also measured clockwise from the north.]

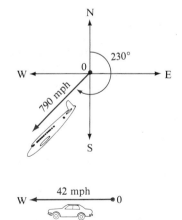

　　Magnitude:　790

　　Direction:　230° measured from the north

b. A car travels at 42 mph due west.

　　Magnitude:　42

　　Direction:　west

---

* Optional section.

c. The wind blows at 46.1 mph in the direction of S63°W.

> Magnitude: 46.1
>
> Direction: S63°W

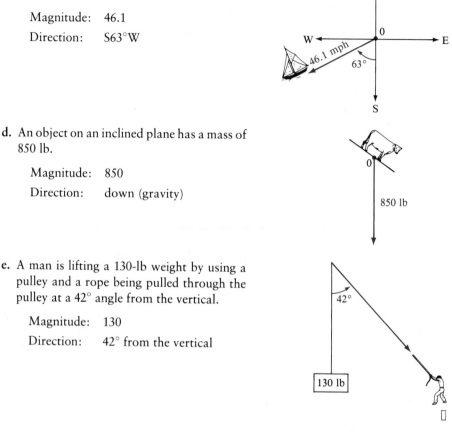

d. An object on an inclined plane has a mass of 850 lb.

> Magnitude: 850
>
> Direction: down (gravity)

e. A man is lifting a 130-lb weight by using a pulley and a rope being pulled through the pulley at a 42° angle from the vertical.

> Magnitude: 130
>
> Direction: 42° from the vertical

Vectors are used to describe quantities having both magnitude and direction. That is, a **vector** is a directed line segment specifying both a magnitude and a direction. The *length* of the vector represents the **magnitude** of the quantity being represented, and the *direction* of the vector represents the **direction** of the quantity. Two vectors are *equal* if they have the same magnitude and direction.

Suppose we choose a point O in the plane and call it the origin. A vector may be represented by a directed line segment from O to a point $P(x, y)$ in the plane. This vector is denoted by $\overrightarrow{OP}$ or **v**. In the text we use **v**; in your work you will write $\vec{v}$. The magnitude of $\overrightarrow{OP}$ is denoted by $|\mathbf{OP}|$ or $|\mathbf{v}|$. The vector from O to O is called the zero vector, **0**.

If **v** and **w** represent any two vectors having different (but not opposite) directions, they can be drawn so that they have the same base point. If they are represented in this fashion, they determine a geometrical figure called a *parallelogram* by drawing two other sides parallel to the given vectors forming a quadrilateral with the opposite sides parallel and equal in length, as shown in Figure 5.35.

FIGURE 5.35   A parallelogram formed by vectors **v** and **w**

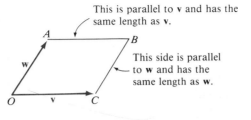

This is parallel to **v** and has the same length as **v**.

This side is parallel to **w** and has the same length as **w**.

Segments $\overline{OB}$ and $\overline{AC}$ are called the *diagonals* of the parallelogram, and angles having a common side are called *adjacent angles*. It is easy to show that the sum of two adjacent angles in a parallelogram is 180°.

Sum of Vectors (Resultant)

FIGURE 5.36   Resultant of vectors **v** and **w**

The **sum** or **resultant** of two vectors **v** and **w** is the vector that is the diagonal of the parallelogram having **v** and **w** as the adjacent sides and O as its base, as shown in Figure 5.36. The vectors **v** and **w** are called **components**.

There are basically two types of vector problems dealing with addition of vectors. The first is to find the resultant vector. To do this you use a right triangle, Law of Sines, or Law of Cosines. The second problem is to **resolve** a given vector into two component vectors. If the two vectors form a right angle, they are called **rectangular components**. You will usually resolve a vector into rectangular components.

Example 2

Consider two forces, one with magnitude 3.0 in a N20°W direction and the other with magnitude 7.0 in a S50°W direction. Find the resultant vector.

Solution

Sketch the given vectors and draw the parallelogram formed by these vectors. The diagonal is the resultant vector, as shown in Figure 5.37. You can easily find $\theta = 110°$, but you really need an angle inside the shaded triangle. Use the property from geometry that adjacent angles in a parallelogram are supplementary; that is, they add up to 180°. This tells you that $\phi = 70°$. Thus, you know SAS, so use the Law of Cosines to find the magnitude, $|\mathbf{v}|$.

$$|\mathbf{v}|^2 = 3^2 + 7^2 - 2(3)(7)\cos 70°$$
$$|\mathbf{v}| \approx 6.60569103$$

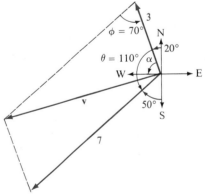

FIGURE 5.37   Finding a resultant vector

The direction of **v** can be found using the Law of Sines to find α:

$$\frac{\sin 70°}{|\mathbf{v}|} = \frac{\sin \alpha}{7}$$

$$\sin \alpha = \frac{7}{|\mathbf{v}|} \sin 70° \approx .9957850459$$

Thus, $\alpha \approx 84.737565°$.

Since $20° + 85° = 105°$, you can see that the direction of $\mathbf{v}$ should be measured from the south. Thus, the magnitude of $\mathbf{v}$ is 6.6 and the direction is S75°W.   ☐

Example 3

Suppose a vector, $\mathbf{v}$, has a magnitude of 5.00 and a direction given by $\theta = 30.0°$, where θ is the angle the vector makes with the positive x-axis. Resolve this vector into horizontal and vertical components.

Solution

Let $\mathbf{v}_x$ be the horizontal component and $\mathbf{v}_y$ the vertical component, as shown in Figure 5.38. Then

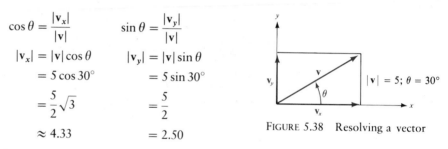

$$\cos \theta = \frac{|\mathbf{v}_x|}{|\mathbf{v}|} \qquad \sin \theta = \frac{|\mathbf{v}_y|}{|\mathbf{v}|}$$

$$|\mathbf{v}_x| = |\mathbf{v}| \cos \theta \qquad |\mathbf{v}_y| = |\mathbf{v}| \sin \theta$$

$$= 5 \cos 30° \qquad = 5 \sin 30°$$

$$= \frac{5}{2}\sqrt{3} \qquad = \frac{5}{2}$$

$$\approx 4.33 \qquad = 2.50$$

$|\mathbf{v}| = 5; \theta = 30°$

FIGURE 5.38   Resolving a vector

Thus, $\mathbf{v}_x$ is a horizontal vector with magnitude 4.33 and $\mathbf{v}_y$ is a vertical vector with magnitude 2.50.   ☐

Inclined Plane

One application involves forces acting on an object that is resting on an inclined plane. If the object is resting on a horizontal plane, the mass of the object has the same magnitude as the force pressing against the plane, as shown in Figure 5.39a.

FIGURE 5.39   Inclined plane application

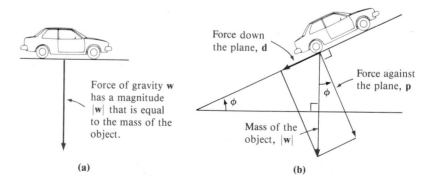

Force of gravity $\mathbf{w}$ has a magnitude $|\mathbf{w}|$ that is equal to the mass of the object.

Force down the plane, $\mathbf{d}$

Force against the plane, $\mathbf{p}$

Mass of the object, $|\mathbf{w}|$

(a)   (b)

However, if the object is on an inclined plane, the mass, |**w**|, of the object is the resultant of two forces—the downward pull, **d**, along the inclined plane and the force pressing against the plane, **p**, as shown in Figure 5.39b. Since **w** and **p** are perpendicular to the sides making up the angle ϕ, it follows that the angle of inclination, ϕ, is equal to the angle between the vectors representing the weight of the object and the force against the plane. You can use this information as shown in Example 4.

Example 4

What is the smallest force necessary to keep a 3,420-lb car from sliding down a hill that makes an angle of 5.25° with the horizontal? Also, what is the force against the hill? Assume that friction is ignored.

Solution

The mass (or force) of the car, |**OC**|, acts vertically downward. This vector can be resolved into two vectors: **OH**, the force against the hill, and **OD**, the force pushing down the hill, as shown in Figure 5.40.

FIGURE 5.40

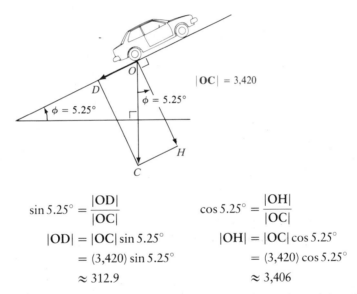

$$\sin 5.25° = \frac{|OD|}{|OC|} \qquad \cos 5.25° = \frac{|OH|}{|OC|}$$

$$|OD| = |OC| \sin 5.25° \qquad |OH| = |OC| \cos 5.25°$$

$$= (3,420) \sin 5.25° \qquad = (3,420) \cos 5.25°$$

$$\approx 312.9 \qquad\qquad \approx 3,406$$

Thus a force of 312.9 lb is necessary to keep the car from sliding down the hill, and the force against the hill is 3,406 lb. □

Problem Set 5.6

A *Find the resultant vector in Problems 1–3.*

1. **v** with direction west and magnitude 3.0 and **w** with direction north and magnitude 4.0.

2. **v** with direction south and magnitude 4.0 and **w** with direction east and magnitude 3.0.

3. **v** with direction east and magnitude 14 and **w** with direction north and magnitude 18.

Resolve the vectors given in Problems 4–6 into horizontal and vertical components.

4. $|\mathbf{v}| = 10, \theta = 30°$ 5. $|\mathbf{v}| = 23, \theta = 18°$ 6. $|\mathbf{v}| = 120, \theta = 27°$

7. *Navigation.* A woman sets out in a rowboat on a due-west heading and rows at 4.8 mph. The current is carrying the boat due south at 12 mph. What is the true course of the rowboat, and how fast is the boat traveling relative to the ground?

8. *Aviation.* An airplane is headed due west at 240 mph. The wind is blowing due south at 43 mph. What is the true course of the plane, and how fast is it traveling across the ground?

9. *Physics.* A cannon is fired at an angle of 10.0° above the horizontal, with the initial speed of the cannonball 2,120 fps. Find the magnitude of the vertical and horizontal components of the initial velocity.

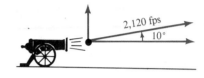

10. *Physics.* Answer Problem 9 for an angle of 17.8°.

11. *Aviation.* An airplane is heading 215° with a velocity of 723 mph. How far south has it traveled in one hour?

12. *Aviation.* An airplane is heading 43.0° with a velocity of 248 mph. How far east has it traveled in two hours?

B   13. *Navigation.* A boat moving at 18.0 knots heads in the direction of S38.2°W across a large river whose current is traveling at 5.00 knots in the direction of S13.1°W. Give the true course of the boat. How fast is it traveling as a result of these forces?

14. *Navigation.* A boat moving at 25.0 knots heads in the direction of N68.0°W across a river whose current is traveling at 12.0 knots in the direction of N16.0°W. Give the true course of the boat. How fast is it traveling as a result of these forces?

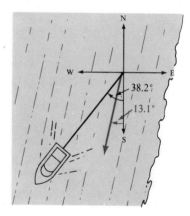

15. *Physics.* Two forces of 220 lb and 180 lb are acting on the same point in directions that differ by 52°. What is the magnitude of the resultant, and what is the angle that it makes with each of the given forces?

16. *Geometry.* If two sides of a parallelogram are 50 cm and 70 cm, and the diagonal connecting the ends of those sides is 40 cm, find the angles of the parallelogram.

17. *Physics.* If the resultant of two forces of 30 kilograms (kg) and 50 kg is a force of 60 kg, find the angle the resultant makes with each component force.

18. *Navigation.* Two boats leave a dock at the same time; one travels S15°W at 8 mph, and the other travels N28°W at 12 mph. How far apart are they in three hours?

19. *Aviation.* Two airplanes leave an airport at the same time; one travels at 180 mph with a course of

280°, while the other travels at 260 mph with a course of 35°. How far apart are the planes in one hour?

20. *Navigation*. A boat traveling at a constant rate in the direction N25°W passes a 3-mile marker from a lighthouse whose direction from the marker is due west. Twenty minutes after the boat passes the marker, it is N58°E from the lighthouse. How fast is the boat traveling?

21. *Navigation*. A boat leaves a dock at noon and travels N53°W at 10 kph. When will the boat be 12 km from a lighthouse located 15 km due west of the dock?

22. *Aviation*. At noon a plane leaves an airport and flies with a course of 61° at 200 mph. When will the plane be 35 miles from a point located 50 miles due east of the airport?

23. *Navigation*. Two boats leave port at the same time. The first travels 24.1 mph in the direction of S58.5°W, while the other travels 9.80 mph in the direction of N42.1°E. How far apart are the boats after two hours?

24. *Navigation*. A sailboat is in a 5.30-mph current in the direction of S43.2°W and is being blown by a 3.20-mph wind in a direction of S25.3°W. Find the course and speed of the sailboat.

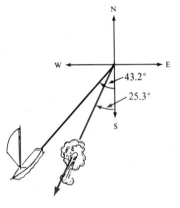

25. *Physics*. The world's steepest standard-gauge railroad grade measures 5.2° and is located between the Samala River Bridge and Zunil in Guatemala. What is the minimum force necessary to keep a boxcar weighing 52.0 tons from sliding down this incline? What is the force against the hill? Assume that friction is ignored.

26. *Physics*. What is the minimum force necessary to keep a 250-lb barrel from sliding down an inclined plane making an angle of 12° with the horizontal? What is the force against the plane? Assume that friction is ignored.

27. *Physics*. A force of 253 lb is necessary to keep a weight of exactly 400 lb from sliding down an inclined plane. What is the angle of inclination of the plane? Assume that friction is ignored.

28. *Physics*. A force of 486 lb is necessary to keep a weight of exactly 800 lb from sliding down an inclined plane. What is the angle of inclination of the plane? Assume that friction is ignored.

29. *Physics*. A cable that can withstand 5,250 lb is used to pull cargo up an inclined ramp for storage. If the inclination of the ramp is 25.5°, find the heaviest piece of cargo that can be pulled up the ramp. Assume that friction is ignored.

30. *Physics*. Answer the question posed in Problem 29 for a ramp whose angle of inclination is 18.2°.

C   31. *Space science.* The weight of astronauts on the moon is about $\frac{1}{6}$ of their weight on earth. This fact has a marked effect on such simple acts as walking, running, and jumping. To study these effects and to train astronauts for working under lunar-gravity conditions, scientists at NASA's Langley Research Center have designed an inclined-plane apparatus to simulate reduced gravity (see Figure 5.41). The apparatus consists of a sling that holds the astronaut in a position perpendicular to an inclined plane. The sling is attached to one end of a long cable that runs parallel to the inclined plane. The other end of the cable is attached to a trolley that runs along an overhead track. This device allows the astronaut to move freely in a plane perpendicular to the inclined plane. Let W be the weight of the astronaut and θ the angle between the inclined plane and the ground. Make a vector diagram showing the tension in the cable and the force exerted by the inclined plane against the feet of the astronaut.

FIGURE 5.41
Apparatus to simulate reduced gravity

32. *Space science.* From the point of view of the astronaut in Problem 31 the inclined plane is the ground and the astronaut's simulated mass (that is, the downward force against the inclined plane) is $W \cos \theta$. What value of θ is required in order to simulate lunar gravity?

*5.7   VECTOR OPERATIONS

The process of resolving a vector can be simplified with a definition of scalar multiplication and two very special vectors. **Scalar multiplication** is the multiplication of a vector by a real number. It is called scalar multiplication because real numbers are sometimes called scalars. If c is a positive real number and $\mathbf{v}$ is a vector, then the vector $c\mathbf{v}$ is a vector representing the scalar multiplication of c and $\mathbf{v}$. It is defined geometrically as a vector in the same direction as $\mathbf{v}$ but with a magnitude c times the original magnitude of $\mathbf{v}$, as shown in Figure 5.42. If c is a negative real number, then the scalar multiplication results in a vector in exactly the opposite direction as $\mathbf{v}$ with length c times as long, as shown in Figure 5.42. If $c = 0$, then $c\mathbf{v}$ is the **0** vector.

FIGURE 5.42   Examples of scalar multiplication

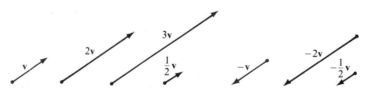

_____
* Optional section.

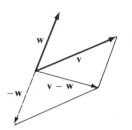

FIGURE 5.43 Vector subtraction

Scalar multiplication allows us to define **subtraction** for vectors:

$$\mathbf{v} - \mathbf{w} = \mathbf{v} + (-\mathbf{w})$$

Geometrically, subtraction is shown in Figure 5.43.

Algebraic Representation of a Vector

Two very special vectors help us to treat vectors algebraically as well as geometrically.

THE i AND j VECTORS

> **i** is the vector of unit length in the direction of the positive x-axis.
> **j** is the vector of unit length in the direction of the positive y-axis.

Any vector **v** can be written as

$$\mathbf{v} = a\mathbf{i} + b\mathbf{j}$$

where a and b are the magnitude of the horizontal and vertical components, respectively. This is called the **algebraic representation of a vector** (see Figure 5.44).

FIGURE 5.44 Algebraic representation of a vector

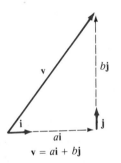

$$\mathbf{v} = a\mathbf{i} + b\mathbf{j}$$

Example 1

Find the algebraic representation for a vector **v** with magnitude 10 making an angle of 60° with the positive x-axis.

Solution

Figure 5.45 shows the general procedure for writing the algebraic representation of a vector when given the magnitude and direction of that vector.

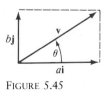

FIGURE 5.45

$$|b\mathbf{j}| = b$$
$$|a\mathbf{i}| = a$$
$$\cos\theta = \frac{a}{|\mathbf{v}|} \qquad a = |\mathbf{v}|\cos\theta$$
$$\sin\theta = \frac{b}{|\mathbf{v}|} \qquad b = |\mathbf{v}|\sin\theta$$
$$\mathbf{v} = |\mathbf{v}|\cos\theta\mathbf{i} + |\mathbf{v}|\sin\theta\mathbf{j}$$

From Figure 5.45,

$$a = |\mathbf{v}|\cos\theta \qquad b = |\mathbf{v}|\sin\theta$$
$$= 10\cos 60° \qquad = 10\sin 60°$$
$$= 5 \qquad\qquad \approx 8.7$$

Therefore,

$$\mathbf{v} = 5\mathbf{i} + 8.7\mathbf{j}$$

□

Example 2

Find the algebraic representation for a vector $\mathbf{v}$ with initial point $(4, -3)$ and endpoint $(-2, 4)$.

Solution

Figure 5.46 shows the general procedure for writing the algebraic representation of a vector when given the endpoints of that vector.

FIGURE 5.46

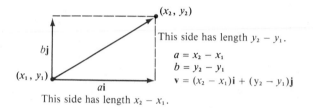

This side has length $y_2 - y_1$.

$$a = x_2 - x_1$$
$$b = y_2 - y_1$$
$$\mathbf{v} = (x_2 - x_1)\mathbf{i} + (y_2 - y_1)\mathbf{j}$$

This side has length $x_2 - x_1$.

From Figure 5.46,

$$a = x_2 - x_1 \qquad b = y_2 - y_1$$
$$= -2 - 4 \qquad = 4 - (-3)$$
$$= -6 \qquad\qquad = 7$$

Thus,

$$\mathbf{v} = -6\mathbf{i} + 7\mathbf{j}$$

□

Magnitude of a Vector

Example 3

Find the magnitude of the vector in Example 2.

Solution

$\mathbf{v} = -6\mathbf{i} + 7\mathbf{j}$. Thus,

$$|\mathbf{v}| = \sqrt{(-6)^2 + (7)^2} \quad \textit{Pythagorean theorem}$$
$$= \sqrt{36 + 49}$$
$$= \sqrt{85}$$

□

Example 3 leads to the following general result.

MAGNITUDE OF A
VECTOR

The **magnitude** of a vector $\mathbf{v} = a\mathbf{i} + b\mathbf{j}$ is given by

$$|\mathbf{v}| = \sqrt{a^2 + b^2}$$

The operations of addition, subtraction, and scalar multiplication can also be stated algebraically. Let $\mathbf{v} = a\mathbf{i} + b\mathbf{j}$ and $\mathbf{w} = c\mathbf{i} + d\mathbf{j}$. Then

$$\mathbf{v} + \mathbf{w} = (a + c)\mathbf{i} + (b + d)\mathbf{j}$$
$$\mathbf{v} - \mathbf{w} = (a - c)\mathbf{i} + (b - d)\mathbf{j}$$
$$c\mathbf{v} = ca\mathbf{i} + cb\mathbf{j}$$

Example 4

Let $\mathbf{v} = 6\mathbf{i} + 4\mathbf{j}$ and $\mathbf{w} = -2\mathbf{i} + 3\mathbf{j}$.

a. $|\mathbf{v}| = \sqrt{6^2 + 4^2}$
   $= \sqrt{36 + 16}$
   $= 2\sqrt{13}$

b. $|\mathbf{w}| = \sqrt{(-2)^2 + 3^2}$
   $= \sqrt{4 + 9}$
   $= \sqrt{13}$

c. $\mathbf{v} + \mathbf{w} = (6 - 2)\mathbf{i} + (4 + 3)\mathbf{j}$
   $= 4\mathbf{i} + 7\mathbf{j}$

d. $\mathbf{v} - \mathbf{w} = (6 + 2)\mathbf{i} + (4 - 3)\mathbf{j}$
   $= 8\mathbf{i} + \mathbf{j}$

e. $-\mathbf{v} = (-1)6\mathbf{i} + (-1)4\mathbf{j}$
   $= -6\mathbf{i} - 4\mathbf{j}$

f. $-2\mathbf{w} = (-2)(-2)\mathbf{i} + (-2)(3)\mathbf{j}$
   $= 4\mathbf{i} - 6\mathbf{j}$ □

You can see from Example 4 that the algebraic representation of a vector makes it easy to handle vectors and their operations. Another advantage of the algebraic representation is that it specifies a direction and a magnitude, but not a particular location. The directed line segment in Example 2 defined a given vector. However, there are infinitely many other vectors represented by the form $-6\mathbf{i} + 7\mathbf{j}$. Two of these are shown in Figure 5.47.

FIGURE 5.47 Some vectors represented by $\mathbf{v} = -6\mathbf{i} + 7\mathbf{j}$

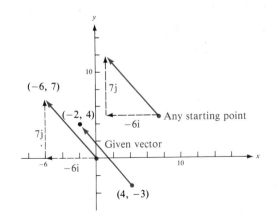

Scalar Product

In arithmetic you are used to using the words *multiplication* and *product* to mean the same thing. When working with vectors these words are used to denote different ideas. Recall that scalar multiplication does not tell us how to multiply two vectors; instead, it tells us how to multiply a scalar and a vector. Now we define an operation called **scalar product** in order to multiply two vectors to obtain a number. It is called *scalar product* because we obtain a number, or scalar, as an answer. Sometimes scalar product is called *dot product* or *inner product*. In subsequent courses you will define another vector multiplication, called **vector product**, in which you obtain a vector as an answer.

SCALAR PRODUCT

> Let $\mathbf{v} = a\mathbf{i} + b\mathbf{j}$ and $\mathbf{w} = c\mathbf{i} + d\mathbf{j}$. Then the **scalar product**, written $\mathbf{v} \cdot \mathbf{w}$, is defined by
>
> $$\mathbf{v} \cdot \mathbf{w} = ac + bd$$

Example 5

Find the scalar product of the given vectors.

a. If $\mathbf{v} = 2\mathbf{i} + 5\mathbf{j}$ and $\mathbf{w} = 6\mathbf{i} - 3\mathbf{j}$, then

$$\begin{aligned}
\mathbf{v} \cdot \mathbf{w} &= 2(6) + 5(-3) \\
&= 12 - 15 \\
&= -3
\end{aligned}$$

b. If $\mathbf{v} = \cos 30° \, \mathbf{i} + \sin 30° \, \mathbf{j}$ and $\mathbf{w} = \cos 60° \, \mathbf{i} - \sin 60° \, \mathbf{j}$, then

$$\begin{aligned}
\mathbf{v} \cdot \mathbf{w} &= \cos 30° \cos 60° - \sin 30° \sin 60° \\
&= \cos(30° + 60°) \\
&= 0
\end{aligned}$$

c. If $\mathbf{v} = -\sqrt{3}\,\mathbf{i} + \sqrt{2}\,\mathbf{j}$ and $\mathbf{w} = 3\sqrt{3}\,\mathbf{i} + 5\sqrt{2}\,\mathbf{j}$, then

$$\begin{aligned}
\mathbf{v} \cdot \mathbf{w} &= (-\sqrt{3})(3\sqrt{3}) + \sqrt{2}(5\sqrt{2}) \\
&= -9 + 10 \\
&= 1
\end{aligned}$$

d. If $\mathbf{v} = 2\mathbf{i} - 3\mathbf{j}$ and $\mathbf{w} = 4\mathbf{i} + a\mathbf{j}$, then

$$\mathbf{v} \cdot \mathbf{w} = 8 - 3a$$

         ☐

Angle Between Vectors

There is a very useful geometric property for scalar product that is apparent if we find an expression for the angle between two vectors.

ANGLE BETWEEN
VECTORS

The angle θ between nonzero vectors $\mathbf{v}$ and $\mathbf{w}$ is found by

$$\cos \theta = \frac{\mathbf{v} \cdot \mathbf{w}}{|\mathbf{v}||\mathbf{w}|}$$

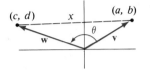

FIGURE 5.48   Finding the angle
between two vectors

Proof: Let $\mathbf{v} = a\mathbf{i} + b\mathbf{j}$ and $\mathbf{w} = c\mathbf{i} + d\mathbf{j}$ be drawn with their bases at the origin, as shown in Figure 5.48. Let x be the distance between the endpoints of the vectors. Then, by the Law of Cosines,

$$\cos \theta = \frac{|\mathbf{v}|^2 + |\mathbf{w}|^2 - x^2}{2|\mathbf{v}||\mathbf{w}|}$$

$$= \frac{(\sqrt{a^2 + b^2})^2 + (\sqrt{c^2 + d^2})^2 - (\sqrt{(a-c)^2 + (b-d)^2})^2}{2|\mathbf{v}||\mathbf{w}|}$$

$$= \frac{a^2 + b^2 + c^2 + d^2 - (a^2 - 2ac + c^2 + b^2 - 2bd + d^2)}{2|\mathbf{v}||\mathbf{w}|}$$

$$= \frac{2ac + 2bd}{2|\mathbf{v}||\mathbf{w}|}$$

$$= \frac{ac + bd}{|\mathbf{v}||\mathbf{w}|}$$

$$= \frac{\mathbf{v} \cdot \mathbf{w}}{|\mathbf{v}||\mathbf{w}|}$$

Orthogonal Vectors

There is a useful geometric property of vectors whose directions differ by $90°$. If you are dealing with lines, they are called **perpendicular lines**, but if they are vectors forming a $90°$ angle, they are called **orthogonal**. Notice, if $\theta = 90°$, then $\cos 90° = 0$; therefore, from the angle between vectors formula,

$$0 = \frac{\mathbf{v} \cdot \mathbf{w}}{|\mathbf{v}||\mathbf{w}|}$$

If you multiply both sides by $|\mathbf{v}||\mathbf{w}|$, the result is the following important condition.

ORTHOGONAL VECTORS

Nonzero vectors $\mathbf{v}$ and $\mathbf{w}$ are **orthogonal** if and only if $\mathbf{v} \cdot \mathbf{w} = 0$.

Example 6

Show that $\mathbf{v} = 3\mathbf{i} - 2\mathbf{j}$ and $\mathbf{w} = 6\mathbf{i} + 9\mathbf{j}$ are orthogonal.

Solution

$$\mathbf{v} \cdot \mathbf{w} = 3(6) + (-2)(9)$$
$$= 18 - 18 = 0$$

Since the scalar product is 0, the vectors are orthogonal.

Example 7   Find a so that $\mathbf{v} = 3\mathbf{i} + a\mathbf{j}$ and $\mathbf{w} = \mathbf{i} - 2\mathbf{j}$ are orthogonal.

Solution $\mathbf{v} \cdot \mathbf{w} = 3 - 2a$

                   If they are orthogonal, then

                        $3 - 2a = 0$

                        $a = \dfrac{3}{2}$ □

Problem Set 5.7

A   *Find the algebraic representation for each vector given in Problems 1–6. Approximate answers should be rounded to four decimal places. $|\mathbf{v}|$ is the magnitude of the vector $\mathbf{v}$, and θ is the angle the vector makes with the positive x-axis. A and B are the endpoints of the vector $\mathbf{v}$, and A is the base point. Draw each vector.*

   1. $|\mathbf{v}| = 12,\ \theta = 60°$ 2. $|\mathbf{v}| = 8,\ \theta = 30°$ 3. $|\mathbf{v}| = 4,\ \theta = 112°$
   4. $|\mathbf{v}| = 10,\ \theta = 214°$ 5. $A(4, 1),\ B(2, 3);\ \overrightarrow{AB}$ 6. $A(-1, -3),\ B(4, 5);\ \overrightarrow{AB}$

Find the magnitude of each of the vectors given in Problems 7–12.

   7. $\mathbf{v} = 3\mathbf{i} + 4\mathbf{j}$ 8. $\mathbf{v} = 5\mathbf{i} - 12\mathbf{j}$ 9. $\mathbf{v} = 6\mathbf{i} - 7\mathbf{j}$
   10. $\mathbf{v} = -3\mathbf{i} + 5\mathbf{j}$ 11. $\mathbf{v} = -2\mathbf{i} + 2\mathbf{j}$ 12. $\mathbf{v} = 5\mathbf{i} - 8\mathbf{j}$

State whether the given pairs of vectors in Problems 13–16 are orthogonal.

   13. $\mathbf{v} = 3\mathbf{i} - 2\mathbf{j};\ \mathbf{w} = 6\mathbf{i} + 9\mathbf{j}$ 14. $\mathbf{v} = 2\mathbf{i} + 3\mathbf{j};\ \mathbf{w} = 6\mathbf{i} - 9\mathbf{j}$
   15. $\mathbf{v} = 4\mathbf{i} - 5\mathbf{j};\ \mathbf{w} = 8\mathbf{i} + 10\mathbf{j}$ 16. $\mathbf{v} = 5\mathbf{i} + 4\mathbf{j};\ \mathbf{w} = 8\mathbf{i} - 10\mathbf{j}$

B   *In Problems 17–25, find $\mathbf{v} \cdot \mathbf{w}$, $|\mathbf{v}|$, $|\mathbf{w}|$, and $\cos\theta$, where θ is the angle between $\mathbf{v}$ and $\mathbf{w}$.*

   17. $\mathbf{v} = 3\mathbf{i} + 4\mathbf{j}$ 18. $\mathbf{v} = 8\mathbf{i} - 6\mathbf{j}$ 19. $\mathbf{v} = 2\mathbf{i} + \sqrt{5}\,\mathbf{j}$
       $\mathbf{w} = 5\mathbf{i} + 12\mathbf{j}$ $\mathbf{w} = -5\mathbf{i} + 12\mathbf{j}$ $\mathbf{w} = 3\sqrt{5}\,\mathbf{i} - 3\mathbf{j}$
   20. $\mathbf{v} = 7\mathbf{i} - \sqrt{15}\,\mathbf{j}$ 21. $\mathbf{v} = 2\mathbf{i} + 3\mathbf{j}$ 22. $\mathbf{v} = 3\mathbf{i} + 9\mathbf{j}$
       $\mathbf{w} = 2\sqrt{15}\,\mathbf{i} + 14\mathbf{j}$ $\mathbf{w} = 6\mathbf{i} + 5\mathbf{j}$ $\mathbf{w} = 2\mathbf{i} - 5\mathbf{j}$
   23. $\mathbf{v} = \mathbf{i}$ 24. $\mathbf{v} = \mathbf{j}$ 25. $\mathbf{v} = \mathbf{i}$
       $\mathbf{w} = \mathbf{i}$ $\mathbf{w} = \mathbf{j}$ $\mathbf{w} = -\mathbf{j}$

In Problems 26–31, find the angle θ to the nearest degree, $0° \le \theta \le 180°$, between the vectors $\mathbf{v}$ and $\mathbf{w}$.

   26. $\mathbf{v} = \frac{1}{2}\mathbf{i} + \frac{\sqrt{3}}{2}\mathbf{j}$ 27. $\mathbf{v} = \sqrt{2}\,\mathbf{i} - \sqrt{2}\,\mathbf{j}$ 28. $\mathbf{v} = \mathbf{j}$
       $\mathbf{w} = \frac{1}{2}\mathbf{i} + \frac{1}{2}\mathbf{j}$ $\mathbf{w} = \frac{\sqrt{3}}{2}\mathbf{i} + \frac{1}{2}\mathbf{j}$ $\mathbf{w} = \frac{1}{2}\mathbf{i} - \frac{\sqrt{3}}{2}\mathbf{j}$
   29. $\mathbf{v} = -\mathbf{i}$ 30. $\mathbf{v} = 2\mathbf{i} + 3\mathbf{j}$ 31. $\mathbf{v} = -3\mathbf{i} + 2\mathbf{j}$
       $\mathbf{w} = -2\sqrt{2}\,\mathbf{i} + 2\sqrt{2}\,\mathbf{j}$ $\mathbf{w} = -\mathbf{i} + 4\mathbf{j}$ $\mathbf{w} = 6\mathbf{i} + 9\mathbf{j}$

In Problems 32–34, find a number a so that the given vectors are orthogonal.

32. $\mathbf{v} = 2\mathbf{i} + 3\mathbf{j}$
$\mathbf{w} = 5\mathbf{i} + a\mathbf{j}$

33. $\mathbf{v} = 4\mathbf{i} - a\mathbf{j}$
$\mathbf{w} = -2\mathbf{i} + 5\mathbf{j}$

34. $\mathbf{v} = a\mathbf{i} + 5\mathbf{j}$
$\mathbf{w} = a\mathbf{i} - 15\mathbf{j}$

C   **35.** If $a = |\mathbf{v}|$ and $b = |\mathbf{w}|$, show that $a\mathbf{w} + b\mathbf{v}$ and $b\mathbf{v} - a\mathbf{w}$ are orthogonal.

5.8   Summary and Review

	OBJECTIVES	PAGES/EXAMPLES
5.1 Right Triangles	1. State the right-triangle definition of the trigonometric functions.	p. 190.
	2. Solve right triangles.	pp. 190–192; Examples 1–3; Problems 1–18.
	3. Solve applied right-triangle problems.	pp. 192–195; Examples 4–6; Problems 19–32.
5.2 Law of Cosines	4. State the Law of Cosines and tell when it is used.	p. 200.
	5. Solve SSS triangles using the Law of Cosines.	p. 201; Example 1; Problems 1–6, 13–15, 22, 23.
	6. Solve SAS triangles using the Law of Cosines.	p. 202; Example 2; Problems 7–12, 16–21.
	7. Solve applied problems using the Law of Cosines.	p. 203; Problems 25–35.
5.3 Law of Sines	8. State the Law of Sines and tell when it is used.	p. 205.
	9. Solve ASA or AAS triangles using the Law of Sines.	p. 205; Example 1; Problems 1–24.
	10. Solve applied problems using the Law of Sines.	pp. 206–207; Example 2; Problems 25–30.
5.4 The Ambiguous Case: SSA	11. Solve SSA triangles.	p. 210; Examples 1–3; Problems 1–24.
	12. Solve triangles by deciding the method of solution.	p. 213; Example 4; Problems 25–33.
5.5 Areas and Volumes	13. Find the area of a triangle given a. two sides and an included angle; b. two angles and an included side; c. three sides.	p. 218; Examples 1–3; Problems 1–24; 28.

14. Find the area of a sector of a circle.

p. 220; Examples 4–5; Problems 25–27.

15. Find the volume of a cone when the height is not known.

p. 221; Example 6; Problems 29–30.

*5.6 VECTOR TRIANGLES

16. Given two nonzero vectors, find the resultant vector.

p. 225; Example 2; Problems 1–3.

17. Solve applied problems by finding the resultant vector.

p. 225; Example 2; Problems 7–24.

18. Resolve a vector into horizontal and vertical components.

p. 226; Example 3; Problems 4–6.

19. Solve applied problems involving inclined planes.

pp. 226–227; Example 4; Problems 25–30.

*5.7 VECTOR OPERATIONS

20. Write the algebraic representation of a vector.

p. 231; Examples 1–2; Problems 1–6.

21. Find the magnitude of a vector.

p. 232; Examples 3–4; Problems 7–12.

22. Find the scalar product of two vectors.

p. 234; Example 5; Problems 17–25.

23. Find the angle between vectors.

p. 235; Problems 17–31.

24. Determine whether two vectors are orthogonal.

p. 235; Examples 6–7; Problems 13–16, 32–34.

Terms

*Addition of vectors [5.6]
Adjacent side [5.1, 5.4]
*Algebraic representation [5.7]
Ambiguous case [5.4]
*Angle between vectors [5.7]
Area [5.5]
Bearing [5.1]
*Components [5.6]
Cone [5.5]
Depression, angle of [5.1]
Elevation, angle of [5.1]

*Heading [5.6]
Hypotenuse [5.1]
*$\mathbf{i}$, [5.7]
Inclination, angle of [5.1]
*Inclined plane [5.6]
*$\mathbf{j}$, [5.7]
Law of Cosines [5.2]
Law of Sines [5.3]
*Magnitude [5.6, 5.7]
Opposite side [5.1, 5.4]
*Orthogonal [5.7]
*Rectangular components [5.6]

*Resolve [5.6]
*Resultant [5.6]
*Scalar multiplication [5.7]
*Scalar product [5.7]
Sector [5.5]
Solved triangle [5.1]
*Sum of vectors [5.6]
Triangle [5.1]
*Vector [5.6]
Volume [5.5]

* Optional sections

Chapter 5 Review of Objectives

The problem numbers correspond to the objectives listed in Section 5.8.

1. Fill in the blanks by referring to the right triangle in Figure 5.49.
   a. $\cos \beta =$ _____
   b. $\sin \beta =$ _____
   c. $\tan \beta =$ _____

2. Solve the triangle shown in Figure 5.49 if

$$a = 5.30 \quad \text{and} \quad b = 12.2$$

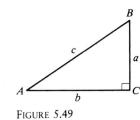

FIGURE 5.49

3. a. The tallest human-built structure in the world is the TV transmitter tower of KTHI-TV in North Dakota. Find its height if the angle of elevation at 501.0 ft away is 76.35°.
   b. To measure the distance across the Rainbow Bridge in Utah, a surveyor selected two points P and Q on either end of the bridge. From point Q the surveyor measured 500 ft in the direction N38.4°E to point R. Point P was then determined to be in the direction S67.5°W. What is the distance across the Rainbow Bridge if all the preceding measurements are in the same plane and angle PQR is a right angle?

4. Fill in the blanks by using the Law of Cosines.
   a. $a^2 =$ _____ b. $\cos \beta =$ _____

5. Solve the triangle where $a = 14$, $b = 27$, and $c = 19$.
6. Solve the triangle where $b = 7.2$, $c = 15$, and $\alpha = 113°$.
7. *Engineering.* A mine shaft is dug into the side of a sloping hill. The shaft is dug horizontally for 485 feet. Next, a turn is made so that the angle of elevation of the second shaft is 58.0°, thus forming a 58° angle between the shafts. The shaft is then continued for 382 feet before exiting, as shown in Figure 5.50. How far is it along a straight line from the entrance to the exit, assuming that all tunnels are in a single plane?

FIGURE 5.50 Determining the exit of a mine shaft

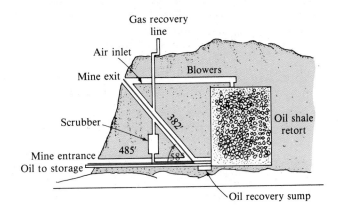

8. State the Law of Sines.
9. Solve the triangle where $a = 92.6$, $\alpha = 18.3°$, $\beta = 112.4°$.
10. A triangularly shaped garden has two angles measuring 46.5° and 105.8°, with the side opposite the 46.5° angle measuring 38.0 m. How much fence is needed to enclose the garden?
11. Ferndale is 7.0 miles N50°W of Fortuna. If I leave Fortuna at noon and travel due west at 2.0 mph, when will I be exactly 6.0 miles from Ferndale?
12. Fill in the blanks for the triangle ABC.

To find	Known	Procedure	Solution
Example:		Solution:	
β	a, b, α	$\dfrac{\sin \alpha}{a} = \dfrac{\sin \beta}{b}$	$\beta = \sin^{-1}\left(\dfrac{b \sin \alpha}{a}\right)$
α	a, b, c	a. _____	b. _____
α	a, b, β	c. _____	d. _____
α	a, β, γ	e. _____	f. _____
b	a, α, β	g. _____	h. _____
b	a, c, β	i. _____	j. _____

13. Find the area of each of the following triangles.
    a. $b = 16$, $c = 43$, $\alpha = 113°$
    b. $\alpha = 40.0°$, $\beta = 51.8°$, $c = 14.3$
    c. $a = 121$, $b = 46$, $c = 92$
14. A farmer plows a circular field with a radius of 1,000 ft into separate equal sections. If one section has a central angle of 15° ($\frac{\pi}{12}$ radians), what is the area of one section?
15. If a conical pile of sawdust has a radius of 135 ft and an angle of elevation of 35.0°, what is the volume of the pile?
*16. *Physics*. Consider two forces, one with a magnitude 5.0 in a S30°E direction and another with a magnitude of 12.0 in a N40°W direction. Find the resultant vector.
*17. *Physics*. An object is hurled from a catapult due east with a velocity of 38 feet per second (fps). If the wind is blowing due south at 15 mph (22 fps), what are the true bearing and the velocity of the object?
*18. Resolve the vector with magnitude 4.5 and $\theta = 51°$ into horizontal and vertical components.
*19. *Physics*. What is the smallest force necessary to keep a 3-ton boxcar from sliding down a hill that makes an angle of 4° with the horizontal? Assume that friction is ignored.

---

* Optional sections

*20.   Find the algebraic representation of the given vectors.
       a. $|\mathbf{v}| = \sqrt{2}, \theta = 45°$
       b. $|\mathbf{v}| = 9.3, \theta = 118°$
       c. $A(1, -2), B(-5, -7); \overrightarrow{AB}$
*21.   Find the magnitude of $\mathbf{v} = 3\mathbf{i} - 2\mathbf{j}$ and $\mathbf{w} = \mathbf{i} - \mathbf{j}$.
*22.   Find $\mathbf{v} \cdot \mathbf{w}$ for the vectors in Problem 21.
*23.   Find the angle (to the nearest degree) between the vectors given in Problem 21.
*24.   If $\mathbf{t} = \mathbf{i} - a\mathbf{j}$, find a so that $\mathbf{v}$ (from Problem 21) and $\mathbf{t}$ are orthogonal.

Miscellaneous Problems

Solve the triangles given in Problems 1–30.

1. $a = 24; c = 61; \beta = 58°$
2. $b = 34; c = 21; \gamma = 16°$
3. $a = 30; \beta = 50°; \gamma = 100°$
4. $b = 23; \alpha = 25°; \beta = 110°$
5. $b = 40; \alpha = 50°; \gamma = 60°$
6. $b = 90; \beta = 85°; \gamma = 25°$
7. $a = 15; b = 8.0; \gamma = 38°$
8. $a = 18; c = 11; \beta = 63°$
9. $b = 14; c = 12; \alpha = 82°$
10. $b = 21; c = 35; \alpha = 125°$
11. $a = 7.0; b = 8.0; c = 2.0$
12. $a = 10; b = 4.0; c = 8.0$
13. $a = 6.80; b = 12.2; c = 21.5$
14. $b = 4.6; \alpha = 108°; \gamma = 38°$
15. $a = 14.5; b = 17.2; \alpha = 35.5°$
16. $\alpha = 48.0°; b = 25.5; c = 48.5$
17. $a = 121; b = 315; \gamma = 50.0°$
18. $a = 6.50; b = 8.30; c = 12.6$
19. $a = 11; b = 9.0; c = 8.0$
20. $a = 12; b = 6.0; c = 15$
21. $a = 12.4; b = 32.5; c = 35.1$
22. $a = 2.72; c = 3.57; \alpha = 43.7°$
23. $b = 68.2; \alpha = 145°; \beta = 52.4°$
24. $a = 151; b = 235; c = 412$
25. $\beta = 15.0°; \gamma = 18.0°; b = 23.5$
26. $b = 45.7; \alpha = 82.3°; \beta = 61.5°$
27. $b = 82.5; c = 52.2; \gamma = 23.1°$
28. $a = 352; b = 230; c = 418$
29. $a = 10.2; b = 11.8; \alpha = 47.0°$
30. $a = 8.62; b = 11.8; \alpha = 37.0°$

31.   *Surveying.* A surveyor needs to estimate the distance of a tunnel through a hill. The distances to either end are 1,000 ft and 1,236 ft with an angle of 50.6° between these sides. What is the length of the tunnel?

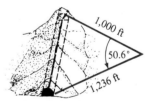

Proposed tunnel

32.   *Surveying.* What is the area of a triangular lot whose sides measure 275 ft, 690 ft, and 735 ft?

33.   *Aviation.* A pilot leaves Santa Rosa and flies 250 miles with a heading of 128° to the Nut Tree. He then leaves the Nut Tree at noon and flies north at 160 mph.

---

* Optional section.

When he is exactly 200 miles from Santa Rosa after leaving the Nut Tree, he must log his time. What time (to the nearest minute) should his log show?

34. *Aviation.* San Francisco and Los Angeles are approximately 450 miles apart. A pilot flying from San Francisco finds that she is 15.0° off course after 150 miles. How far is she from Los Angeles at this time?

35. A fire is simultaneously spotted from two observation towers that are located 10,520 ft apart. The first observation post reports the fire at N54.8°E and the second post reports the fire at N48.2°W. How far is the fire from each observation location? Assume that all the measurements are in the same plane and that the towers are on an east–west line.

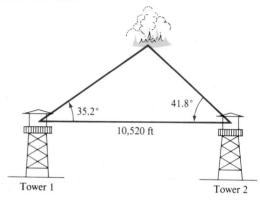

36. A large advertising sign stands on top of a tall building. The angle of elevation from a point on level ground 275 ft from the base of the building up to the bottom of the sign is 18.5°, and to the top of the sign it is 21.5°. How tall is the sign?

APPLICATION FOR FURTHER STUDY

PARKING-LOT DESIGN

Courtesy of Wide World Photos, Inc.

A city engineer must design the parking along a certain street. She can use parallel or angle parking. Angle parking requires greater street width, as shown in Figure 5.51, but it allows more spaces along the curb. Let us assume that city code calls for a 10-ft width per parking space. This means that 100 ft of curb could accommodate ten parking spaces if $\theta = 90°$.

FIGURE 5.51   Street parking

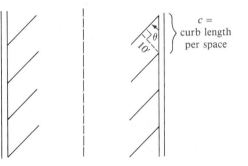

$c =$ curb length per space

Example 1

Find the curb length required for ten spaces if $\theta = 45°$.

Solution

From Figure 5.51, $\sin \theta = \dfrac{10}{c}$, so $c = \dfrac{10}{\sin \theta}$. For $\theta = 45°$,

$$c = \frac{10}{\sin 45°} \approx 14.14213562$$

This means that at $45°$ the curb length required for ten spaces is a little less than 142 ft. This is an increase in curb space of about 42% over the amount necessary for $\theta = 90°$.

▯

243

For this example, you can see that changing θ from $90°$ to $45°$ increased the amount of curb space needed by 42%. Now consider the change in the amount of street space needed. Would you guess it is larger or smaller? Suppose that city code calls for parking spaces to be 21 ft long, as shown in Figure 5.52.

FIGURE 5.52   Parking-space angle

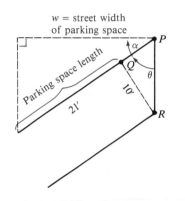

w = street width
of parking space

Parking space length

21'

The parking-space length cannot be measured from P, so the 21-ft requirement is measured from a point Q, where Q is drawn so that PQ is perpendicular to QR. We see from Figure 5.52 that, for $\theta = 90°$, $w = 21$. Let $\alpha = 90° - \theta$. Then

$$\cos \alpha = \frac{w}{21 + |PQ|}$$
$$w = (21 + |PQ|) \cos \alpha$$

Notice that $\tan \theta = \dfrac{10}{|PQ|}$, so

$$|PQ| = \frac{10}{\tan \theta} = 10 \cot \theta$$

which can be substituted into the equation for w:

$$w = (21 + 10 \cot \theta) \cos \alpha$$
$$= (21 + 10 \cot \theta) \sin \theta \qquad \textit{Cofunctions of complementary angles are equal.}$$
$$w = 21 \sin \theta + 10 \cot \theta \sin \theta$$
$$= 21 \sin \theta + 10 \cos \theta$$

For $\theta = 45°$,

$$w = 21 \sin 45° + 10 \cos 45°$$
$$\approx 21.92$$

This result means that for an angle of $45°$ the width that the parking space extends into the street is *increased* by about 4%. Does this match your guess?

Example 2   How does the street width change if θ is changed from $90°$ to $30°$?

Solution

$$w = 21 \sin 30° + 10 \cos 30°$$
$$\approx 19.16$$

This result is a decrease of about 1.8 ft, or about 9%, in the street width of the parking space. □

Problems for Further Study: Parking-Lot Design

θ	c	w
$10°$		
$20°$		
$30°$		
$40°$		
$50°$		
$60°$		
$70°$		
$80°$		
$90°$		

1. Complete the table in the margin for the parking-lot design problem presented in this section.
2. *Engineering.* You can see from the table in Problem 1 that there appear to be some physical limitations. For example, if c exceeds 21 ft, parallel parking appears to be more advantageous (why?). Find the angle θ for which $c = 21$, and then specify what you think would be a reasonable domain for θ for angle parking.
3. *Engineering.* Suppose you measured the parking width in Figure 5.52 from P instead of Q, as shown in Figure 5.53. For example, if $\theta = 50°$, then $\alpha = 40°$ and

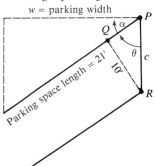

$$\cos \alpha = \frac{w}{21}$$
$$w = 21 \cos 40°$$
$$\approx 16 \text{ ft}$$

But this would make the usable length of the space 21 ft $- |QP|$ or

FIGURE 5.53   Angle parking

$$21 - 10 \cot \theta = 21 - 10 \cot 50°$$
$$\approx 13 \text{ ft}$$

Complete a table similar to the one shown in Problem 1 for usable length of various parking spaces. Calculate values of θ for $30°, 40°, 50°, \ldots, 80°$, and $90°$.
4. *Engineering.* If street size were of primary importance, what type of parking would you recommend, based on your observations from Problems 1–3?
5. *Engineering.* If curb size were of primary importance, what type of parking would you recommend, based on your observations from Problems 1–3?

245

CHAPTER 6

COMPLEX NUMBERS*

CHAPTER OVERVIEW

In mathematics, there are several applications that give rise to equations such as $x^2 + 1 = 0$, which cannot be solved in the set of real numbers. A *set of complex numbers* has been defined that includes not only the real numbers, but also square roots of negative numbers. The number *i*, defined in Section 6.1, is called the *imaginary unit*, although it is not more "imaginary" than the number 5. When working with complex and imaginary numbers, you must keep in mind that *all* numbers are concepts and that the number *i* is just as much a number as any of the "real" numbers. The trigonometric form of a complex number, introduced in Section 6.2, is a very useful form you will use not only in this course, but also in your future mathematical work.

6.1 THE IMAGINARY UNIT AND COMPLEX NUMBERS†

Definition of Complex Numbers

To find the roots of certain equations, you must sometimes consider the square roots of negative numbers. Since the set of real numbers does not allow for such a possibility, a number that is *not a real number* is defined. This number is denoted by the symbol *i*.

THE IMAGINARY UNIT

> The number *i*, called the **imaginary unit**, is defined as a number with the following properties:
>
> $i^2 = -1$, and if $a > 0$, $\sqrt{-a} = i\sqrt{a}$

With this number, you can write the square roots of any negative numbers as the product of a real number and the number *i*. Thus, $\sqrt{-9} = i\sqrt{9} = 3i$. For any positive real number *b*, $\sqrt{-b} = \sqrt{b}i$. We now consider a new set of numbers.

COMPLEX NUMBERS

> The set of all numbers that can be written in the form
>
> $a + bi$
>
> with real numbers *a* and *b* and the imaginary unit *i* is called the set of **complex numbers**.

---

* The material from this chapter is optional.

† If you have recently had an intermediate algebra course, you can skip or briefly review this section.

If $b = 0$, then $a + bi = a + 0i = a$, which is a **real number**; thus, the real numbers form a subset of the complex numbers. If $a = 0$ and $b \neq 0$, then $a + bi = 0 + bi = bi$ is called a **pure imaginary number**. If $a = 0$ and $b = 1$, then $a + bi = 0 + 1 \cdot i = i$, which is the imaginary unit. Sometimes a single symbol, usually z, is used to represent a complex number $a + bi$ (see Problems 32–40).

Operations with Complex Numbers

A complex number is **simplified** if it is written in the form $a + bi$, where a and b are real numbers in simplified form, and i is the imaginary unit. In order to work with complex numbers, you will need definitions of equality, along with the usual arithmetic operations. Let $a + bi$ and $c + di$ be complex numbers. Values that could cause division by zero are excluded.

OPERATIONS WITH
COMPLEX NUMBERS

> **Equality:** $a + bi = c + di$ if and only if $a = c$ and $b = d$
>
> **Addition:** $(a + bi) + (c + di) = (a + c) + (b + d)i$
>
> **Subtraction:** $(a + bi) - (c + di) = (a - c) + (b - d)i$
>
> **Multiplication:** $(a + bi)(c + di) = (ac - bd) + (ad + bc)i$
>
> **Division:** $\dfrac{a + bi}{c + di} = \dfrac{(ac + bd) + (bc - ad)i}{c^2 + d^2} = \dfrac{ac + bd}{c^2 + d^2} + \dfrac{bc - ad}{c^2 + d^2}i$

It is not necessary to memorize these definitions because you can deal with two complex numbers as you would any binomials as long as you remember that $i^2 = -1$.

Example 1

Simplify each expression.

a. $(4 + 5i) + (3 + 4i) = (4 + 3) + (5 + 4)i = 7 + 9i$
b. $(2 - i) - (3 - 5i) = (2 - 3) + (-1 + 5)i = -1 + 4i$
c. $(5 - 2i) + (3 + 2i) = 8 + 0i$ or 8
d. $(4 + 3i) - (4 + 2i) = 0 + i$ or i
e. $(2 + 3i)(4 + 2i) = 8 + 16i + 6i^2$
$\qquad\qquad\qquad\quad = 8 + 16i - 6$
$\qquad\qquad\qquad\quad = 2 + 16i$
f. $i^{95} = i^{92+3}$
$\quad\ = (i^4)^{23}\, i^3$
$\quad\ = (1)^{23}\, i^2 i$ *Since $i^4 = 1$*
$\quad\ = (1)(-1)i$
$\quad\ = -i$ $\qquad\qquad\qquad\qquad\qquad\qquad$ □

Example 2

Verify that multiplying $(a + bi)(c + di)$ in the usual algebraic way gives the same result as that shown in the definition of multiplication of complex numbers.

Solution

$$(a + bi)(c + di) = ac + adi + bci + bdi^2$$
$$= ac + (ad + bc)i - bd$$
$$= (ac - bd) + (ad + bc)i$$

☐

Example 3

Simplify $(4 - 3i)(4 + 3i)$.

Solution

$$(4 - 3i)(4 + 3i) = 16 - 9i^2$$
$$= 16 + 9$$
$$= 25$$

☐

Example 3 gives a clue for dividing complex numbers. The definition would be difficult to remember, so instead we use the idea of *conjugates*. The complex numbers $a + bi$ and $a - bi$ are called **complex conjugates**, and each is the conjugate of the other:

$$(a + bi)(a - bi) = a^2 - b^2i^2$$
$$= a^2 + b^2$$

which is a real number. Thus, simplify a quotient by using the conjugate of the denominator, as illustrated by Examples 4–6. This process is similar to an earlier technique used for simplifying trigonometric identities.

Example 4

Simplify $\dfrac{15 - 5i}{2 - i}$.

Solution

Multiply by 1.

$$\frac{15 - 5i}{2 - i} = \frac{15 - 5i}{2 - i} \cdot \frac{2 + i}{2 + i}$$

Conjugates

$$= \frac{30 + 5i - 5i^2}{4 - i^2}$$

$$= \frac{35 + 5i}{5}$$

$$= 7 + i$$

You can check this division by multiplying $(2 - i)(7 + i)$:

$$(2 - i)(7 + i) = 14 - 5i - i^2$$
$$= 15 - 5i$$

☐

Example 5

$$\frac{6 + 5i}{2 + 3i} = \frac{6 + 5i}{2 + 3i} \cdot \frac{2 - 3i}{2 - 3i}$$

$$= \frac{12 - 8i - 15i^2}{4 - 9i^2}$$

$$= \frac{27}{13} - \frac{8}{13}i$$ ☐

Example 6   Verify that the conjugate method gives the same result as the definition of division of complex numbers.

Solution

$$\frac{a + bi}{c + di} = \frac{a + bi}{c + di} \cdot \frac{c - di}{c - di}$$

$$= \frac{ac - adi + bci - bdi^2}{c^2 - d^2i^2}$$

$$= \frac{(ac + bd) + (bc - ad)i}{c^2 + d^2}$$

$$= \frac{ac + bd}{c^2 + d^2} + \frac{bc - ad}{c^2 + d^2}i$$ ☐

Problem Set 6.1

A   *Simplify the expressions in Problems 1–14.*

1. **a.** $(3 + 3i) + (5 + 4i)$ **b.** $(6 - 2i) + (5 + 3i)$ **c.** $(4 - 2i) - (3 + 4i)$

2. **a.** $5 - (2 - 3i)$ **b.** $5i - (5 + 5i)$ **c.** $(5 - 3i) - (5 + 2i)$

3. **a.** $(3 + 4i) - (7 + 4i)$ **b.** $-2(-4 + 5i)$ **c.** $4(2 - i) - 3(-1 - i)$

4. **a.** $6(3 + 2i) + 4(-2 - 3i)$ **b.** $i(2 + 3i)$ **c.** $i(5 - 2i)$

5. **a.** $(3 - i)(2 + i)$ **b.** $(4 - i)(2 + i)$ **c.** $(5 - 2i)(5 + 2i)$

6. **a.** $(8 - 5i)(8 + 5i)$ **b.** $(3 - 5i)(3 + 5i)$ **c.** $(7 - 9i)(7 + 9i)$

7. **a.** i^2 **b.** $-i^2$ **c.** $(-i)^2$

8. **a.** i^3 **b.** $-i^3$ **c.** $(-i)^3$

9. **a.** i^4 **b.** $-i^4$ **c.** $(-i)^4$

10. **a.** i^5 **b.** $-i^5$ **c.** $(-i)^5$

11. **a.** i^6 **b.** i^8 **c.** i^{11}

12. **a.** i^{26} **b.** i^{25} **c.** i^{27}

13. **a.** i^{144} **b.** i^{917} **c.** i^{236}

14. **a.** i^{2001} **b.** $-i^{1980}$ **c.** i^{1984}

B   *Simplify the expressions in Problems 15–29.*

15. $(6 - 2i)^2$ 16. $(3 + 3i)^2$ 17. $(4 + 5i)^2$ 18. $(3 - 5i)^3$ 19. $\dfrac{-3}{1 + i}$

20. $\dfrac{5}{4 - i}$ 21. $\dfrac{2}{i}$ 22. $\dfrac{5}{i}$ 23. $\dfrac{3i}{5 - 2i}$ 24. $\dfrac{-2i}{3 + i}$

25. $\dfrac{5 + 3i}{4 - i}$ 26. $\dfrac{4 - 2i}{3 + i}$ 27. $\dfrac{1 - 6i}{1 + 6i}$ 28. $\dfrac{2 + 7i}{2 - 7i}$ 29. $\dfrac{3 - 2i}{5 + i}$

C   30. Simplify $(1.9319 + .5176i)(2.5981 + 1.5i)$.   31. Simplify $\dfrac{-3.2253 + 8.4022i}{3.4985 + 1.9392i}$.

Let $z_1 = (1 + \sqrt{3}) + (2 + \sqrt{3})i$ and $z_2 = (2 + \sqrt{12}) + \sqrt{3}\,i$. Perform the indicated operation in Problems 32–37.

32. $z_1 + z_2$ 33. $z_1 - z_2$ 34. $z_1 z_2$ 35. $\dfrac{z_1}{z_2}$ 36. $(z_1)^2$ 37. $(z_2)^2$

For each of Problems 38–40, let $z_1 = a + bi$, $z_2 = c + di$, and $z_3 = e + fi$.

38. Prove the commutative laws for complex numbers. That is, prove that

$$z_1 + z_2 = z_2 + z_1$$
$$z_1 z_2 = z_2 z_1$$

39. Prove the associative laws for complex numbers. That is, prove that

$$z_1 + (z_2 + z_3) = (z_1 + z_2) + z_3$$
$$z_1(z_2 z_3) = (z_1 z_2) z_3$$

40. Prove the distributive law for complex numbers. That is, prove that

$$z_1(z_2 + z_3) = z_1 z_2 + z_1 z_3$$

6.2 TRIGONOMETRIC FORM OF COMPLEX NUMBERS

Plotting Complex Numbers

Consider a graphical representation of a complex number. To give a graphical representation of complex numbers such as

$$2 + 3i, \quad -i, \quad -3 - 4i, \quad 3i, \quad -2 + \sqrt{2}\,i, \quad \frac{3}{2} - \frac{5}{2}i$$

a two-dimensional coordinate system is used: the horizontal axis represents the **real axis** and the vertical axis is the **imaginary axis**, so that $a + bi$ is represented by the ordered pair (a, b). Remember that a and b represent real numbers, so we plot (a, b) in the usual manner, as shown in Figure 6.1. The coordinate system in Figure 6.1 is called the **complex plane**, or the **Gaussian plane**, in honor of Karl Friedrich Gauss (1777–1855).

FIGURE 6.1 Complex plane

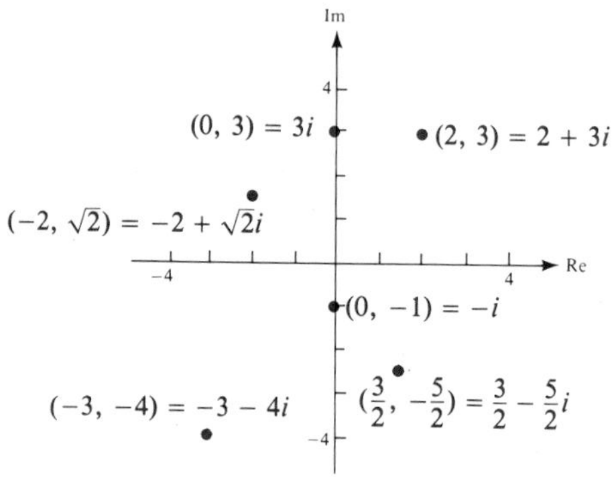

The **absolute value** of a complex number z is, graphically, the distance between z and the origin (just as it is for real numbers). The absolute value of a complex number is also called the **modulus**. The distance formula leads to the following definition.

ABSOLUTE VALUE, OR MODULUS, OF A COMPLEX NUMBER

> If $z = a + bi$, then the *absolute value*, or *modulus*, of z is denoted by $|z|$ and defined by
>
> $$|z| = \sqrt{a^2 + b^2}$$

Example 1

Find the absolute value.

a. $3 + 4i$

absolute value:

$$\begin{aligned} |3 + 4i| &= \sqrt{3^2 + 4^2} \\ &= \sqrt{25} \\ &= 5 \end{aligned}$$

b. $-2 + \sqrt{2}\,i$

absolute value:

$$\begin{aligned} |-2 + \sqrt{2}\,i| &= \sqrt{4 + 2} \\ &= \sqrt{6} \end{aligned}$$

c. -3

absolute value:

$$\begin{aligned} |-3 + 0i| &= \sqrt{9 + 0} \\ &= 3 \end{aligned}$$

This example shows that the definition of absolute value for complex numbers is consistent with the definition of absolute value given for real numbers.

d. $4i$

absolute value:

$$\begin{aligned} |4i| &= \sqrt{0^2 + 4^2} \\ &= 4 \end{aligned}$$

□

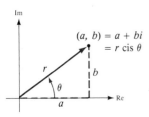

FIGURE 6.2   Trigonometric form and rectangular form of a complex number

Trigonometric Form of a Complex Number

The form $a + bi$ is called the **rectangular form** of a complex number, but another very useful representation uses trigonometry. Consider the graphical representation of a complex number $a + bi$, as shown in Figure 6.2.

Let r be the distance from the origin to (a, b) and let θ be the angle the segment makes with the real axis. Then

$$r = \sqrt{a^2 + b^2}$$

and θ, called the **argument**, is chosen so that it is the smallest nonnegative angle the terminal side makes with the positive real axis. From the definition of the trigonometric functions,

$$\cos \theta = \frac{a}{r} \qquad \sin \theta = \frac{b}{r} \qquad \tan \theta = \frac{b}{a}$$

$$a = r \cos \theta \qquad b = r \sin \theta$$

Therefore,

$$a + bi = r \cos \theta + ir \sin \theta$$
$$= r(\cos \theta + i \sin \theta)$$

Sometimes, $r(\cos \theta + i \sin \theta)$ is abbreviated by

$$r(\cos \theta + i \sin \theta)$$

$r(c \quad\quad is \quad \theta)$ or $r \operatorname{cis} \theta$ (see Figure 6.2)

TRIGONOMETRIC FORM
OF A COMPLEX NUMBER

> The **trigonometric form** of a complex number $z = a + bi$ is
>
> $$r(\cos \theta + i \sin \theta) = r \operatorname{cis} \theta$$
>
> where $r = \sqrt{a^2 + b^2}$; $\tan \theta = \frac{b}{a}$ if $a \neq 0$; $a = r \cos \theta$; $b = r \sin \theta$. This representation is unique for $0 \leq \theta < 360°$ for all z except $0 + 0i$.

The placement of θ in the proper quadrant is an important consideration because there are two values of $0 \leq \theta < 360°$ that will satisfy the relationship

$$\tan \theta = \frac{b}{a}$$

For example, compare the following:

1. $-1 + i$; $a = -1, b = 1$; $\tan \theta = \dfrac{1}{-1}$ or $\tan \theta = -1$

2. $1 - i$; $a = 1, b = -1$; $\tan \theta = \dfrac{-1}{1}$ or $\tan \theta = -1$

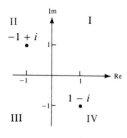

Notice the same trigonometric equation for both complex numbers, even though $-1 + i$ is in Quadrant II and $1 - i$ is in Quadrant IV. This consideration of quadrants is even more important when you are doing the problem on a calculator, since the proper sequence of steps for this example is

giving the result $-45°$, which is not true for either example since $0 \leq \theta < 360°$. The entire process can be taken care of quite simply if you let θ' be the reference angle for θ. Then find the reference angle

$$\theta' = \tan^{-1}\left|\frac{b}{a}\right|$$

After you know the reference angle and the quadrant, it is easy to find θ. For these examples,

$$\theta' = \tan^{-1}|-1| \qquad \text{On a calculator:} \quad \boxed{1}\ \boxed{\text{INV}}\ \boxed{\text{tan}}$$
$$= 45°$$

For Quadrant II, $\theta = 135°$; for Quadrant IV, $\theta = 315°$.

Example 2

Change the complex numbers to trigonometric form.

a. $1 - \sqrt{3}\,i$

$a = 1$ and $b = -\sqrt{3}$; the number is in Quadrant IV.

$$r = \sqrt{1^2 + (-\sqrt{3})^2} \qquad \theta' = \tan^{-1}\left|\frac{-\sqrt{3}}{1}\right|$$
$$= \sqrt{4} \qquad\qquad\qquad = 60°$$
$$= 2 \qquad\qquad\qquad \textit{The reference angle is } 60°;$$
$$\textit{in Quadrant IV: } \theta = 300°.$$

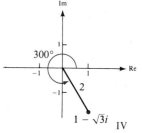

Thus, $1 - \sqrt{3}\,i = 2\,\text{cis}\,300°$.

b. $6i$

$a = 0$ and $b = 6$; notice that $\tan \theta$ is not defined for $\theta = 90°$. By inspection, $6i = 6\,\text{cis}\,90°$.

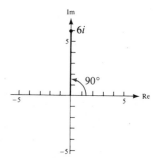

c. $4.310 + 5.516i$

$a = 4.310$ and $b = 5.516$; the number is in Quadrant I.

$$r = \sqrt{(4.310)^2 + (5.516)^2} \qquad \theta = \tan^{-1}\left|\frac{5.516}{4.310}\right|$$

$$\approx \sqrt{49} \qquad\qquad \approx \tan^{-1}(1.2798)$$

$$= 7 \qquad\qquad \approx 52°$$

Thus, $4.310 + 5.516i \approx 7\operatorname{cis} 52°$.

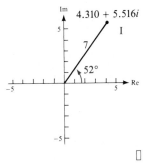

Example 3

Change the complex numbers to rectangular form.

a. $4\operatorname{cis} 150°$

$r = 4$ and $\theta = 150°$.

$$a = 4\cos 150° \qquad b = 4\sin 150°$$

$$= 4\left(-\frac{\sqrt{3}}{2}\right) \qquad = 4\left(\frac{1}{2}\right)$$

$$= -2\sqrt{3} \qquad\qquad = 2$$

Thus, $4\operatorname{cis} 150° = -2\sqrt{3} + 2i$.

b. $5(\cos 218° + i\sin 218°)$

$r = 5$ and $\theta = 218°$

$$a = 5\cos 218° \qquad b = 5\sin 218°$$

$$\approx -3.94 \qquad\qquad \approx -3.08$$

Thus, $5(\cos 218° + i\sin 218°) \approx -3.94 - 3.08i$.

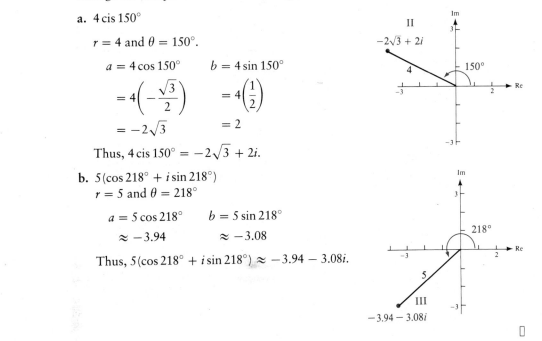

Problem Set 6.2

A   *Plot each pair of numbers in Problems 1–9, and find the absolute value of each number.*

1. $3 + i, 5 + 5i$

2. $7 - i, 2 - 5i$

3. $3 + 2i, -2 + 5i$

4. $-1 + 3i, 4 - i$

5. $2 + 2i, -1 + i$

6. $3 + 5i, -3 + i$

7. $2, -4i$

8. $-5, -i$

9. $-2i, 5$

B   *Plot and then change each of the numbers in Problems 10–21 to trigonometric form. Write the angle to the nearest degree.*

10. $1 + i$ 11. $1 - i$ 12. $\sqrt{3} - i$

13. $\sqrt{3} + i$ 14. $-1 - \sqrt{3}\,i$ 15. 1

16. 5 17. $-i$ 18. $4i$

19. $1.7207 + 2.4575i$ 20. $5.7956 - 1.5529i$ 21. $-.6946 + 3.9392i$

Plot and then change each of the numbers in Problems 22–31 to rectangular form. Use exact values whenever possible; otherwise, give answers with four decimal place accuracy.

22. $2(\cos 45° + i \sin 45°)$ 23. $3(\cos 60° + i \sin 60°)$ 24. $5(\cos \frac{4\pi}{3} + i \sin \frac{4\pi}{3})$

25. $\cos \frac{5\pi}{6} + i \sin \frac{5\pi}{6}$ 26. $\operatorname{cis} 0$ 27. $4 \operatorname{cis} 30°$

28. $7 \operatorname{cis} 135°$ 29. $6 \operatorname{cis} 247°$ 30. $9 \operatorname{cis} 190°$

31. $10 \operatorname{cis} 371°$

C   32. If $z = a + bi$ and $\bar{z} = a - bi$, show that $|z| = \sqrt{z \cdot \bar{z}}$.

33. If $z_1 = a + bi$ and $z_2 = c + di$, show that $|z_1 + z_2| \le |z_1| + |z_2|$. This is called the triangle inequality.

34. In calculus it is shown (using series) that

$$e^{i\theta} = \cos \theta + i \sin \theta$$

which is one of the most remarkable formulas in all of mathematics. This means, among other things, that $re^{i\theta} = r \operatorname{cis} \theta$. Plot the following complex numbers.

a. $e^{i\pi/4}$ b. $2e^{-i\pi/6}$ c. $e^{-i\pi}$ d. $2e^{2\pi i}$ e. $\sqrt{2}\,e^{5\pi/4}$

> We are number $e^{-i\pi}$.

6.3 DE MOIVRE'S FORMULA

Products and Quotients of Complex Numbers

In Section 6.1, the operations involving complex numbers were fairly simple, so you encountered little difficulty when carrying them out. For example,

$$(1 + 2i)(2 + 3i) = 2 + 7i + 6i^2$$
$$= 2 + 7i + 6(-1)$$
$$= -4 + 7i$$

But consider

$$(1.9319 + .5176i)(2.5981 + 1.5i) = 5.0193 + 2.8979i + 1.3448i + .7764i^2$$
$$= 4.2429 + 4.2427i$$

You almost need a calculator to carry out the operations. If these numbers are changed to trigonometric form, are the calculations easier?

$$1.9319 + \ \ .5176i \approx 2\operatorname{cis}15°$$
$$2.5981 + 1.5i \ \ \ \approx 3\operatorname{cis}30°$$

The product, changed to trigonometric form, is

$$4.2429 + 4.2427i \approx 6\operatorname{cis}45°$$

Product of 2 and 3 ⌐   ⌐ *Sum of 15° and 30°*

A similar result holds for division. For division, consider the number

$$10(\cos 150° + i\sin 150°) \div 2(\cos 120° + i\sin 120°)$$
$$= (-5\sqrt{3} + 5i) \div (-1 + \sqrt{3}\,i)$$

In algebraic (rectangular) form:

$$\frac{-5\sqrt{3} + 5i}{-1 + \sqrt{3}\,i} = \frac{-5\sqrt{3} + 5i}{-1 + \sqrt{3}\,i} \cdot \frac{-1 - \sqrt{3}\,i}{-1 - \sqrt{3}\,i}$$

$$= \frac{5\sqrt{3} + 15i - 5i - 5\sqrt{3}\,i^2}{1 - 3i^2}$$

$$= \frac{10\sqrt{3} + 10i}{4}$$

$$= \frac{5}{2}\sqrt{3} + \frac{5}{2}i$$

$$= 5\operatorname{cis}30° \quad \textit{Can you fill in the details of changing to trigonometric form?}$$

But from the trigonometric forms,

$$\frac{10\operatorname{cis}150°}{2\operatorname{cis}120°} = 5\operatorname{cis}30°$$

would seem to indicate that

$$\frac{10\operatorname{cis}150°}{2\operatorname{cis}120°} = \frac{10}{2}\operatorname{cis}(150° - 120°)$$

Difference of arguments

Quotient of 10 and 2

which is true not only for this example but also in general.

These multiplication and division rules are summarized by the following theorem.

PRODUCTS AND QUOTIENTS OF COMPLEX NUMBERS IN TRIGONOMETRIC FORM

Let $z_1 = r_1 \operatorname{cis} \theta_1$ and $z_2 = r_2 \operatorname{cis} \theta_2$ be nonzero complex numbers. Then

$$z_1 z_2 = r_1 r_2 \operatorname{cis}(\theta_1 + \theta_2) \qquad \frac{z_1}{z_2} = \frac{r_1}{r_2} \operatorname{cis}(\theta_1 - \theta_2)$$

Proof of product form:

$$
\begin{aligned}
z_1 z_2 &= (r_1 \operatorname{cis} \theta_1)(r_2 \operatorname{cis} \theta_2) \\
&= [r_1(\cos \theta_1 + i \sin \theta_1)][r_2(\cos \theta_2 + i \sin \theta_2)] \\
&= r_1 r_2 (\cos \theta_1 + i \sin \theta_1)(\cos \theta_2 + i \sin \theta_2) \\
&= r_1 r_2 (\cos \theta_1 \cos \theta_2 + i \cos \theta_1 \sin \theta_2 + i \sin \theta_1 \cos \theta_2 - \sin \theta_1 \sin \theta_2) \\
&= r_1 r_2 [(\cos \theta_1 \cos \theta_2 - \sin \theta_1 \sin \theta_2) + i(\cos \theta_1 \sin \theta_2 + \sin \theta_1 \cos \theta_2)] \\
&= r_1 r_2 [\cos(\theta_1 + \theta_2) + i \sin(\theta_1 + \theta_2)] \\
&= r_1 r_2 \operatorname{cis}(\theta_1 + \theta_2)
\end{aligned}
$$

The proof of the quotient form is similar and is left as a problem.

Example 1   Simplify.

a. $5 \operatorname{cis} 38° \cdot 4 \operatorname{cis} 75° = 5 \cdot 4 \operatorname{cis}(38° + 75°)$
$$= 20 \operatorname{cis} 113°$$

b. $\sqrt{2} \operatorname{cis} 188° \cdot 2\sqrt{2} \operatorname{cis} 310° = 4 \operatorname{cis} 498°$
$$= 4 \operatorname{cis} 138°$$

c. $(2 \operatorname{cis} 48°)^3 = (2 \operatorname{cis} 48°)(2 \operatorname{cis} 48°)^2$ *Notice that this result is the same as*
$$= (2 \operatorname{cis} 48°)(4 \operatorname{cis} 96°) \qquad (2 \operatorname{cis} 48°)^3 = 2^3 \operatorname{cis}(3 \cdot 48°)$$
$$= 8 \operatorname{cis} 144° \qquad\qquad\qquad\quad = 8 \operatorname{cis} 144°$$

Example 2   Find $\dfrac{15(\cos 48° + i \sin 48°)}{5(\cos 125° + i \sin 125°)}$.

Solution $\dfrac{15 \operatorname{cis} 48°}{5 \operatorname{cis} 125°} = 3 \operatorname{cis}(48° - 125°)$

$$= 3 \operatorname{cis}(-77°)$$
$$= 3 \operatorname{cis} 283° \quad \textit{Keep the arguments between } 0° \textit{ and } 360°.$$

Example 3   Simplify $(1 - \sqrt{3} \, i)^5$.

Solution   First change to trigonometric form: $a = 1$; $b = -\sqrt{3}$; Quadrant IV.

$$r = \sqrt{1+3} \qquad \theta' = \tan^{-1}\left|-\frac{\sqrt{3}}{1}\right|$$

$$= 2 \qquad\qquad = 60°$$

$$\theta = 300° \quad \textit{Quadrant} \text{ IV}$$

$$(1 - \sqrt{3}\,i)^5 = (2\operatorname{cis}300°)^5$$

$$= 2^5\operatorname{cis}(5 \cdot 300°)$$

$$= 32\operatorname{cis}1{,}500°$$

$$= 32\operatorname{cis}60°$$

If you want the answer in rectangular form, you can now change back:

$$a = 32\cos 60° \qquad b = 32\sin 60°$$

$$= 32\left(\frac{1}{2}\right) \qquad = 32\left(\frac{1}{2}\sqrt{3}\right)$$

$$= 16 \qquad\qquad = 16\sqrt{3}$$

Thus, $(1 - \sqrt{3}\,i)^5 = 16 + 16\sqrt{3}\,i$.

De Moivre's Formula

As you can see from Example 3, multiplication in trigonometric form extends quite nicely to any positive integral power in a result called **De Moivre's Formula**.

DE MOIVRE'S FORMULA

> If n is a natural number, then
>
> $$(r\operatorname{cis}\theta)^n = r^n\operatorname{cis}n\theta$$
>
> for a complex number $r\operatorname{cis}\theta = r(\cos\theta + i\sin\theta)$.

Although De Moivre's formula is useful for powers, as illustrated by Example 3, its main usefulness is in finding the complex roots of numbers. Recall from algebra that $\sqrt[n]{r} = r^{1/n}$ is used to denote the principal nth root of r. However, $r^{1/n}$ is only *one* of the nth roots of r. How do you find *all* nth roots of r? You can use a calculator or logarithms to find the principal root, and then use the following theorem to find the other roots.

nTH ROOT THEOREM

> If n is any positive integer, then the nth roots of $r\operatorname{cis}\theta$ are given by
>
> $$\sqrt[n]{r}\operatorname{cis}\left(\frac{\theta + 360°k}{n}\right) \quad\text{or}\quad \sqrt[n]{r}\operatorname{cis}\left(\frac{\theta + 2\pi k}{n}\right)$$
>
> for $k = 0, 1, 2, 3, \ldots, n - 1$.

The proof of this theorem follows directly from De Moivre's formula and is left as a problem.

ABRAHAM DE MOIVRE
1667–1754

Historical Note

The naming of results in the history of mathematics has often been a strange process. In this section we introduced De Moivre's theorem, named after Abraham De Moivre (1667–1754), who used his famous formula in *Philosophical Transactions* (1707) and in *Miscellanea analytica* (1730). He was an outstanding mathematician who worked extensively with probability. However, De Moivre's formula was really first used by Roger Cotes (1682–1716). On the other hand, De Moivre was the inventor of a result called Stirling's formula. De Moivre was unable to receive a professorship, so he supported himself by making a London pub his headquarters and solving problems for anyone who came to him. A story about De Moivre's death is told by Howard Eves. De Moivre noticed that each day he required a quarter of an hour more sleep than on the preceding day. De Moivre said he would die when the arithmetic progression reached 24 hours. Finally, after sleeping longer and longer every night, he went to bed and slept for more than 24 hours and then died in his sleep.

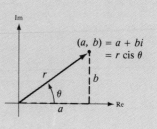

Example 4

Find the square roots of $-\frac{9}{2} + \frac{9}{2}\sqrt{3}\,i$.

Solution

First change to trigonometric form:

$$r = \sqrt{\left(-\frac{9}{2}\right)^2 + \left(\frac{9}{2}\sqrt{3}\right)^2} \qquad \theta' = \tan^{-1}\left|\frac{\frac{9}{2}\sqrt{3}}{-\frac{9}{2}}\right|$$

$$= \sqrt{\frac{81}{4} + \frac{81 \cdot 3}{4}} \qquad\qquad = \tan^{-1}(\sqrt{3})$$

$$\qquad\qquad\qquad\qquad\qquad = 60°$$

$$= \sqrt{81\left(\frac{1}{4} + \frac{3}{4}\right)} \qquad \theta = 120° \quad \textit{Quadrant } \text{II}$$

$$= 9$$

By the *n*th root theorem, the square roots of $9\,\mathrm{cis}\,120°$ are

$$9^{1/2}\,\mathrm{cis}\left(\frac{120° + 360°k}{2}\right) = 3\,\mathrm{cis}(60° + 180°k)$$

$$k = 0: \quad 3\,\mathrm{cis}\,60° = \frac{3}{2} + \frac{3}{2}\sqrt{3}\,i$$

$$k = 1: \quad 3\,\mathrm{cis}\,240° = -\frac{3}{2} - \frac{3}{2}\sqrt{3}\,i$$

All other integral values of k repeat one of the previously found roots. For example,

$$k = 2: \quad 3 \operatorname{cis} 420° = \frac{3}{2} + \frac{3}{2}\sqrt{3}\,i$$

Check:

$$\left(\frac{3}{2} + \frac{3}{2}\sqrt{3}\,i\right)^2 = \frac{9}{4} + \frac{9}{2}\sqrt{3}\,i + \frac{9}{4} \cdot 3i^2$$

$$= -\frac{9}{2} + \frac{9}{2}\sqrt{3}\,i$$

$$\left(-\frac{3}{2} - \frac{3}{2}\sqrt{3}\,i\right)^2 = \frac{9}{4} + \frac{9}{2}\sqrt{3}\,i + \frac{9}{4} \cdot 3i^2$$

$$= -\frac{9}{2} + \frac{9}{2}\sqrt{3}\,i \qquad\qquad \square$$

Example 5

Find the fifth roots of 32.

Solution

You might ask "Why not find these on a calculator?" The reason you cannot is that it will give you only one of the roots—not all five.

Begin by writing 32 in trigonometric form: $32 = 32 \operatorname{cis} 0°$. The fifth roots are found by

$$32^{1/5} \operatorname{cis}\left(\frac{0° + 360°k}{5}\right) = 2 \operatorname{cis} 72°k$$

$k = 0:$ $2 \operatorname{cis} 0°$ $= 2$ ⟵ ────────── *The first root, which is located*
$k = 1:$ $2 \operatorname{cis} 72°$ $= .6180 + 1.9021i$ *so that its argument is θ/n, is*
$k = 2:$ $2 \operatorname{cis} 144° = -1.6180 + 1.1756i$ *called the* **principal nth root.**
$k = 3:$ $2 \operatorname{cis} 216° = -1.6180 - 1.1756i$ *This is the root you can find on*
$k = 4:$ $2 \operatorname{cis} 288° = .6180 - 1.9021i$ *a calculator if the number*
                                        whose root you are finding is
                                        real and positive.

All other integral values for k repeat those listed here. $\square$

If all the fifth roots of 32 are represented graphically, as shown in Figure 6.3, notice that they all lie on a circle of radius 2 and are equally spaced.

FIGURE 6.3 Graphical representation of the fifth roots of 32

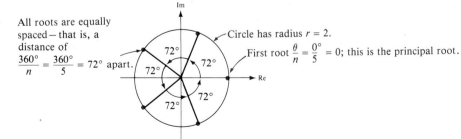

All roots are equally spaced—that is, a distance of $\frac{360°}{n} = \frac{360°}{5} = 72°$ apart.

Circle has radius $r = 2$.

First root $\frac{\theta}{n} = \frac{0°}{5} = 0$; this is the principal root.

If n is a positive integer, then the nth roots of a complex number $a + bi = r \operatorname{cis} \theta$ are equally spaced on the circle of radius r centered at the origin.

Example 6   Solve $x^3 + 1 = 0$.

Solution $x^3 = -1$, so we are looking for the cube roots of -1.

Since $-1 = \operatorname{cis} 180°$ we have $1^{1/3} \operatorname{cis} \dfrac{180° + 360°k}{3}$.

$k = 0:$ $\operatorname{cis} 60° = \dfrac{1}{2} + \dfrac{\sqrt{3}}{2} i$

$k = 1:$ $\operatorname{cis} 180° = -1$

$k = 2:$ $\operatorname{cis} 300° = \dfrac{1}{2} - \dfrac{\sqrt{3}}{2} i$

Thus, the roots of $x^3 + 1 = 0$ are $\frac{1}{2} \pm \frac{\sqrt{3}}{2} i,\ -1$.   ⬚

Problem Set 6.3

A   *Perform the indicated operations in Problems 1–11. Leave your answer in a form that matches the form in the question.*

1. $2 \operatorname{cis} 60° \cdot 3 \operatorname{cis} 150°$ 2. $4(\cos 65° + i \sin 65°) \cdot 12(\cos 87° + i \sin 87°)$

3. $\dfrac{5(\cos 315° + i \sin 315°)}{2(\cos 48° + i \sin 48°)}$ 4. $\dfrac{8 \operatorname{cis} 30°}{4 \operatorname{cis} 15°}$ 5. $\dfrac{12 \operatorname{cis} 120°}{4 \operatorname{cis} 250°}$

6. $(2 \operatorname{cis} 50°)^3$ 7. $(3 \operatorname{cis} 60°)^4$ 8. $(\cos 210° + i \sin 210°)^5$

9. $(2 - 2i)^4$ 10. $(1 + i)^6$ 11. $(\sqrt{3} - i)^8$

B   *Find the indicated roots of the numbers in Problems 12–21. Leave your answers in trigonometric form.*

12.   Square roots of $16 \operatorname{cis} 100°$ 13.   Cube roots of $8 \operatorname{cis} 240°$

14.   Fourth roots of $81 \operatorname{cis} 88°$ 15.   Fifth roots of $32 \operatorname{cis} 200°$

16.   Cube roots of $64 \operatorname{cis} 216°$ 17.   Fifth roots of $32 \operatorname{cis} 160°$

18.   Cube roots of -1 19.   Cube roots of 27

20.   Fourth roots of $1 + i$ 21.   Fourth roots of $-1 - i$

Find the indicated roots of the numbers in Problems 22–27. Leave your answers in trigonometric form.

22.   Sixth roots of -64 23.   Sixth roots of $64i$ 24.   Ninth roots of 1

25.   Ninth roots of $-1 + i$ 26.   Tenth roots of i 27.   Tenth roots of 1

Find the indicated roots of the numbers in Problems 28–33. Leave your answers in rectangular form. Also, show the roots graphically. Use exact values whenever possible.

28. Cube roots of 1

29. Fourth roots of 16

30. Cube roots of -8

31. Cube roots of $4\sqrt{3} - 4i$

32. Square roots of $\dfrac{25}{2} - \dfrac{25\sqrt{3}}{2}i$

33. Fourth roots of $12.2567 + 10.2846i$

Solve the equations in Problems 34–41 for all complex roots. Leave your answers in rectangular form and give approximate answers to four decimal places.

34. $x^3 - 1 = 0$

35. $x^5 - 1 = 0$

36. $x^5 + 1 = 0$

37. $x^6 - 1 = 0$

38. $x^4 + 16 = 0$

39. $x^4 - 16 = 0$

40. $x^4 + x^3 + x^2 + x + 1 = 0$
[*Hint:* Multiply both sides by $(x - 1)$ and compare with Problem 35.]

41. $x^5 + x^4 + x^3 + x^2 + x + 1 = 0$
[*Hint:* Multiply both sides by $(x - 1)$ and compare with Problem 37.]

C

For Problems 42–45, leave your answers in trigonometric form.

42. Find the cube roots of $(4\sqrt{2} + 4\sqrt{2}\,i)^2$.

43. Find the fifth roots of $(-16 + 16\sqrt{3}\,i)^3$.

44. Prove that

$$\frac{r_1 \operatorname{cis} \theta_1}{r_2 \operatorname{cis} \theta_2} = \frac{r_1}{r_2} \operatorname{cis}(\theta_1 - \theta_2)$$

45. Prove that

$$\left[\sqrt[n]{r} \operatorname{cis}\left(\frac{\theta + 360°k}{n} \right) \right]^n = r \operatorname{cis} \theta$$

46. In the previous section (Problem 34) we noted $re^{i\theta} = r \operatorname{cis} \theta$. Use this fact and the laws of exponents to prove De Moivre's theorem.

6.4   SUMMARY AND REVIEW

	OBJECTIVES	PAGES/EXAMPLES
6.1 THE IMAGINARY UNIT AND COMPLEX NUMBERS	**1.** Add and subtract complex numbers.	p. 248; Example 1; Problems 1–3.
	2. Multiply complex numbers.	p. 248; Examples 1–3; Problems 4–18.
	3. Divide complex numbers.	p. 249; Examples 4–6; Problems 19–29.

6.2 TRIGONOMETRIC FORM
OF COMPLEX NUMBERS

4. Find the absolute value of a complex number and plot that number in the Gaussian plane.

p. 251; Figure 6.1 and Example 1; Problems 1–9.

5. Write complex numbers in trigonometric form.

p. 253; Example 2; Problems 10–21.

6. Write complex numbers in rectangular form.

p. 255; Example 3; Problems 22–31.

6.3 DE MOIVRE'S FORMULA

7. Multiply and divide complex numbers in trigonometric form.

p. 256; Examples 1 and 2; Problems 1–5.

8. Raise complex numbers to integral powers.

p. 258; Example 3; Problems 6–11.

9. Find the nth roots of a complex number.

p. 259; Examples 4 and 5; Problems 12–33.

Terms

Absolute value [6.2]	Gaussian plane [6.2]	Pure imaginary number [6.1]
Argument [6.2]	i [6.1]	
Complex conjugates [6.1]	Imaginary axis [6.2]	Real axis [6.2]
Complex number [6.1]	Imaginary unit [6.1]	Real number [6.1]
Complex plane [6.2]	Modulus [6.2]	Rectangular form [6.2]
Conjugate [6.1]	nth root theorem [6.3]	Trigonometric form [6.2]
De Moivre's formula [6.3]	Principal nth root [6.3]	

Chapter 5 Review of Objectives

The problem numbers correspond to the objectives listed in Section 6.4. Simplify the expressions in Problems 1–3.

1. **a.** $(6 + 3i) + (-4 + 2i)$ **b.** $(2 - 5i) - (3 - 2i)$
2. **a.** $2i(-1 + i)(-2 + 2i)$ **b.** $(2 - 5i)^2$
3. **a.** $\dfrac{2 + 3i}{1 - i}$ **b.** $\dfrac{-4}{i}$
4. Plot each number and find its absolute value.
   a. $4 + 2i$ **b.** $1 + 5i$ **c.** $-4 - i$ **d.** $4 - 2i$
5. Change the given complex numbers to trigonometric form.
   a. $7 - 7i$ **b.** $-3i$
6. Change the given complex numbers to rectangular form.
   a. $2 \operatorname{cis} 150°$ **b.** $4(\cos \frac{7\pi}{4} + i \sin \frac{7\pi}{4})$
7. Simplify $\dfrac{2 \operatorname{cis} 158° \cdot 4 \operatorname{cis} 212°}{(2 \operatorname{cis} 312°)^3}$ and leave your answer in trigonometric form.
8. Simplify $(\sqrt{12} - 2i)^4$ and leave your answer in rectangular form.

9. Find the indicated roots and represent them in trigonometric form, in rectangular form, and graphically.

   a. Square roots of $\frac{7}{2}\sqrt{3} - \frac{7}{2}i$ **b.** Fourth roots of 1

Miscellaneous Problems

Change the rectangular-form complex numbers to trigonometric form in Problems 1–8.

1. $9 - 9i$ 2. -6 3. $-\frac{7}{2}\sqrt{3} + \frac{7}{2}i$

4. $2 + 3i$ 5. 9 6. $-1.5321 - 1.2856i$

7. $16i$ 8. $-2 + 3i$

Change the trigonometric-form complex numbers to rectangular form in Problems 9–14.

9. $9(\cos \frac{3\pi}{4} + i \sin \frac{3\pi}{4})$ 10. $2(\cos 3 + i \sin 3)$ 11. $3 \operatorname{cis} 240°$

12. $5 \operatorname{cis} 270°$ 13. $5 \operatorname{cis} 25°$ 14. $3 \operatorname{cis} \frac{11\pi}{6}$

Simplify the given expressions and leave your answer in the form indicated in Problems 15–18.

15. $(\sqrt{12} - 2i)^4$; rectangular   16. $\dfrac{(3 + 3i)(\sqrt{3} - 1)}{1 + i}$; rectangular

17. $\dfrac{2 \operatorname{cis} 185° \cdot 4 \operatorname{cis} 223°}{(2 \operatorname{cis} 300°)^3}$; trigonometric

18. $2i(-1 + i)(-2 + 2i)$; trigonometric

Find the indicated roots in Problems 19–30. Leave your answers in trigonometric form unless otherwise indicated.

19. cube roots of i 20. sixth roots of -64

21. sixth roots of $64i$ 22. ninth roots of 1

23. ninth roots of $-1 + i$ 24. tenth roots of i

25. tenth roots of 1   26. fourth roots of -16

27. cube roots of $-16 + 16\sqrt{3}\,i$ 28. cube roots of $4 \operatorname{cis} \frac{5\pi}{4}$

29. square roots of $-\frac{7}{2}\sqrt{3} + \frac{7}{2}i$ (rectangular form)

30. cube roots of $4\sqrt{2} + 4\sqrt{2}i$ (rectangular form)

APPLICATION FOR FURTHER STUDY

ELECTRIC GENERATOR

COURTESY OF P.G. & E.

In 1831 Michael Faraday (1791–1867) discovered induced electric currents by performing an experiment with a coil of wire and a magnet. He reasoned that if current through a wire would produce a magnetic field, then why not the reverse, in which a magnet is made to produce current in a wire?

Consider the electric generator illustrated in Figure 6.4.

FIGURE 6.4 An electric generator

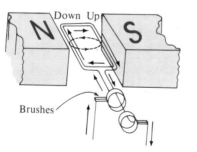

The current will flow first in one direction and then in the other, according to the turning of the armature, because each wire moves up across the field one instant and down at the next. (See Figure 6.5.)

The graph of the current leads us to suspect that the equation for the current is trigonometric, which it is. The current, I, can be found from an equation of the form

$$I = k \sin \omega t$$

for constants k and ω. Although it is beyond the scope of this course, we might mention that the complex numbers introduced in this chapter play an important role in the study of electricity.

FIGURE 6.5   Alternating current
from an electric generator

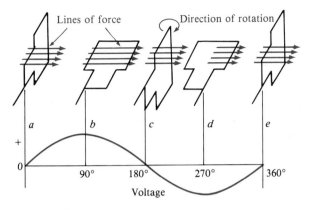

Voltage

For alternating currents, the current sometimes lags behind the voltage, V, which means that the power supplied to any alternating-current circuit is given by

$$P = VI \cos \theta$$

The quantity $\cos \theta$ is called the power factor and it is desirable to have θ as close to 0 as possible because VI should be as close to P as possible. Notice that, since $-1 \le \cos \theta \le 1$ has a maximum value, $\cos \theta = 1$, $VI \le P$ unless $\cos \theta = 1$. A low power factor should be avoided because, for a given supply voltage V, a large current I would be needed to transmit appreciable amounts of electrical energy. (See Figure 6.6.)

FIGURE 6.6   The current is
lagging 45° behind the voltage

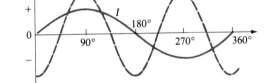

Example 1

If $\theta = 15°$, find the power factor.

Solution

$P = VI \cos 15° \approx VI(.9659)$; the power factor is about 97%.   □

Example 2

If you need to have a power factor of at least 90%, what is the maximum current lag (to the nearest degree)?

Solution

$$\cos \theta = .90$$
$$\theta = 25.8°$$

Since $\cos \theta$ is decreasing in the first quadrant, if $\cos \theta \ge .90$, then $\theta \le 25.8°$. Thus, the maximum current lag to the nearest degree is 25°.   □

267

Problems for Further Study: Electric Generator

1. Graph the current if $k = 15$ and $\omega = 120\pi$.

2. *Engineering*. The voltage is given by

$$V = k \sin \lambda (t - p)$$

   where $k = 10$, $\lambda = 60\pi$, and $p = \frac{1}{2}$. Graph the voltage.

3. *Engineering*. If you need to have a power factor of at least 96%, what will the maximum current lag be?

CHAPTER 7

POLAR CURVES*

CHAPTER OVERVIEW

Up to now, the only coordinate system you have studied in mathematics is a rectangular coordinate system. You have used this system to plot pairs of real numbers (Cartesian coordinate system) as well as to plot single complex numbers (Gaussian plane). This chapter introduces another method of representing a point in a plane: a *polar coordinate system*. It is reported that polar coordinates were used to plot global locations on the floor of the Paris observatory (see Problems 13 and 14 of Problem Set 3.1). The photograph of hurricane Hugo (September 1989) shows a curve called a *spiral*, which is introduced in this chapter. The mathematical equation for a spiral is easily written in polar form.

Courtesy of NOAA, Washington, D.C.

7.1 POLAR COORDINATES

Plotting Points

Up to this point in the book, we have used a rectangular coordinate system. Now, consider a different system, called the **polar coordinate system**. In this system, fix a point O, called the *origin*, or **pole**, and represent a point in the plane by an ordered pair $P(r, \theta)$, where θ measures the angle from the positive x-axis and r represents the directed distance from the pole to the point P. Both r and θ can be any real numbers. When plotting points, consider a **polar axis**, fixed at the pole and coinciding with the x-axis. Now rotate the polar axis through an angle θ as shown in Figure 7.1. You might find it helpful to rotate your pencil as the axis—the tip points in the positive direction. If θ is positive, the angle is measured in a counterclockwise direction, and if θ is negative, it is measured in a clockwise direction.

---

* The material in this chapter is optional.

FIGURE 7.1   Polar-form points

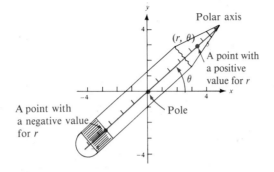

Next, plot r on the polar axis (pencil). Notice that any real number can be plotted on this real number line. Plotting points seems easy, but it is necessary that you completely understand this process. Study each part of Example 1 and make sure you understand how each point is plotted.

Example 1

Plot each of the following polar-form points: $A(4, \frac{\pi}{3})$, $B(-4, \frac{\pi}{3})$, $C(3, -\frac{\pi}{6})$, $D(-3, -\frac{\pi}{6})$, $E(-3, 3)$, $F(-3, -3)$, $G(-4, -2)$, $H(5, \frac{3\pi}{2})$, $I(-5, \frac{\pi}{2})$, $J(5, -\frac{\pi}{2})$.

Solution

Points A–D illustrate the basic ideas of plotting polar-form points.

Plot $A(4, \frac{\pi}{3})$; positive θ Plot $B(-4, \frac{\pi}{3})$; positive θ

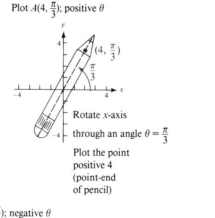

Plot $C(3, -\frac{\pi}{6})$; negative θ Plot $D(-3, -\frac{\pi}{6})$; negative θ

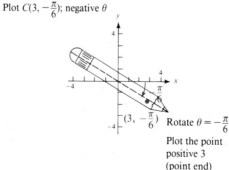

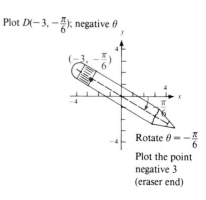

Points *E–H* illustrate common situations that can sometimes be confusing. Make sure you understand each example.

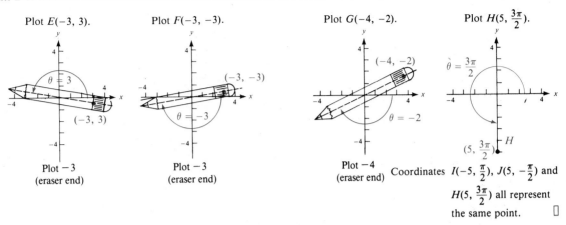

Plot *E*(−3, 3).   Plot *F*(−3, −3).   Plot *G*(−4, −2).   Plot $H(5, \frac{3\pi}{2})$.

Plot −3 (eraser end)   Plot −3 (eraser end)   Plot −4 (eraser end)   Coordinates $I(-5, \frac{\pi}{2})$, $J(5, -\frac{\pi}{2})$ and $H(5, \frac{3\pi}{2})$ all represent the same point.   □

Primary Representations of a Point

One thing you will notice from Example 1 is that ordered pairs in polar form are not associated in a one-to-one fashion with points in the plane. Indeed, given any point in the plane, there are infinitely many ordered pairs of polar coordinates associated with that point. If you are given a point (r, θ) other than the pole in polar coordinates, then $(-r, \theta + \pi)$ also represents the same point. In addition, there are infinitely many others, all of which have the same first component as one of these and have second components that are multiples of 2π added to these angles. We call (r, θ) and $(-r, \theta + \pi)$ the **primary representations of the point** if the angles θ and $\theta + \pi$ are between 0 and 2π.

WARNING! Remember to simplify so the second component represents a nonnegative angle less than one revolution.

PRIMARY REPRESENTATIONS OF A POINT IN POLAR FORM

> Every point in polar form has two primary representations:
>
> (r, θ), where $0 \le \theta < 2\pi$ and $(-r, \pi + \theta)$, where $0 \le \pi + \theta < 2\pi$

Example 2

Give both primary representations for each of the given points.

a. $(3, \frac{\pi}{4})$ has primary representations $(3, \frac{\pi}{4})$ and $(-3, \frac{5\pi}{4})$.

$\frac{\pi}{4} + \pi = \frac{5\pi}{4}$

b. $(5, \frac{5\pi}{4})$ has primary representations $(5, \frac{5\pi}{4})$ and $(-5, \frac{\pi}{4})$.

$\frac{5\pi}{4} + \pi = \frac{9\pi}{4}$, *but* $(-5, \frac{9\pi}{4})$ *is not a primary representation of the point* $(5, \frac{5\pi}{4})$ *since* $\frac{9\pi}{4} > 2\pi$.

c. $(-6, -\frac{2\pi}{3})$ has primary representations $(-6, \frac{4\pi}{3})$ and $(6, \frac{\pi}{3})$.

θ *should be positive.*

d. $(9, 5)$ has primary representations $(9, 5)$ and $(-9, 5 - \pi)$; a point such as $(-9, 5 - \pi)$ is usually

Notice that $(-9, 5 + \pi)$ *is not a primary representation, since* $5 + \pi > 2\pi$.

approximated by writing
$(-9, 1.86)$.

e. $(9, 7)$ has primary representations $(9, 7 - 2\pi)$ or $(9, .72)$ and $(-9, 7 - \pi)$ or $(-9, 3.86)$.

*Add or subtract from θ **whatever** multiple of 2π will force the new angle to be between 0 and approximately 6.28 (2π).*

□

Relationship Between Polar and Rectangular Coordinates

The relationship between the two coordinate systems can easily be found by using the definition of the trigonometric functions (see Figure 7.2).

RELATIONSHIP BETWEEN RECTANGULAR AND POLAR COORDINATES

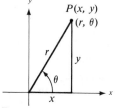

FIGURE 7.2 Relationship between rectangular and polar coordinates

1. To change **from polar to rectangular**:

$$x = r \cos \theta \qquad y = r \sin \theta$$

2. To change **from rectangular to polar**:

$$r = \sqrt{x^2 + y^2} \qquad \theta' = \tan^{-1} \left| \frac{y}{x} \right|, \quad x \neq 0$$

where θ' is the reference angle for θ. Place θ in the proper quadrant by noting the signs of x and y. If $x = 0$, then $\theta' = \frac{\pi}{2}$.

Example 3

Change the polar coordinates $(-3, \frac{5\pi}{4})$ to rectangular coordinates.

Solution

$$x = -3 \cos \frac{5\pi}{4} = -3 \left(-\frac{\sqrt{2}}{2} \right) = \frac{3\sqrt{2}}{2} \qquad y = -3 \sin \frac{5\pi}{4} = -3 \left(-\frac{\sqrt{2}}{2} \right) = \frac{3\sqrt{2}}{2}$$

$$\underbrace{\left(-3, \frac{5\pi}{4} \right)}_{Polar\ form}; \underbrace{\left(\frac{3\sqrt{2}}{2}, \frac{3\sqrt{2}}{2} \right)}_{Rectangular\ form}$$

□

Example 4

Write both primary representations of the polar-form coordinates for the point $(\frac{5\sqrt{3}}{2}, -\frac{5}{2})$.

Solution

$$r = \sqrt{\left(\frac{5\sqrt{3}}{2} \right)^2 + \left(-\frac{5}{2} \right)^2}$$

$$= \sqrt{\frac{75}{4} + \frac{25}{4}}$$

$$= 5$$

$$\theta' = \tan^{-1} \left| \frac{-\frac{5}{2}}{\frac{5\sqrt{3}}{2}} \right| = \tan^{-1} \left(\frac{1}{\sqrt{3}} \right)$$

$$= \frac{\pi}{6} \qquad \theta = \frac{11\pi}{6} \qquad (Quadrant\ IV)$$

$$\underbrace{\left(\frac{5\sqrt{3}}{2}, -\frac{5}{2}\right)}_{\substack{\text{Rectangular}\\ \text{form}}}; \underbrace{\left(5, \frac{11\pi}{6}\right); \left(-5, \frac{5\pi}{6}\right)}_{\text{Polar form}} \quad \frac{11\pi}{6} + \pi = \frac{17\pi}{6}, \text{ and } \frac{17\pi}{6} \text{ is coterminal with } \frac{5\pi}{6}.$$

▯

Points on a Polar Curve

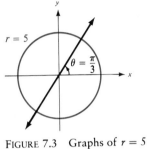

FIGURE 7.3   Graphs of $r = 5$ and $\theta = \frac{\pi}{3}$

If r and θ are related by an equation, we can speak of the *graph of the equation*. For example, $r = 5$ is the equation of a circle with center at the origin and radius 5. Also, $\theta = \frac{\pi}{3}$ is the equation of the line passing through the pole as drawn in Figure 7.3.

As with other equations we have graphed in this book, we will begin graphing polar-form curves by plotting some points. However, you must first be able to recognize whether a point in polar form satisfies a given equation. We will do this in the next section.

Example 5

Show that each of the given points lies on the graph of $r = \dfrac{2}{1 - \cos\theta}$.

  a. $(2, \frac{\pi}{2})$ **b.** $(-2, \frac{3\pi}{2})$ **c.** $(-1, 2\pi)$

Solution

Begin by substituting the given coordinates into the equation.

a. $2 \overset{?}{=} \dfrac{2}{1 - \cos\frac{\pi}{2}} = \dfrac{2}{1 - 0} = 2$

Thus, the point $(2, \frac{\pi}{2})$ is on the curve, since it satisfies the equation.

b. $-2 \overset{?}{=} \dfrac{2}{1 - \cos\frac{3\pi}{2}} = \dfrac{2}{1 - 0} = 2$ *This equation is **not** true.*

The equation is not satisfied, but we *cannot* say that the point is not on the curve. Indeed, we see from part a that it is on the curve, since $(-2, \frac{3\pi}{2})$ and $(2, \frac{\pi}{2})$ name the same point! So even if one primary representation of a point does not satisfy the equation, we must still check the other primary representation of the point.

c. $-1 \overset{?}{=} \dfrac{2}{1 - \cos 2\pi} = \dfrac{2}{1 - 1}$,   which is undefined.

Check the other representation of the same point—namely, $(1, \pi)$:

$$1 \overset{?}{=} \dfrac{2}{1 - \cos\pi} = \dfrac{2}{1 - (-1)} = 1$$

Thus, the point is on the curve.   ▯

Example 5 leads us to the following definition of what is necessary in order for a polar-form point to **satisfy** an equation.

A POINT SATISFYING AN
EQUATION

> An ordered pair representing a polar-form point (other than the pole) **satisfies an equation** involving a function of $n\theta$ (for an integer n) if and only if at least one of its primary representations satisfies the given equation.

We turn our attention to graphing polar-form curves in the next section.

Problem Set 7.1

A   *In Problems 1–6 plot each of the given polar-form points. Give both primary representations, and give the rectangular coordinates of the point.*

1. $(4, \frac{\pi}{4})$ **2.** $(6, \frac{\pi}{3})$ **3.** $(5, \frac{2\pi}{3})$ **4.** $(3, -\frac{\pi}{6})$ **5.** $(\frac{3}{2}, -\frac{5\pi}{6})$ **6.** $(-4, 4)$

In each of Problems 7–12, plot the given rectangular-form points and give both primary representations in polar form.

7. $(5, 5)$ **8.** $(-1, \sqrt{3})$ **9.** $(2, -2\sqrt{3})$

10. $(-2, -2)$ **11.** $(3, -3)$ **12.** $(3, 7)$

13. *Geography.* Name the country in Figure 7.4 in which each of the named points is located.
   a. $(5.5, 260°)$
   b. $(7.5, 78°)$
   c. $(-2, 140°)$
   d. $(-5, 80°)$

14. *Geography.* Name possible coordinates for each of the following cities (see Figure 7.4).
   a. Miami, Florida
   b. Los Angeles, California
   c. Mexico City, Mexico
   d. London, England

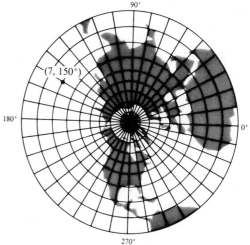

FIGURE 7.4   This is reproduced from the floor of the Paris Observatory.

B   *In Problems 15–19 tell whether each of the given points lies on the curve $r = \dfrac{5}{1 - \sin\theta}$.*

15. $(10, \frac{\pi}{6})$ **16.** $(5, \frac{\pi}{2})$ **17.** $(-10, \frac{5\pi}{6})$

18. $(-\frac{10}{3}, \frac{5\pi}{6})$ **19.** $(20 + 10\sqrt{3}, \frac{\pi}{3})$

In Problems 20–24 tell whether each of the given points lies on the curve $r = 2(1 - \cos\theta)$.

20. $(1, \frac{\pi}{3})$ **21.** $(1, -\frac{\pi}{3})$ **22.** $(-1, \frac{\pi}{3})$

23. $(-2, \frac{\pi}{2})$

24. $(-2 - \sqrt{2}, \frac{\pi}{4})$

Find three ordered pairs satisfying each of the equations in Problems 25–36. Give both primary representations for each point.

25. $r^2 = 9 \cos \theta$ **26.** $r^2 = 9 \cos 2\theta$ **27.** $r = 3\theta$ **28.** $r = 5\theta$

29. $r\theta = 4$ **30.** $r = 2 - 3 \sin \theta$ **31.** $r = 2(1 + \cos \theta)$ **32.** $r = 2 \cos \theta$

33. $r = 6 \cos 3\theta$ **34.** $r = \tan \theta$ **35.** $\dfrac{r}{1 - \sin \theta} = 2$ **36.** $r = \dfrac{8}{1 - 2 \cos \theta}$

C

37. Derive the equations for changing from polar coordinates to rectangular coordinates.

38. Derive the equations for changing from rectangular coordinates to polar coordinates.

39. What is the distance between $(3, \frac{\pi}{3})$ and $(7, \frac{\pi}{4})$? Explain why you cannot use the distance formula for these ordered pairs.

40. What is the distance between the polar-form points (r, θ) and (c, α)?

7.2 GRAPHING POLAR-FORM EQUATIONS

The general process for graphing a new type of curve is to plot points, make generalizations, and then graph future similar curves using those generalizations along with plotting only a few key points.

Example 1

Graph $r - \theta = 0$ for $\theta \geq 0$.

Solution

Set up a table of values.

θ	r
0	0
1	1
2	2
3	3
4	4
5	5
6	6

Choose a θ, and then find a corresponding r so that (r, θ) satisfies the equation. Plot each of these points and connect them as shown in Figure 7.5.

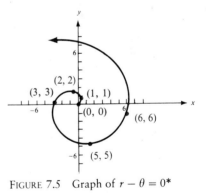

FIGURE 7.5   Graph of $r - \theta = 0$*

Notice that as θ increases, r must also increase. □

---

* Many books use what is called **polar graph paper**, but such paper is not really necessary. It also obscures the fact that polar curves and rectangular curves are both plotted on a Cartesian coordinate system only with a different meaning attached to the ordered pairs. You can estimate the angles as necessary without polar graph paper. Remember, just as when graphing rectangular curves, the key is not in plotting many points, but in recognizing the curve and then plotting a few key points.

Example 2

Graph $r = \cos\theta$ by plotting points.

Solution

	Press	Display
If $\theta = 0$, then $r = 1$; plot $(1, 0)$.	0 cos	1.
If $\theta = 1$, then $r \approx .54$; plot $(.54, 1)$.	1 cos	.5403023059
If $\theta = 2$, then $r \approx -.42$; plot $(-.42, 2)$.	2 cos	−.4161468365
If $\theta = 3$, then $r \approx -.99$; plot $(-.99, 3)$.	3 cos	−.9899924966
If $\theta = 4$, then $r \approx -.65$; plot $(-.65, 4)$.	4 cos	−.6536436209
If $\theta = 5$, then $r \approx .28$; plot $(.28, 5)$.	5 cos	.2836621855
If $\theta = 6$, then $r \approx .96$; plot $(.96, 6)$.	6 cos	.9601702867
If $\theta = 7$, then $r \approx .75$; plot $(.75, 7)$.	7 cos	.7539022543
If $\theta = 8$, then $r \approx -.15$; plot $(-.15, 8)$.	8 cos	−.1455000338

Plot these points as shown in Figure 7.6. Notice that, if these points are connected, they form a circle with center at $(0, .5)$ and radius $.5$.

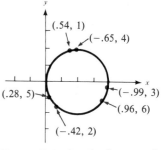

FIGURE 7.6   Graph of $r = \cos\theta$

Cardioids

Example 3

Graph $r = 2(1 - \cos\theta)$ by plotting points.

Solution

Construct a table of values by choosing values for θ and approximating the corresponding values for r.

θ	0	$\dfrac{\pi}{6}$	$\dfrac{\pi}{3}$	$\dfrac{\pi}{2}$	$\dfrac{2\pi}{3}$	$\dfrac{5\pi}{6}$	π	$\dfrac{7\pi}{6}$	$\dfrac{4\pi}{3}$	$\dfrac{3\pi}{2}$	$\dfrac{5\pi}{3}$	$\dfrac{11\pi}{6}$
r (APPROX. VALUE)	0	.27	1	2	3	3.7	4	3.7	3	2	1	.27

These points are plotted and then connected, as shown in Figure 7.7.

Jacob Bernoulli
1654–1705

Polar coordinates were introduced by Jacob Bernoulli in 1691 but were not used to any extent until the close of the 18th century. Jacob was a member of one of the most distinguished families in mathematics. He was one of 12 related Bernoullis, all of whom achieved some fame and recognition as mathematicians. In regard to polar coordinates, Jacob was so moved by the way the equiangular spiral reproduces itself that he asked that it be engraved on his tombstone with the inscription

Eadem mutata resurgo
("I shall arise the same, though changed").

FIGURE 7.7   Graph of $r = 2(1 - \cos\theta)$

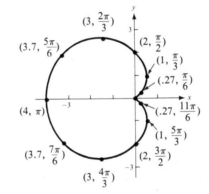

The curve in Example 3 is called a **cardioid** because it is heart-shaped. Compare the curve graphed in Example 3 with the general curve $r = a(1 - \cos\theta)$, which is called the **standard-position cardioid**. Consider the following table of values:

NOTE: This table presents the key points for a cardioid.

θ	0	$\dfrac{\pi}{2}$	π	$\dfrac{3\pi}{2}$
r	0	a	$2a$	a

These values for θ should be included whenever you are making a graph in polar coordinates and are using the method of plotting points.

These reference points are all that is necessary to plot for other standard-position cardioids, because they will all have the same shape as the one shown in Figure 7.7.

Rotations of Polar-Form Curves

What about cardioids that are not in standard position? In Chapter 2 we translated the graphs to new locations, but the translation of a polar-form curve is not so easy because points are not labeled in a rectangular fashion. We can, however, easily rotate the polar-form curve.

ROTATION OF
POLAR-FORM GRAPHS

> The polar graph of $r = f(\theta - \alpha)$ is the same as the polar graph of $r = f(\theta)$ that has been rotated through an angle α.

Example 4   Graph $r = 3 - 3\cos(\theta - \frac{\pi}{6})$.

Solution   Recognize this as a cardioid with $a = 3$ and a rotation of $\frac{\pi}{6}$. Plot the four points shown in Figure 7.8 and draw the cardioid.

FIGURE 7.8   Graph of
$r = 3 - 3\cos(\theta - \frac{\pi}{6})$

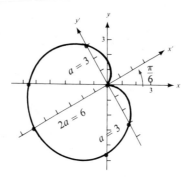

If the rotation is 180°, the equation simplifies considerably. Consider

$$
\begin{aligned}
r &= 3 - 3\cos(\theta - \pi) && \textit{Cardioid with a 180° rotation} \\
&= 3 - 3[\cos\theta\cos\pi + \sin\theta\sin\pi] && \textit{Identity 16} \\
&= 3 - 3[\cos\theta(-1) + \sin\theta(0)] && \textit{Exact values} \\
&= 3 - 3(-1)\cos\theta \\
&= 3 + 3\cos\theta
\end{aligned}
$$

Compare this with Example 4 and you will see that the only difference is a 180° rotation instead of a 30° rotation. This means that, whenever you graph an equation of the form

$$r = a(1 + \cos\theta) \quad \textit{Note the plus sign.}$$

instead of $r = a(1 - \cos\theta)$, it is a standard-form cardioid with a 180° rotation.

Example 5 — Graph $r = 4(1 - \sin \theta)$.

Solution — Notice that

$$\sin \theta = \cos\left(\frac{\pi}{2} - \theta\right) = \cos\left(\theta - \frac{\pi}{2}\right)$$

This is a cardioid with $a = 4$ that has been rotated $\frac{\pi}{2}$. The graph is shown in Figure 7.9. A curve of the form $r = a(1 - \sin \theta)$ is a cardioid that has been rotated $\frac{\pi}{2}$.

FIGURE 7.9   Graph of $r = 4(1 - \sin \theta)$

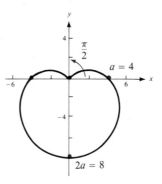

The cardioid is only one of the interesting polar-form curves. It is a special case of a curve called a **limaçon**, which is developed in Problem 34 of Problem Set 7.2. Polar-form curves with multiple angles are discussed in the next section.

Problem Set 7.2

A   *Sketch each of the equations in Problems 1–18 by relating it to the standard-position cardioid. Use a rotation if necessary.*

1. $r = 2 - 2\cos\theta$
2. $r = 3(1 - \cos\theta)$
3. $r = 4 - 4\cos\theta$
4. $r - 3 = 3\cos\theta$
5. $r - 5 = 5\cos\theta$
6. $r - 4 = 4\cos\theta$
7. $r = 2 - 2\sin\theta$
8. $r = 1 - \sin\theta$
9. $4r = 1 - \sin\theta$
10. $r = 2 + 2\sin\theta$
11. $r - 3 = 3\sin\theta$
12. $r - 4\sin\theta = 4$
13. $r = 3 + 3\cos(\theta - \frac{\pi}{3})$
14. $r = 2 - 2\cos(\theta - \frac{\pi}{4})$
15. $r = 2 + 2\cos(\theta - \frac{\pi}{6})$
16. $r = 3 + 3\sin(\theta - \frac{\pi}{6})$
17. $r = 2 - 2\sin(\theta - \frac{\pi}{3})$
18. $r = 2 - 2\sin(\theta - \frac{\pi}{6})$

B   *Sketch each of the equations in Problems 19–33 by plotting polar-form points. For your convenience, plot positive values of θ only.*

19. $r = \theta$
20. $r - 2\theta = 0$
21. $r = 3\theta$
22. $r = 5\sin\frac{\pi}{6}$
23. $r = 4\cos\frac{\pi}{3}$
24. $r = 3\tan\frac{\pi}{4}$
25. $\theta = 2$
26. $\theta = 3$
27. $\theta = \frac{1}{2}$
28. $r = 2\sin\theta$
29. $r = 3\cos\theta$
30. $r = \tan\theta$
31. $r = \sec\theta$
32. $r = \csc\theta$
33. $r = \cot\theta$

C

*34. The limaçon is a curve of the form $r = b \pm a\cos\theta$ or $r = b \pm a\sin\theta$, where $a > 0$ and $b > 0$. There are four types of limaçons:

a. $\frac{b}{a} < 1$ (limaçon with inner loop); graph $r = 2 - 3\cos\theta$ by plotting points.

b. $\frac{b}{a} = 1$ (cardioid); graph $r = 3 - 3\cos\theta$.

c. $1 < \frac{b}{a} < 2$ (limaçon with a dimple); graph $r = 3 - 2\cos\theta$ by plotting points.

d. $\frac{b}{a} \geq 2$ (convex limaçon); graph $r = 3 - \cos\theta$ by plotting points.

*35. Graph the following limaçons (see Problem 34).

a. $r = 1 - 2\cos\theta$ **b.** $r = 2 - \cos\theta$ **c.** $r = 2 + 3\cos\theta$

d. $r = 2 - 3\sin\theta$ **e.** $r = 2 + 3\sin\theta$ **f.** $r = 3 - 2\sin\theta$

36. *Spirals* are interesting mathematical curves. There are three general types of spirals:

a. A spiral of Archimedes has the form $r = a\theta$; graph $r = 2\theta$ by plotting points (let $\theta > 0$).

b. A hyperbolic spiral has the form $r\theta = a$; graph $r\theta = 2$ by plotting points (let $\theta > 0$).

c. A logarithmic spiral has the form $r = a^{k\theta}$; graph $r = 2^\theta$ by plotting points (let $\theta > 0$).

37. Identify and graph the following spirals (see Problem 36).

a. $r = \theta$ **b.** $r = -\theta$ **c.** $r\theta = 1$ **d.** $r\theta = -1$ **e.** $r = 2^{2\theta}$ **f.** $r = 3^\theta$

7.3 SPECIAL POLAR-FORM CURVES

Rose Curves

The cardioid discussed in the last section involved a cosine or a sine of a single angle. The first example of this section is a function of a double angle and is called a **rose curve**.

Example 1

Graph $r = 4\cos 2\theta$.

Solution

When presented with a polar-form curve that you do not recognize, you will graph the curve by plotting points. You can use tables, exact values, or a calculator. A calculator was used to form the following table.

θ	r (APPROX. VALUE)	θ	r (APPROX. VALUE)	θ	r (APPROX. VALUE)	θ	r (APPROX. VALUE)
0	4	$\frac{\pi}{2}$	-4	π	4	$\frac{3\pi}{2}$	-4
$\frac{\pi}{12}$	3.5	$\frac{7\pi}{12}$	-3.5	$\frac{13\pi}{12}$	3.5	$\frac{19\pi}{12}$	-3.5
$\frac{\pi}{6}$	2	$\frac{2\pi}{3}$	-2	$\frac{7\pi}{6}$	2	$\frac{5\pi}{3}$	-2
$\frac{\pi}{4}$	0	$\frac{3\pi}{4}$	0	$\frac{5\pi}{4}$	0	$\frac{7\pi}{4}$	0
$\frac{\pi}{3}$	-2	$\frac{5\pi}{6}$	2	$\frac{4\pi}{3}$	-2	$\frac{11\pi}{6}$	2
$\frac{5\pi}{12}$	-3.5	$\frac{11\pi}{12}$	3.5	$\frac{17\pi}{12}$	-3.5	$\frac{23\pi}{12}$	3.5

---

* Problems 34 and 35 are required for optional Section 7.5.

For each θ you choose, press:

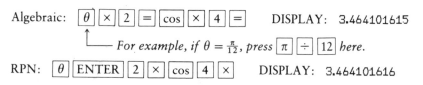

Algebraic: $\boxed{\theta}\ \boxed{\times}\ \boxed{2}\ \boxed{=}\ \boxed{\cos}\ \boxed{\times}\ \boxed{4}\ \boxed{=}$ DISPLAY: 3.464101615

For example, if $\theta = \frac{\pi}{12}$, press $\boxed{\pi}\ \boxed{\div}\ \boxed{12}$ here.

RPN: $\boxed{\theta}\ \boxed{\text{ENTER}}\ \boxed{2}\ \boxed{\times}\ \boxed{\cos}\ \boxed{4}\ \boxed{\times}$ DISPLAY: 3.464101616

The graph is shown in Figure 7.10. Notice that as θ increases from the starting point 0 to $\frac{\pi}{4}$, r decreases from 4 to 0. As θ increases from $\frac{\pi}{4}$ to $\frac{\pi}{2}$, you can see that r is negative and decreases from 0 to -4. Study the table at the left as it relates to the graph in Figure 7.10.

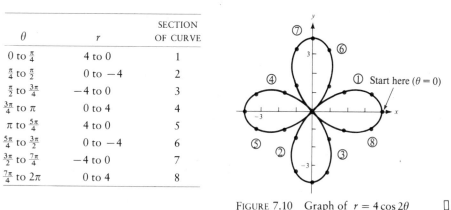

θ	r	SECTION OF CURVE
0 to $\frac{\pi}{4}$	4 to 0	1
$\frac{\pi}{4}$ to $\frac{\pi}{2}$	0 to -4	2
$\frac{\pi}{2}$ to $\frac{3\pi}{4}$	-4 to 0	3
$\frac{3\pi}{4}$ to π	0 to 4	4
π to $\frac{5\pi}{4}$	4 to 0	5
$\frac{5\pi}{4}$ to $\frac{3\pi}{2}$	0 to -4	6
$\frac{3\pi}{2}$ to $\frac{7\pi}{4}$	-4 to 0	7
$\frac{7\pi}{4}$ to 2π	0 to 4	8

FIGURE 7.10   Graph of $r = 4\cos 2\theta$ ☐

The equation

$$r = a \cos n\theta$$

is a four-leaved rose if $n = 2$, and the length of the leaves is a. In general, if n is an even number, the curve has $2n$ leaves; if n is odd, the number of leaves is n. These leaves are equally spaced on a radius of a.

Example 2

Graph $r = 4\cos 2(\theta - \frac{\pi}{4})$.

Solution

This is a rose curve with four leaves of length 4 equally spaced on a circle. However, this curve has been rotated $\frac{\pi}{4}$, as shown in Figure 7.11.

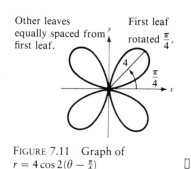

FIGURE 7.11   Graph of $r = 4\cos 2(\theta - \frac{\pi}{4})$ ☐

Notice that Example 2 can be rewritten as a sine curve:

$$r = 4\cos 2(\theta - \tfrac{\pi}{4})$$
$$= 4\cos(2\theta - \tfrac{\pi}{2})$$
$$= 4\cos(\tfrac{\pi}{2} - 2\theta)$$
$$= 4\sin 2\theta$$

These steps can be reversed in order to graph a rose curve written in terms of a sine function.

Example 3

Graph $r = 5\sin 4\theta$.

Solution

$$r = 5\sin 4\theta$$
$$= 5\cos(\tfrac{\pi}{2} - 4\theta)$$
$$= 5\cos(4\theta - \tfrac{\pi}{2})$$
$$= 5\cos 4(\theta - \tfrac{\pi}{8})$$

Recognize this as a rose curve rotated $\tfrac{\pi}{8}$. There are eight leaves of length 5. The leaves are a distance of $\tfrac{\pi}{4}$ apart, as shown in Figure 7.12.

FIGURE 7.12   Graph of $r = 5\sin 4\theta$

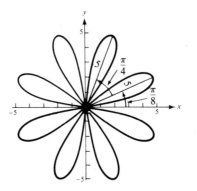

Lemniscates

The third and last general type of polar-form curve we will consider is called a **lemniscate** and has the general form

$$r^2 = a^2 \cos 2\theta$$

Example 4

Graph $r^2 = 9\cos 2\theta$.

Solution

As before, when graphing a curve for the first time, begin by plotting points. For this example, be sure to obtain two values for r when solving this quadratic equation. For example, if $\theta = 0$, then $\cos 2\theta = 1$ and $r^2 = 9$, so $r = 3$ or -3.

θ	0	$\dfrac{\pi}{12}$	$\dfrac{\pi}{6}$	$\dfrac{\pi}{4}$	$\dfrac{\pi}{4}$ to $\dfrac{3\pi}{4}$	$\dfrac{5\pi}{6}$	$\dfrac{11\pi}{12}$	π
r (APPROX. VALUE)	± 3	± 2.8	± 2.1	0	undefined	± 2.1	± 2.8	± 3

Notice that for $\frac{\pi}{4} < \theta < \frac{3\pi}{4}$ there are no values for r since $\cos 2\theta$ is negative. For $\pi \le \theta \le 2\pi$, the values repeat the sequence given above, so these points are plotted and then connected, as shown in Figure 7.13.

θ	r	SECTION OF CURVE
0 to $\frac{\pi}{4}$	3 to 0	1 (two values) Quadrants I and III
$\frac{\pi}{4}$ to $\frac{3\pi}{4}$	No value	
$\frac{3\pi}{4}$ to π	0 to 3	2 (two values) Quadrants II and IV
π to 2π	$\begin{cases} \text{Repeat of} \\ \text{values from} \\ 0 \text{ to } \pi \end{cases}$	

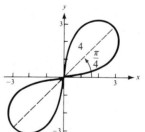

FIGURE 7.13   Graph of $r^2 = 9\cos 2\theta$

The graphs of $r^2 = a^2 \cos 2\theta$ and $r^2 = a^2 \sin 2\theta$ are called *lemniscates*. There are always two leaves to a lemniscate and the length of the leaves is a. The procedure for graphing a lemniscate given in the form $r^2 = a^2 \sin 2\theta$ is illustrated by Example 5.

Example 5

Graph $r^2 = 16\sin 2\theta$.

Solution

$$\begin{aligned} r^2 &= 16\sin 2\theta \\ &= 16\cos(\tfrac{\pi}{2} - 2\theta) \\ &= 16\cos(2\theta - \tfrac{\pi}{2}) \\ &= 16\cos 2(\theta - \tfrac{\pi}{4}) \end{aligned}$$

This is a lemniscate whose leaf has length $\sqrt{16} = 4$, rotated $\frac{\pi}{4}$ as shown in Figure 7.14.

FIGURE 7.14   Graph of $r^2 = 16\sin 2\theta$

Summary of Polar Curves

We conclude this section by summarizing the special types of polar-form curves we have examined. There are many others, some of which are presented in the problems, but these three special types are the most common.

Cardioid: $r = a(1 \pm \cos \theta)$ or $r = a(1 \pm \sin \theta)$

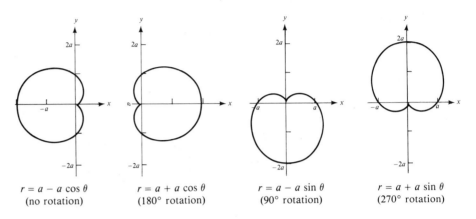

| $r = a - a \cos \theta$ | $r = a + a \cos \theta$ | $r = a - a \sin \theta$ | $r = a + a \sin \theta$ |
| (no rotation) | (180° rotation) | (90° rotation) | (270° rotation) |

Rose curve: $r = a \cos n\theta$ or $r = a \sin n\theta$, where n is a positive integer.

a. If n is odd, the rose is n-leaved.   **b.** If n is even, the rose is $2n$-leaved.

One leaf: If $n = 1$, the rose is a curve with one petal and is circular.

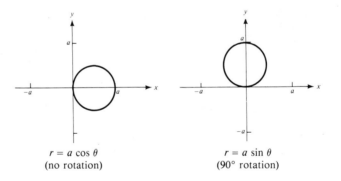

| $r = a \cos \theta$ | $r = a \sin \theta$ |
| (no rotation) | (90° rotation) |

Two leaves: See the lemniscate described on page 285.

Three leaves: $n = 3$ *Four leaves:* $n = 2$

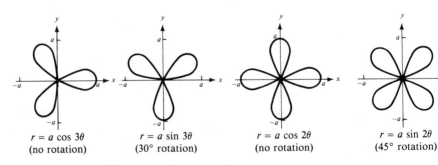

| $r = a \cos 3\theta$ | $r = a \sin 3\theta$ | $r = a \cos 2\theta$ | $r = a \sin 2\theta$ |
| (no rotation) | (30° rotation) | (no rotation) | (45° rotation) |

Lemniscate: $r^2 = a^2 \cos 2\theta$ or $r^2 = a^2 \sin 2\theta$

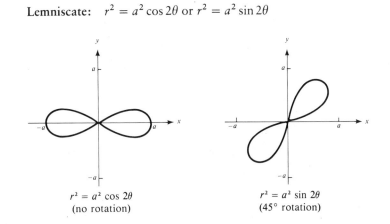

$r^2 = a^2 \cos 2\theta$
(no rotation)

$r^2 = a^2 \sin 2\theta$
(45° rotation)

Problem Set 7.3

A   *Identify each of the curves in Problems 1–15 as a cardioid, a rose curve (state number of leaves), a lemniscate, or none of the above.*

1. $r^2 = 9 \cos 2\theta$ 2. $r = 2 \sin 2\theta$ 3. $r = 3 \sin 3\theta$ 4. $r^2 = 2 \cos 2\theta$

5. $r = 2 - 2 \cos \theta$ 6. $r = 3 + 3 \sin \theta$ 7. $r^2 = \sin 3\theta$ 8. $r = 4 \sin 30°$

9. $r = 5 \cos 60°$ 10. $r = 5 \sin 8\theta$ 11. $r = 3\theta$ 12. $r\theta = 3$

13. $\theta = \tan \frac{\pi}{4}$ 14. $r = 5(1 - \sin \theta)$ 15. $\cos \theta = 1 - r$

B   *Sketch each of the equations in Problems 16–30 by relating it to a standard rose curve or lemniscate. Use a rotation if necessary.*

16. $r = 2 \cos 2\theta$ 17. $r = 3 \cos 3\theta$ 18. $r = 5 \sin 3\theta$

19. $r^2 = 9 \cos 2\theta$ 20. $r^2 = 16 \cos 2\theta$ 21. $r^2 = 9 \sin 2\theta$

22. $r = 3 \cos 3(\theta - \frac{\pi}{3})$ 23. $r = 2 \cos 2(\theta + \frac{\pi}{3})$ 24. $r = 5 \cos 3(\theta - \frac{\pi}{4})$

25. $r = \sin 3(\theta + \frac{\pi}{6})$ 26. $r = \sin(2\theta + \frac{\pi}{3})$ 27. $r = \cos(2\theta + \frac{\pi}{3})$

28. $r^2 = 16 \cos 2(\theta - \frac{\pi}{6})$ 29. $r^2 = 9 \cos(2\theta - \frac{\pi}{3})$ 30. $r^2 = 9 \cos(2\theta - \frac{\pi}{4})$

C   31. The *strophoid* is a curve of the form $r = a \cos 2\theta \sec \theta$; graph this curve where $a = 2$ by plotting points.

32. The *bifolium* has the form $r = a \sin \theta \cos^2\theta$; graph this curve where $a = 1$ by plotting points.

33. The *folium of Descartes* has the form

$$r = \frac{3a \sin \theta \cos \theta}{\sin^3\theta + \cos^3\theta}$$

Graph this curve where $a = 2$ by plotting points.

*7.4   PARAMETRIC EQUATIONS

It is sometimes convenient to graph curves that consist of points (x, y) where *each* of the variables x and y is defined as a function. For example, suppose

$$x = f(t) \quad \text{and} \quad y = g(t)$$

for functions f and g, where the domain of these functions is some interval I. If we let

$$f(t) = 1 + 3t \quad \text{and} \quad y = g(t) = 2t \qquad (\text{for } 0 \le t \le 5)$$

then, if $t = 1$,

$$x = f(1) = 1 + 3(1) = 4$$
$$y = g(1) = 2(1) = 2$$

Plot the point $(x, y) = (4, 2)$ as one point on the curve. Other values are shown in the following table and are plotted in Figure 7.15.

t	0	1	2	3	4	5
x	1	4	7	10	13	16
y	0	2	4	6	8	10

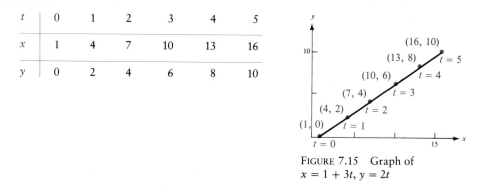

FIGURE 7.15   Graph of $x = 1 + 3t$, $y = 2t$

The variable t is called a **parameter** and the equations $x = 1 + 3t$ and $y = 2t$ are called **parametric equations** for the line segment shown in Figure 7.15.

PARAMETER AND
PARAMETRIC EQUATIONS

> Let t be a number in an interval I. Consider the curve defined by the set of ordered pairs (x, y), where
>
> $$x = f(t) \quad \text{and} \quad y = g(t)$$
>
> for functions f and g defined on I. Then the variable t is called a **parameter** and the equations $x = f(t)$ and $y = g(t)$ are called the **parametric equations** for the curve defined by (x, y).

* Optional section.

Example 1

Plot the curve represented by the parametric equations

$$x = \cos \theta$$
$$y = \sin \theta$$

Solution

The parameter is θ and you can generate a table of values by using Appendix D Table, exact values, or a calculator:

θ	0°	15°	30°	45°	60°	75°	90°	120°	⋯
x	1.00	.97	.87	.71	.50	.26	.00	−.50	⋯
y	.00	.26	.50	.71	.87	.97	1.00	.87	⋯

These points are plotted in Figure 7.16. If the plotted points are connected, you can see that the curve is a circle.

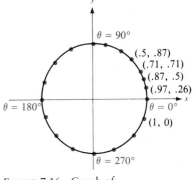

FIGURE 7.16 Graph of $x = \cos \theta, y = \sin \theta$

It is possible to recognize the parametric equations in Example 1 as a unit circle if you square both sides of the equation and add:

$$x^2 = \cos^2\theta$$
$$y^2 = \sin^2\theta$$
$$x^2 + y^2 = \cos^2\theta + \sin^2\theta$$

So

$$x^2 + y^2 = 1$$

This process is called **eliminating the parameter**.

Suppose a light is attached to the edge of a bike wheel. The path of light is shown in Figure 7.17, and the curve traced by the light is called a **cycloid**. If the radius of the wheel is a, we can find the parametric equations for the cycloid. Let $P(x, y)$ be any

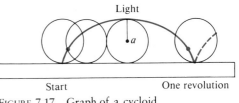

FIGURE 7.17 Graph of a cycloid

COMPUTER APPLICATION

TRIGPAK, which has been mentioned before, does an excellent job of graphing parametric equations. For example, the curve shown at the right is the TRIGPAK plot of the curve wiith the following parametric equation:

$$x = 4\sin\theta + 3\cos\theta, \quad y = 4\cos\theta + 3\sin\theta$$

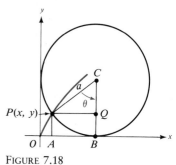

point of the cycloid, and let θ be the amount of rotation of the circle, as shown in Figure 7.18. Then,

$$x = |OA|$$
$$= |OB| - |AB|$$
$$= \overset{\frown}{PB} - |PQ| \quad \text{where } \overset{\frown}{PB} \text{ is the arc length.}$$
$$\text{Also notice } \overset{\frown}{PB} = a\theta \text{ by the}$$
$$\text{arc length formula.}$$
$$= a\theta - a\sin\theta, \quad \text{since } \sin\theta = \frac{|PQ|}{a}$$

FIGURE 7.18

Also,

$$y = |PA|$$
$$= |QB|$$
$$= |CB| - |CQ|$$
$$= a - a\cos\theta, \quad \text{since } \cos\theta = \frac{|CQ|}{a}$$

Therefore,

$$y = a\theta - a\sin\theta = a(\theta - \sin\theta)$$
$$y = a - a\cos\theta = a(1 - \cos\theta)$$

COMPUTER APPLICATION

Two special cases of cycloids involve the path traced by a fixed point P on a circle. If it rolls *inside* a larger circle, the trace is called a **hypocycloid** and as it rolls around the *outside* of a fixed circle, the trace is called an **epicycloid**. Without computer assistance, these interesting curves are beyond the scope of this course. However, if you have access to a computer you might look at an article by Florence and Sheldon Gordon entitled "Mathematics Discovery via Computer Graphics: Hypocycloids and Epicycloids," in *The Two-Year College Mathematics Journal*, November 1984, p. 441.

If R is the radius of the fixed circle and r is the radius of the rolling circle, then the parametric equations are:

Hypocycloid: $x = (R - r) \cos t + r \cos\left(\dfrac{R - r}{r}\right)t;$

$$y = (R - r) \sin t - r \sin\left(\dfrac{R - r}{r}\right)t$$

Epicycloid: $x = (R + r) \cos t - r \cos\left(\dfrac{R + r}{r}\right)t;$

$$y = (R + r) \sin t - r \sin\left(\dfrac{R + r}{r}\right)t$$

Some variations of these curves, which were drawn by computer, are shown below.

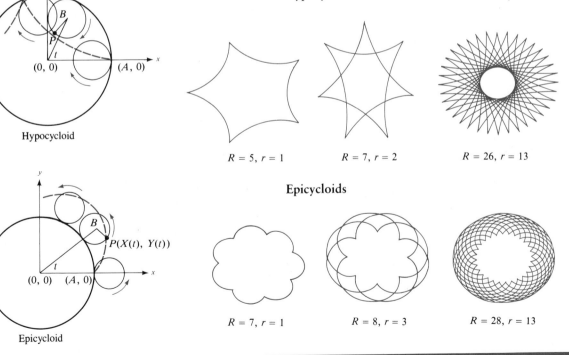

Hypocycloid

Hypocycloids

$R = 5, r = 1$ $R = 7, r = 2$ $R = 26, r = 13$

Epicycloid

Epicycloids

$R = 7, r = 1$ $R = 8, r = 3$ $R = 28, r = 13$

Example 2   Graph the cycloid for $a = 2$ and $0 \leq \theta \leq 2\pi$.

θ	0	$\frac{\pi}{6}$	$\frac{\pi}{3}$	$\frac{\pi}{2}$	$\frac{2\pi}{3}$	$\frac{5\pi}{6}$	π	$\frac{7\pi}{6}$	$\frac{4\pi}{3}$	$\frac{3\pi}{2}$	$\frac{5\pi}{3}$	$\frac{11\pi}{6}$	2π
x	0	.05	.36	1.14	2.46	4.24	6.28	8.33	10.11	11.42	12.20	12.52	12.57
y	0	.27	1.00	2.00	3.00	3.73	4.00	3.73	3.00	2.00	1.00	.27	0

These points are plotted in Figure 7.19.

FIGURE 7.19   Graph of a cycloid

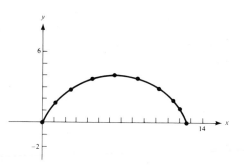

Problem Set 7.4

A   *Graph the curves whose parametric equations are given in Problems 1–12.*

1. $x = 4t$, $y = -2t$
2. $x = t + 1$, $y = 2t$
3. $x = t$, $y = 2 + \frac{2}{3}(t - 1)$
4. $x = t$, $y = 3 - \frac{3}{5}(t + 2)$
5. $x = 2t$, $y = t^2 + t + 1$
6. $x = 3t$, $y = t^2 - t + 6$
7. $x = t$, $y = t^2 + 2t + 3$
8. $x = t$, $y = 2t^2 - 5t + 6$
9. $x = 3\cos\theta$, $y = 3\sin\theta$
10. $x = 2\cos\theta$, $y = 2\sin\theta$
11. $x = 4\cos\theta$, $y = 3\sin\theta$
12. $x = 5\cos\theta$, $y = 2\sin\theta$

B   *Graph the curves whose parametric equations are given in Problems 13–26.*

13. $x = 10\cos t$, $y = 10\sin t$
14. $x = 8\sin t$, $y = 8\cos t$
15. $x = 5\cos\theta$, $y = 3\sin\theta$
16. $x = 4\cos\theta$, $y = 2\sin\theta$
17. $x = \theta - \sin\theta$, $y = 1 - \cos\theta$
18. $x = \theta + \sin\theta$, $y = 1 - \cos\theta$
19. $x = 4\tan 2t$, $y = 3\sec 2t$
20. $x = 2\tan 2t$, $y = 4\sec 2t$
21. $x = 1 + \cos t$, $y = 3 - \sin t$
22. $x = 2 - \sin t$, $y = -3 + \cos t$
23. $x = 3\cos\theta + \cos 3\theta$, $y = 3\sin\theta - \sin 3\theta$
24. $x = \cos t + t\sin t$, $y = \sin t - t\cos t$
25. $x = \sin t$, $y = \csc t$
26. $x = \cos\theta$, $y = \sec\theta$

*27. Graph the hypocycloid where $R = 4$ and $r = 1$.

*28. Graph the hypocycloid where $R = 7$ and $r = 1$.

---

* See Computer Application on page 289. You do not need a computer to work these problems.

*29. Graph the epicycloid where $R = 4$ and $r = 1$.

*30. Graph the epicycloid where $R = 6$ and $r = 1$.

C   **31.** Suppose a string is wound around a circle of radius a. The string is then unwound in the plane of the circle while it is held tight, as shown in Figure 7.20. Find the equation for this curve, called the *involute of a circle*.

FIGURE 7.20 Graph of the involute of a circle

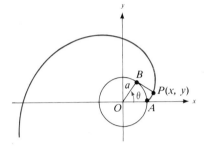

Hint for Problem 31:   Consider Figure 7.21 and find the coordinates of $P(x, y)$. Notice that

$$x = |OB| \qquad y = |PB|$$

FIGURE 7.21

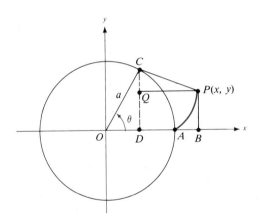

Find x and y in terms of θ, the amount of rotation in radians.

*32. Derive the parametric equations for the hypocycloid.

*33. Derive the parametric equations for the epicycloid.

34. *Computer application.* Write a paper describing what happens if the radius of the inner rolling circle is larger than the radius of the outer fixed circle.

35. *Computer application.* Write a paper describing what happens if the radius of the outer rolling circle is larger than the radius of the inner fixed circle.

---

* See Computer Application on page 289. You do not need a computer to work these problems.

*7.5   INTERSECTION OF POLAR-FORM CURVES

In order to find the points of intersection of graphs in rectangular form, you need only find the simultaneous solution of the equations that define those graphs. It is not always necessary to draw the graphs. This follows because in a rectangular coordinate system there is a one-to-one correspondence between ordered pairs satisfying an equation and points on its graph.

However, as you saw in Section 7.1, this one-to-one property is lost when working with polar coordinates. This means that the simultaneous solution of two equations in polar form may introduce extraneous points of intersection or may even fail to yield all points of intersection. For this reason, our method for finding the intersection of polar-form curves will include sketching the graphs.

Graphical Solution

Example 1

Find the points of intersection of the curves $r = 2\cos\theta$ and $r = 2\sin\theta$.

Solution

First, consider the simultaneous solution of the system of equations.

$$\begin{cases} r = 2\cos\theta \\ r = 2\sin\theta \end{cases}$$

By substitution,

$$2\sin\theta = 2\cos\theta$$
$$\sin\theta = \cos\theta$$
$$\theta = \frac{\pi}{4} + 2n\pi, \frac{5\pi}{4} + 2n\pi \qquad (n \text{ an integer})$$

Then find r using the primary representation for θ:

$$r = 2\cos\frac{\pi}{4} \qquad r = 2\cos\frac{5\pi}{4}$$
$$= 2\left(\frac{\sqrt{2}}{2}\right) \qquad = 2\left(\frac{-\sqrt{2}}{2}\right)$$
$$= \sqrt{2} \qquad = -\sqrt{2}$$

This gives the points $(\sqrt{2}, \frac{\pi}{4})$ and $(-\sqrt{2}, \frac{5\pi}{4})$. Writing the primary representations for these points, we see they represent the same point. Thus, the simultaneous solution yields one point of intersection. Next, consider the graphs of these curves, shown in Figure 7.22.

_____

* Optional section.

FIGURE 7.22 Graphs of
$r = 2\cos\theta$ and $r = 2\sin\theta$

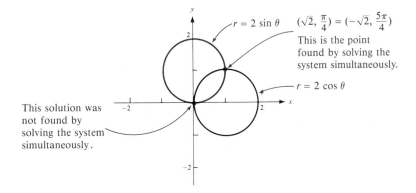

$r = 2\sin\theta$ $(\sqrt{2}, \frac{\pi}{4}) = (-\sqrt{2}, \frac{5\pi}{4})$

This is the point
found by solving the
system simultaneously.

$r = 2\cos\theta$

This solution was
not found by
solving the system
simultaneously.

It looks as though $(0,0)$ is also a point of intersection. Check this point in each of the given equations:

$$r = 2\cos\theta: \quad \text{If } r = 0, \text{ then } \theta = \frac{\pi}{2}.$$

$$r = 2\sin\theta: \quad \text{If } r = 0, \text{ then } \theta = \pi.$$

At first it does not seem that $(0,0)$ satisfies the equation since $r = 0$ gives $(0, \frac{\pi}{2})$ and $(0, \pi)$, respectively. Notice that these coordinates are different and do not satisfy the equations simultaneously. But, if you plot these coordinates, you will see that $(0,0)$, $(0, \frac{\pi}{2})$, and $(0, \pi)$ are all the same point. Points of intersection for the given curves are $\{(0,0), (\sqrt{2}, \frac{\pi}{4})\}$.   □

The pole is often a solution for a system of equations, even though it may not satisfy the equations simultaneously. This is because when $r = 0$, all values of θ will yield the same point—namely, the pole. For this reason, it is necessary to check separately to see if the pole lies on the given graph.

GRAPHICAL SOLUTION
OF THE INTERSECTION
OF POLAR CURVES

> 1. Find the simultaneous solution of the given system of equations.
> 2. Determine whether the pole lies on the two graphs.
> 3. Graph the curves to look for other points of intersection.

Example 2

Find the points of intersection of the curves $r = 2 + 4\cos\theta$ and $r = 6\cos\theta$.

Solution

Step 1: Solve the equations simultaneously.

$$\begin{cases} r = 2 + 4\cos\theta \\ r = 6\cos\theta \end{cases}$$

By substitution,

$$6\cos\theta = 2 + 4\cos\theta$$
$$2\cos\theta = 2$$
$$\cos\theta = 1$$
$$\theta = 0$$

If $\theta = 0$, then $r = 6$, so an intersection point is $(6, 0)$.

Step 2:   Determine whether the pole lies on the graphs.

a. If $r = 0$, then

$$0 = 2 + 4\cos\theta$$
$$\cos\theta = -\tfrac{1}{2}$$
$$\theta = \tfrac{2\pi}{3}, \tfrac{4\pi}{3}$$

Thus, $(0, \tfrac{2\pi}{3})$ and $(0, \tfrac{4\pi}{3})$ satisfy the first equation, and the pole lies on this graph.

b. If $r = 0$, then

$$0 = 6\cos\theta$$
$$\theta = \tfrac{\pi}{2}, \tfrac{3\pi}{2}$$

Thus, $(0, \tfrac{\pi}{2})$ and $(0, \tfrac{3\pi}{2})$ satisfy the second equation, so the pole also lies on this graph.

Step 3:   Graph the curves to look for other points of intersection. The first curve is a limaçon, and the second is a circle as shown in Figure 7.23.

Figure 7.23   Graphs of $r = 2 + 4\cos\theta$ and $r = 6\cos\theta$

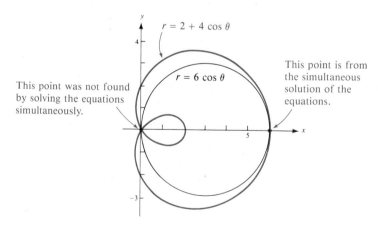

This point was not found by solving the equations simultaneously.

This point is from the simultaneous solution of the equations.

$r = 2 + 4\cos\theta$

$r = 6\cos\theta$

Notice that there are no other points of intersection, so the intersection points are $(6, 0)$ and $(0, 0)$.

Example 3

Find the points of intersection of the curves $r = \frac{3}{2} - \cos\theta$ and $\theta = \frac{2\pi}{3}$.

Solution

Solve $\begin{cases} r = \frac{3}{2} - \cos\theta \\ \theta = \frac{2\pi}{3} \end{cases}$

By substitution,

$$r = \frac{3}{2} - \cos\frac{2\pi}{3}$$
$$= \frac{3}{2} - \left(-\frac{1}{2}\right)$$
$$= 2$$

The solution is $(2, \frac{2\pi}{3})$. Also, if $r = 0$, the first equation has no solution since

$$0 = \frac{3}{2} - \cos\theta$$
$$\cos\theta = \frac{3}{2}$$

and $\cos\theta$ cannot be larger than 1. Now look at the graph, as shown in Figure 7.24.

FIGURE 7.24 Graphs of $r = \frac{3}{2} - \cos\theta$ and $\theta = \frac{2\pi}{3}$

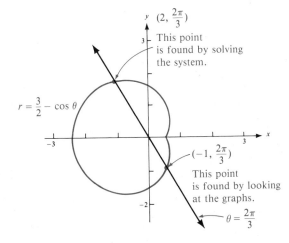

From the graph we see that $(-1, \frac{2\pi}{3})$ looks like a point of intersection. It satisfies the equation $\theta = \frac{2\pi}{3}$, but what about $r = \frac{3}{2}\cos\theta$? Check $(-1, \frac{2\pi}{3})$:

$$-1 \overset{?}{=} \frac{3}{2} - \cos\left(\frac{2\pi}{3}\right)$$
$$= \frac{3}{2} - \left(-\frac{1}{2}\right)$$
$$= 2 \quad \textit{Not satisfied}$$

However, if you check the other representation $(-1, \frac{2\pi}{3}) = (1, \frac{5\pi}{3})$:

$$1 \overset{?}{=} \frac{3}{2} - \cos\left(\frac{5\pi}{3}\right)$$
$$= \frac{3}{2} - \left(\frac{1}{2}\right)$$
$$= 1 \quad \textit{Satisfied}$$

You would not have found this other point without checking the graph. □

Analytic Solution

Now you might ask if there is a procedure you can use that will yield all of the points of intersection without relying on the graph. The answer is yes, if you realize that polar-form equations can have different representations just as we found for polar-form points. Recall, a polar-form point has two primary representations

$$(r, \theta) \quad \text{and} \quad (-r, \theta + \pi)$$

where the angle (θ or $\theta + \pi$, respectively) is between 0 and 2π. For every polar-form equation $r = f(\theta)$, there are equations

$$r = (-1)^n f(\theta + n\pi)$$

for any integer n that yield exactly the same curve.

ANALYTIC SOLUTION FOR
THE INTERSECTION OF
POLAR CURVES

> 1. Determine if the pole lies on the two graphs.
> 2. Solve each equation of one graph simultaneously with each equation of the other graph. If $r = f(\theta)$ is the given equation, then the other equations of the same graph are
>
> $$r = (-1)^n f(\theta + n\pi) .$$

Example 4

Find the points of intersection of the curves

$$r = 1 - 2\cos\theta \quad \text{and} \quad r = 1$$

Solution

Check pole: If $r = 0$, then the second equation ($r = 1$) is not satisfied. That is, the graph of the second equation does not pass through the origin.

Next, solve the equations simultaneously; write out the alternate forms of the equations.

$r = 1 - 2\cos\theta$:

$$r = (-1)^n [1 - 2\cos(\theta + n\pi)]$$

If $n = 0$, $\quad r = 1 - 2\cos\theta$

If $n = 1$, $\quad r = -[1 - 2\cos(\theta + \pi)]$

$$= -[1 + 2\cos\theta]$$

$$= -1 - 2\cos\theta$$

For other integral values, one of these two equations is repeated.

$r = 1$:

$$r = (-1)^n \cdot 1$$

If $n = 0$, $\quad r = 1$

If $n = 1$, $\quad r = -1$

For other integral values, one of these two equations is repeated.
Solve the systems:

$$\begin{cases} r = 1 - 2\cos\theta \\ r = 1 \end{cases} \qquad \begin{cases} r = 1 - 2\cos\theta \\ r = -1 \end{cases} \qquad \begin{cases} r = -1 - 2\cos\theta \\ r = 1 \end{cases} \qquad \begin{cases} r = -1 - 2\cos\theta \\ r = -1 \end{cases}$$

$$\begin{array}{llll} 1 = 1 - 2\cos\theta & -1 = 1 - 2\cos\theta & 1 = -1 - 2\cos\theta & -1 = -1 - 2\cos\theta \\ 0 = \cos\theta & 1 = \cos\theta & -1 = \cos\theta & 0 = \cos\theta \\ \theta = \dfrac{\pi}{2}, \dfrac{3\pi}{2} & \theta = 0 & \theta = \pi & \text{Same as first equation} \end{array}$$

Points: $(1, \frac{\pi}{2})$, $(1, \frac{3\pi}{2})$, $(1, \pi)$

The graphs of these curves are shown in Figure 7.25.

FIGURE 7.25 Graphs of
$r = 1 - 2\cos\theta$ and $r = 1$

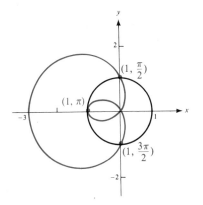

Problem Set 7.5

Find the points of intersection of the curves given by the equations in Problems 1–30. Sketch each of the graphs as a check of your work since you may have introduced extraneous solutions. You need to give only one primary representation for each point of intersection; give the one with a positive r or with $0 \le \theta \le 2\pi$.

A

1. $\begin{cases} r = 4\cos\theta \\ r = 4\sin\theta \end{cases}$

2. $\begin{cases} r = 8\cos\theta \\ r = 8\sin\theta \end{cases}$

3. $\begin{cases} r = 2\cos\theta \\ r = 1 \end{cases}$

4. $\begin{cases} r = 4\cos\theta \\ r = 2 \end{cases}$

5. $\begin{cases} r^2 = 9\cos 2\theta \\ r = 3 \end{cases}$

6. $\begin{cases} r^2 = 4\sin 2\theta \\ r = 2 \end{cases}$

7. $\begin{cases} r = 2(1 + \cos\theta) \\ r = 2(1 - \cos\theta) \end{cases}$

8. $\begin{cases} r = 2(1 + \sin\theta) \\ r = 2(1 - \sin\theta) \end{cases}$

9. $\begin{cases} r^2 = 4\cos 2\theta \\ r = 2 \end{cases}$

10. $\begin{cases} r^2 = 9\sin 2\theta \\ r = 3 \end{cases}$

11. $\begin{cases} r^2 = 4\sin 2\theta \\ r = 2\sqrt{2}\cos\theta \end{cases}$

12. $\begin{cases} r^2 = 9\cos 2\theta \\ r = 3\sqrt{2}\sin\theta \end{cases}$

13. $\begin{cases} r^2 = \cos 2(\theta - 90°) \\ r^2 = \cos 2\theta \end{cases}$

14. $\begin{cases} r = 1 + \sin\theta \\ r = 1 + \cos\theta \end{cases}$

15. $\begin{cases} r = a(1 + \sin\theta) \\ r = a(1 - \sin\theta) \end{cases}$

B

16. $\begin{cases} r = 2\theta \\ \theta = \dfrac{\pi}{6} \end{cases}$

17. $\begin{cases} r = 3\theta \\ \theta = \dfrac{\pi}{3} \end{cases}$

18. $\begin{cases} r^2 = \cos 2\theta \\ r = \sqrt{2}\sin\theta \end{cases}$

19. $\begin{cases} r = 2(1 - \cos\theta) \\ r = 4\sin\theta \end{cases}$

20. $\begin{cases} r = 2(1 + \cos\theta) \\ r = -4\sin\theta \end{cases}$

21. $\begin{cases} r = 2(1 - \sin\theta) \\ r = 4\cos\theta \end{cases}$

22. $\begin{cases} r = 2\cos\theta + 1 \\ r = \sin\theta \end{cases}$

23. $\begin{cases} r = 2\sin\theta + 1 \\ r = \cos\theta \end{cases}$

24. $\begin{cases} r = \dfrac{5}{3 - \cos\theta} \\ r = 2 \end{cases}$

25. $\begin{cases} r = \dfrac{2}{1 + \cos\theta} \\ r = 2 \end{cases}$

26. $\begin{cases} r = \dfrac{4}{1 - \cos\theta} \\ r = 2\cos\theta \end{cases}$

27. $\begin{cases} r = \dfrac{1}{1 + \cos\theta} \\ r = 2(1 - \cos\theta) \end{cases}$

28. $\begin{cases} r = 2\cos\theta \\ r = 2\sec\theta \end{cases}$

29. $\begin{cases} r = 2\sin\theta \\ r = 2\csc\theta \end{cases}$

30. $\begin{cases} r\sin\theta = 1 \\ r = 4\sin\theta \end{cases}$

7.6   SUMMARY AND REVIEW

	OBJECTIVES	PAGES/EXAMPLES
7.1 POLAR COORDINATES	1. Plot polar-form points.	p. 270; Example 1; Problems 1–6; 13–14.
	2. Write the rectangular form of a point, given the polar form.	p. 272; Example 3; Problems 1–6.
	3. Write the two primary representations in polar form, given a point in either rectangular form or polar form.	p. 271; Examples 2 and 4; Problems 1–12.
	4. Determine whether a given point lies on a graph defined by a polar-form equation.	p. 273; Example 5; Problems 15–36.
7.2 GRAPHING POLAR-FORM EQUATIONS	5. Graph polar-form curves by plotting points.	p. 275; Examples 1–3; Problems 19–33.
	6. Graph cardioids by rotating and comparing to a standard-position cardioid.	p. 280; Examples 4 and 5; Problems 1–18.
7.3 SPECIAL POLAR-FORM CURVES	7. Recognize and graph rose curves.	p. 280; Examples 1–3; Problems 16–18, 22–27.

8. Recognize and graph lemniscates.

p. 282; Examples 4 and 5; Problems 19–21, 28–30.

9. Classify and identify the following polar-form curves by inspection of their equations: cardioid, rose curve (number of leaves), and lemniscate. Also be able to determine the rotation, if any.

p. 284; Problems 1–15.

*7.4 PARAMETRIC EQUATIONS

10. Sketch a curve represented by parametric equations.

p. 286; Examples 1–2; Problems 1–26.

*7.5 INTERSECTION OF POLAR-FORM CURVES

11. Find the intersection of polar-form curves.

p. 292; Examples 1–4; Problems 1–30.

Terms

Cardioid [7.2, 7.3]
Lemniscate [7.3]
Limaçon [7.2]
Parameter [7.4]
Parametric equations [7.4]

Polar axis [7.1]
Polar coordinate system [7.1]
Pole [7.1]
Primary representations of a point [7.1]

Rose curve [7.3]
Rotation [7.2]
Standard-position cardioid [7.2]

Chapter 7 Review of Objectives

The problems numbers correspond to the objectives listed in Section 7.6.

1. Plot the given polar-form points.
   a. $(3, -\frac{2\pi}{3})$ **b.** $(-5, -\frac{\pi}{10})$ **c.** $(-2, 2)$ **d.** $(5, \sqrt{75})$

2. Give the rectangular form for the given points.
   a. $(3, -\frac{2\pi}{3})$ **b.** $(-5, -\frac{\pi}{10})$

3. Give both primary representations in polar form for the given points.
   a. $(3, -\frac{2\pi}{3})$ is in polar form.   **b.** $(-5, -\frac{\pi}{10})$ is in polar form.
   c. $(-3, 3\sqrt{3})$ is in rectangular form.   **d.** $(2, 5)$ is in rectangular form.

4. Determine whether the given points lie on the curve defined by the equation

$$r = \frac{10}{2 + \cos\theta}$$

   a. $(4, \frac{\pi}{6})$ **b.** $(-4, \frac{4\pi}{3})$ **c.** $(5, \frac{3\pi}{2})$ **d.** $(5, \pi)$

5. Sketch the given curves by plotting points.
   a. $r - 3\theta = 0, \theta > 0$ **b.** $\theta - 2 = 0$ **c.** $r - 2\tan\frac{3\pi}{4} = 0$

Sketch the curves in Problems 6–8 by comparing them with standard forms.

6. $r + 3\sin\theta = 3$ 7. $r = 6\cos 3(\theta - \frac{\pi}{10})$ 8. $r^2 = 4\sin 2\theta$

---

* Optional sections.

9. Identify each of the given curves.
   a. $r^2 = 9\cos 2\theta$ b. $r = 2 + 2\sin\theta$ c. $r = 4\cos 5\theta$
   d. $r = 5\sin\frac{\pi}{4}$

*10. Graph the curve given by the parametric equations

$$x = 5\cos\theta \qquad y = 3\sin\theta$$

*11. Find the intersection of the curves $r = 4 - 4\sin\theta$ and $r = 4 - 8\sin\theta$.

Miscellaneous Problems

Identify each of the curves in Problems 1–16.

1. $r = \cos 2\theta$ 2. $r = 2\cos\theta$ 3. $r = \cos 2$
4. $r^2 = \cos 2\theta$ 5. $r = 2 - 2\cos\theta$ 6. $r = 2 + 2\cos\theta$
7. $r = 4\sin 3\theta$ 8. $r = 4\sin 3$ 9. $r^2 = 4\sin 2\theta$
10. $r = 4 + 4\sin\theta$ 11. $r = \cos 3(\theta - \frac{\pi}{6})$ 12. $r^2 = \cos 2(\theta - \frac{\pi}{6})$
13. $r = \sin 3(\theta - \frac{\pi}{6})$ 14. $r^2 = 3\sin 2(\theta - \frac{\pi}{6})$
15. $r - 2 = 2\sin(\theta - 3)$ 16. $r - 3 = 3\cos(\theta - 2)$

Graph each of the curves in Problems 17–32 by comparing them with standard forms.

17. $r = \cos 2\theta$ 18. $r = 2\cos\theta$ 19. $r = \cos 2$
20. $r^2 = \cos 2\theta$ 21. $r = 2 - 2\cos\theta$ 22. $r = 2 + 2\cos\theta$
23. $r = 4\sin 3\theta$ 24. $r = 4\sin 3$ 25. $r^2 = 3\sin 2\theta$
26. $r = 4 + 4\sin\theta$ 27. $r = 3\cos 3(\theta - \frac{\pi}{6})$ 28. $r^2 = \cos 2(\theta - \frac{\pi}{6})$
29. $r = \sin 3(\theta - \frac{\pi}{6})$ 30. $r^2 = 3\sin 2(\theta - \frac{\pi}{6})$
31. $r - 2 = 2\sin(3 - \theta)$ 32. $r - 3 = 3\cos(2 - \theta)$

Graph the curves in Problems 33–40 by plotting points.

33. $r = \tan 2\theta$ 34. $r - 2\theta = 0$ 35. $r\theta = 1$

36. $r = \dfrac{8}{4 - 2\cos\theta}$ 37. $x = 5\cos\theta, y = 3\sin\theta$

38. $x = 4\cos\theta, y = 4\sin\theta$ 39. $x = 1 + 4\cos\theta, y = 2 - 4\sin\theta$
40. $x = 2 - 5\cos\theta, y = 1 + 5\sin\theta$

---

* Optional sections.

APPLICATION FOR FURTHER STUDY

COMMUNICATION OF BEES

Courtesy of David Cavagnaro/Dandelet Interlinks.

Polar coordinates can be applied to the manner in which bees communicate the distance and source of a food supply.* Scout bees looking for pollen keep track of their position relative to the hive by referencing it to the sun's position. When a scout finds a food source, the bee returns to the hive, gives away samples of the food, and then performs a dance inside the hive.

The dance is done on the vertical hanging honeycomb inside the hive (see Figure 7.26). The vertical direction represents the sun's direction, and the angle θ represents the direction of the food source.

FIGURE 7.26 Communication of bees. The bee runs straight ahead for a short distance (about a centimeter), returns to the starting point through a 360° turn to the left, again runs through the straight stretch returning through a 360° turn to the right, and so on in an alternating fashion. The straight part of the run is emphasized by lateral wagging of the abdomen.

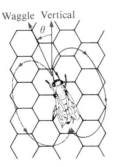

Waggle Vertical

The direction of the straight waggle-run gives an index of the distance, ρ (see Figure 7.27), from the hive, H, to the food source, F. The bee's estimate of ρ depends not on optical perception of the terrain, nor on the duration of flight time, but rather

---

* From Karl von Frisch, *The Dance Language and Orientation of Bees* (Cambridge, Mass.: Belknap Press, 1967). I obtained the idea for this application from Warren Page, New York Technical College.

301

on the bee's expenditure of energy to get there. Notice that the bee is using polar co-ordinates to fix the location of the food source. See Problem 4 for a development of this idea.

FIGURE 7.27   Bee using polar coordinates

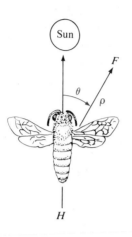

Problems for Further Study: Communication of Bees

1. *Entomology.* Suppose the equation $r = 1 - \sin\theta$ is used to describe the dance shown by the bee in Figure 7.26. Graph this equation. What is the direction of the food source relative to the sun?

2. *Entomology.* Modify the equation in Problem 1 to indicate a food source that is in a direction of $30°$ away from the direction of the sun.

3. *Entomology.* Suppose the equation $r = 2 - 2\sin(\theta - 50°)$ is used to describe the dance shown by the bee in Figure 7.26. Graph this equation. Compare this food source to the one described in Problem 1.

4. *Entomology.* In the bee's dance, a straight run of three seconds indicates a distance of about $\frac{1}{2}$ mile, and a straight run of five seconds means about 2 miles. If the pollen is $20°$ to the right of the sun, we could use the polar coordinates (ρ, θ) to indicate the location of the pollen as $(3, 20°)$ or $(5, 20°)$. What do you think would be the coordinates for a food source about 1 mile from the hive in the same direction? What do you suppose $(6, 20°)$ means in terms of the time of the straight portion of the dance and the distance from the hive?

CUMULATIVE REVIEW FOR CHAPTERS 5–7

SAMPLE TEST I: OBJECTIVE FORM

1. State whether you would use the Law of Cosines, Law of Sines, or a right-triangle solution for each part. If the solution involves the ambiguous case, so state. You do *not* need to solve these triangles.

	a	b	c	α	β	γ
a.	38	—	42	—	108°	—
b.	14.6	28.1	19.4	—	—	—
c.	—	52	49	—	—	63°
d.	83	52	—	121°	—	—
e.	418	—	385	—	218.0°	—
f.	281	318	412	—	—	—
g.	3.0	6.0	—	30°	—	—
h.	18	15	—	—	106°	—
i.	55	61	—	57°	—	—
j.	—	14.1	18.1	—	—	27.4°

2. Identify each of the given curves.
   a. $r = 3 \sin 2\theta$
   b. $r^2 = 3 \sin 2\theta$
   c. $r = 4 \cos 3\theta$
   d. $r = 3(1 + \sin \theta)$

   e. $r = 2 \sin \frac{\pi}{2}$
   f. $r - 5 = 5 \cos \theta$
   g. $r^2 = 5 \cos 2\theta$
   h. $\theta = 3 \sin \frac{\pi}{6}$
   i. $r = 6(1 - \sin \theta)$
   j. $r = 2 + 2 \cos \theta$

3. Solve the triangle given in Problem 1d.

4. Solve the triangle given in Problem 1i.

5. A draftsman drew to scale (10 ft = 10 in.) a triangularly shaped lot with sides 45 ft, 82 ft, and 60 ft. On the plot plan, what are the measures of the angles?

6. For each given polar-form point, give both primary representations in polar form, and then change it to rectangular form.
   a. $(3, -\frac{\pi}{4})$ b. $(-5, -\frac{\pi}{6})$
   c. $(-3, 9.4248)$

7. Write the cube roots (in trigonometric form) of $-\frac{5}{2}\sqrt{3} + \frac{5}{2}i$.

8. Write the square roots (in trigonometric form) of $-8i$.

9. Graph $r = 3 \cos 3\theta$.

10. Graph $r^2 = 4 \sin 2\theta$.

SAMPLE TEST II: FILL IN AND MULTIPLE CHOICE

Fill in Problems 1–5 with the word or words to make the statements complete and correct.

1. The law of cosines for a correctly labeled triangle states $c^2 = $ _____ and $\cos \beta = $ _____.

2. The ambiguous case is checked when given SSA and θ _____ 90° where θ is the given angle. Two solutions will be obtained if _____ where $h = $ (adj) $\sin \theta$.

3. To change a complex number $r \operatorname{cis} \theta$ to rectangular form we use $a = $ _____ and $b = $ _____.

4. If $z_1 = r_1 \operatorname{cis} \theta_1$ and $z_2 = r_2 \operatorname{cis} \theta_2$, then $z_1 z_2 = $ _____.

5. The polar graph $r = f(\theta - \alpha)$ is the same as the polar graph of _____ that has been _____.

Choose the BEST answer from the choices given.

6. If you are given $a = 38$, $c = 42$, and $\beta = 108°$, the correct method of solution is
   A. law of sines—one solution
   B. law of sines—ambiguous case
   C. law of cosines
   D. right-triangle solution
   E. no solution

7. If you are given $a = 55$, $b = 61$, and $\alpha = 57°$, the correct method of solution is
   A. law of sines—one solution
   B. law of sines—ambiguous case
   C. law of cosines
   D. right-triangle solution
   E. no solution

8. If you are given $b = 26$, $\beta = 47°$, and $\gamma = 76°$, the correct method of solution is
   A. law of sines—one solution
   B. law of sines—ambiguous case
   C. law of cosines
   D. right-triangle solution
   E. no solution

9. Which of the following pairs of vectors are orthogonal?
   A. $v = \cos 20°i + \sin 20°j$,
      $w = \cos 70°i + \sin 70°j$
   B. $v = \frac{1}{\sqrt{3}}i - \frac{2}{\sqrt{5}}j$,   C. $v = i$, $w = j$
      $w = 2i - \frac{\sqrt{5}}{5\sqrt{3}}j$
   D. all are orthogonal
   E. none is orthogonal

10. Robin, in the airport's control tower, observes Batman in the Batplane at a distance of 80 km on a bearing of 120° and the Joker in his Learjet 90 km from the airport on a bearing of 140°. How far apart are Batman and the Joker?
    A. 31 km   B. 93 km
    C. 147 km   D. 968 km
    E. none of these

11. $(3 + 5i)^2$ simplified is
    A. -16 B. $-16 + 15i$
    C. $-16 + 30i$ D. $9 + 30i + 25i^2$
    E. none of these

12. $\dfrac{(5 - 5i)(\sqrt{3} - i)}{1 - i}$ simplified is
    A. $5\sqrt{3} - 5i$ B. $10(2 - i)$
    C. $20 \operatorname{cis} 330°$ D. $5(1 - i)^2(\sqrt{3} - i)$
    E. none of these

13. $\dfrac{2 \operatorname{cis} 128° \cdot 9 \operatorname{cis} 285°}{(3 \operatorname{cis} 4°)^3}$ is
    A. $6 \operatorname{cis} 409°$ B. $6 \operatorname{cis} 49°$
    C. $6 \operatorname{cis} 401°$ D. $6 \operatorname{cis} 41°$
    E. none of these

14. One of the cube roots of $\frac{3}{2} - \frac{3}{2}\sqrt{3}\,i$ is
    A. $27 \operatorname{cis} 180°$ B. $\sqrt[3]{3} \operatorname{cis} 220°$
    C. $\sqrt[3]{3} \operatorname{cis} 20°$ D. $1.5 - 2.5981i$
    E. none of these

15. The other primary representation of $(3, \frac{\pi}{10})$ is
    A. $(-3, \frac{\pi}{10})$ B. $(-3, -\frac{\pi}{10})$
    C. $(-3, \frac{9\pi}{10})$ D. $(-3, \frac{11\pi}{10})$
    E. none of these

16. The curve $\theta =$ constant is a
    A. cardioid   B. rose curve
    C. lemniscate   D. circle
    E. line

17. The graph of $r = 3 \sin 2\theta$ is a
    A. cardioid   B. rose curve
    C. lemniscate   D. circle
    E. line

18. What is the rotation of the curve $r^2 = \sin 2\theta$ when compared to the standard-position curve $r^2 = \cos 2\theta$?
    A. $90°$ B. $60°$
    C. $45°$ D. $30°$
    E. none of these

19. The graph of $r = 3 \sin 60°$ is a
    A. cardioid   B. rose curve
    C. lemniscate   D. circle
    E. line

20. A point satisfying the equation $r = \dfrac{3}{1 + \sin \theta}$ is
    A. $(\frac{\pi}{3}, 12 - 6\sqrt{3})$ B. $(\frac{1}{2}, \frac{\pi}{3})$
    C. $(-6, \frac{\pi}{6})$ D. $(12 - 6\sqrt{3}, \frac{\pi}{6})$
    E. none of these

CONIC SECTIONS*

CHAPTER OVERVIEW

Although many calculus books include coverage of analytic geometry, it is becoming more and more necessary for students to have an earlier knowledge of the conic sections. For that reason, a thorough introduction to this topic is presented in this chapter. By a *conic section*, we mean a curve that can be formed by intersecting a cone and a plane; these curves are called *parabolas, ellipses,* and *hyperbolas.* After introducing each of these curves and their equations, we will then translate each, and finally we will use trigonometry to look at their rotated forms. In the last section of this chapter we look at the conic sections in polar form.

8.1 PARABOLAS

Conic Sections

In elementary algebra you learned about graphing the general **first-degree** (or **linear**) **equation**

$$Ax + By + C = 0$$

In this chapter, we consider graphing the general **second-degree** (or **quadratic**) **equation**

$$Ax^2 + Bxy + Cy^2 + Dx + Ey + F = 0$$

Historically, second-degree equations in two variables were first considered in a geometric context and were called **conic sections** because the curves they represent can be described as the intersections of a double-napped right circular cone and a plane. There are three general ways a plane can intersect a cone, as shown in Figure 8.1 on page 306. (Several special cases are discussed later.)

Definition and Terminology

You have probably considered parabolas in your previous algebra courses, but here we will give a geometric definition.

PARABOLA

> A **parabola** is a set of all points in the plane equidistant from a given point (called the **focus**) and a given line (called the **directrix**).

---

* The material in this chapter is optional.

FIGURE 8.1   Conic sections

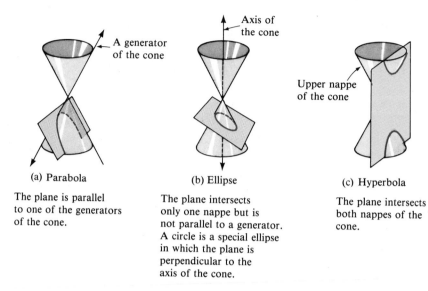

(a) Parabola

The plane is parallel
to one of the generators
of the cone.

(b) Ellipse

The plane intersects
only one nappe but is
not parallel to a generator.
A circle is a special ellipse
in which the plane is
perpendicular to the
axis of the cone.

(c) Hyperbola

The plane intersects
both nappes of the
cone.

To obtain the graph of a parabola from this definition, you can use the special type
of graph paper shown in Figure 8.2a, where F is the focus and L is the directrix.

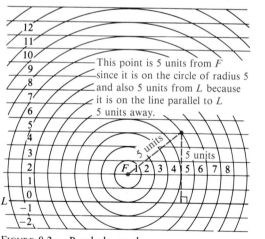

FIGURE 8.2a   Parabola graph paper

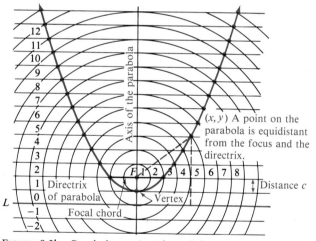

FIGURE 8.2b   Parabola graphed from definition

To sketch a parabola using the definition, let F be any point and let L be any line, as
shown in Figure 8.2a. Plot points in the plane equidistant from the focus and the
directrix. Draw a line through the focus and perpendicular to the directrix. This line is
called the **axis** of the parabola. Let V be the point on this line halfway between the
focus and the directrix. This is the point of the parabola nearest to both the focus and
the directrix. It is called the **vertex** of the parabola. Plot other points equidistant from
F and L, as shown in Figure 8.2b.

In Figure 8.2b, let c be the distance from the vertex to the focus. Notice that the distance from the vertex to the directrix is also c. Consider the segment that passes through the focus perpendicular to the axis and with endpoints on the parabola. This segment has length $4c$ and is called the **focal chord**.

Standard-Form Parabolas

To obtain the equation of a parabola, first consider a special case—a parabola with focus $F(0, c)$ and directrix $y = -c$, where c is any positive number. This parabola must have its vertex at the origin (remember that the vertex is halfway between the focus and the directrix) and must open upward, as shown in Figure 8.3.

FIGURE 8.3   Graph of the parabola $x^2 = 4cy$

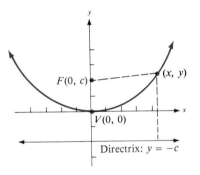

Let (x, y) be any point on the parabola. Then, from the definition of a parabola,

Distance from (x, y) to $(0, c)$ = distance from (x, y) to directrix

or $\sqrt{(x - 0)^2 + (y - c)^2} = y + c$

Square both sides:

$$x^2 + (y - c)^2 = (y + c)^2$$
$$x^2 + y^2 - 2cy + c^2 = y^2 + 2cy + c^2$$
$$x^2 = 4cy$$

This is the equation of the parabola with vertex $(0, 0)$ and directrix $y = -c$.

You can repeat this argument for parabolas that have vertex at the origin and open downward, to the left, and to the right to obtain the results summarized below. A positive number c, the distance from the focus to the vertex, is assumed given. These are called the **standard-form parabolic equations** with vertex $(0, 0)$.

STANDARD-FORM
EQUATIONS FOR
PARABOLAS WITH
VERTEX $(0, 0)$

PARABOLA	FOCUS	DIRECTRIX	VERTEX	EQUATION
Opens *upward*	$(0, c)$	$y = -c$	$(0, 0)$	$x^2 = 4cy$
Opens *downward*	$(0, -c)$	$y = c$	$(0, 0)$	$x^2 = -4cy$
Opens *right*	$(c, 0)$	$x = -c$	$(0, 0)$	$y^2 = 4cx$
Opens *left*	$(-c, 0)$	$x = c$	$(0, 0)$	$y^2 = -4cx$

Given equation, find graph

Example 1

Graph $x^2 = 8y$.

Solution

This equation represents a parabola that opens upward. The vertex is $(0, 0)$, and notice by inspection that

$$4c = 8$$
$$c = 2$$

Thus the focus is $(0, 2)$. After plotting the vertex $V(0, 0)$ and the focus $F(0, 2)$, the only question is the width of the parabola. We determine the width of the parabola by using the focal chord. Remember that **the length of the focal chord is 4c,** so that in this case it is 8. Do you see that in each case $4c$ is the absolute value of the coefficient of the first-degree term in the standard-form equations? Since a parabola is symmetric with respect to its axis, draw a segment of length 8 with the midpoint at F. Using these three points (the vertex and the endpoints of the focal chord), sketch the parabola, as shown in Figure 8.4.

FIGURE 8.4   Graph of the parabola $x^2 = 8y$

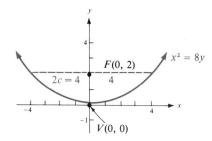

Example 2

Graph $y^2 = -12x$.

Solution

This equation represents a parabola that opens left. The vertex is $(0, 0)$ and

$$4c = 12$$
$$c = 3$$

(recall that c is positive), so the focus is $(-3, 0)$. The length of the focal chord is 12, and the parabola is drawn as in Figure 8.5.

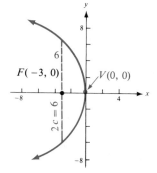

FIGURE 8.5   Graph of the parabola $y^2 = -12x$

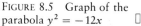

Example 3

Graph $2y^2 - 5x = 0$.

Solution

You must first put the equation into standard form by solving for the second-degree term:

$$y^2 = \frac{5}{2}x$$

The vertex is $(0, 0)$, and

$$4c = \frac{5}{2}$$

$$c = \frac{5}{8}$$

Thus the parabola opens to the right, the focus is $(\frac{5}{8}, 0)$, and the length of the focal chord is $\frac{5}{2}$, as shown in Figure 8.6.

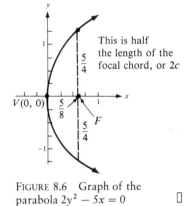

This is half the length of the focal chord, or $2c$

FIGURE 8.6   Graph of the parabola $2y^2 - 5x = 0$

Given graph, find equation

There are two basic types of problems you will need to solve concerning each curve in analytic geometry:

1. Given the equation, draw the graph; this is what you did in Examples 1–3.
2. Given the graph (or information about the graph), write the equation. Example 4 is a problem of this type.

Example 4

Find the equation of the parabola with directrix $y = 4$ and focus at $F(0, -4)$.

Solution

This curve is a parabola that opens downward with vertex at the origin, as shown in Figure 8.7. The value for c is found by inspection: $c = 4$. Thus $4c = 16$. Since the equation is of the form $x^2 = -4cy$, the desired equation is (by substitution)

$$x^2 = -16y$$

FIGURE 8.7   Graph of the parabola with focus at $(0, -4)$ and directrix $y = 4$

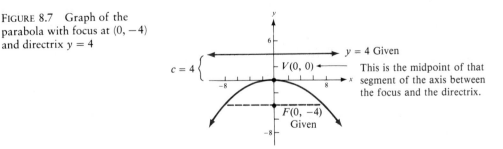

$c = 4$

$y = 4$ Given

$V(0, 0)$

This is the midpoint of that segment of the axis between the focus and the directrix.

$F(0, -4)$
Given

Translated Parabolas

The types of parabolas we have been considering are quite limited, since we have assumed that the vertex is at the origin and the directrix is parallel to one of the coordinate axes. Suppose, however, that you are given a parabola with vertex at (h, k) and a directrix parallel to one of the coordinate axes. In Section 2.2 we showed that you can translate the axes to (h, k) by a substitution:

$$x' = x - h$$
$$y' = y - k$$

Therefore the **standard-form parabolic equations** with vertex (h, k) can be summarized by the following box.

STANDARD-FORM
EQUATIONS FOR
TRANSLATED PARABOLAS

PARABOLA	EQUATION
Opens upward	$(x - h)^2 = 4c(y - k)$
Opens downward	$(x - h)^2 = -4c(y - k)$
Opens right	$(y - k)^2 = 4c(x - h)$
Opens left	$(y - k)^2 = -4c(x - h)$

Example 5

Sketch $(x + 2)^2 = -8(y - 4)$.

Solution

The procedure is identical to that used with the standard parabolas, except you count from the vertex point $(h, k) = (-2, 4)$ rather than from the origin. By inspection, $4c = 8$, $c = 2$, and the parabola opens downward from the vertex, as shown by Figure 8.8.

FIGURE 8.8   Graph of
$(x + 2)^2 = -8(y - 4)$

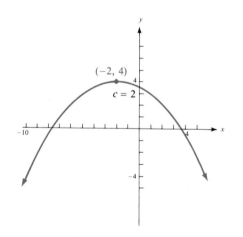

Example 6

Find the equation of the parabola with focus at $(4, -3)$ and directrix the line $x + 2 = 0$.

Solution

Sketch the given information, as shown in Figure 8.9. The vertex is $(1, -3)$, since it must be equidistant from F and the directrix. Note that $c = 3$. Thus, substitute into the equation

$$(y - k)^2 = 4c(x - h)$$

since the parabola opens to the right.
The desired equation is

$$(y + 3)^2 = 12(x - 1)$$

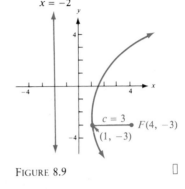

FIGURE 8.9

By sketching the focus and directrix, it is easy to find c and the vertex by inspection. The sketch of the curve is not necessary in order to find this equation.

Problem Set 8.1

A

Using the definition of a parabola, in Problems 1–6 sketch a parabola similar to the one in Figure 8.2a, such that the distance between the focus and the directrix is the given number.

1. 4   2. 6   3. 8   4. 10   5. 3   6. 7

Sketch the curves given by the equations in Problems 7–24. Label the focus F and the vertex V.

7. $y^2 = 8x$

8. $y^2 = -12x$

9. $y^2 = -20x$

10. $x^2 = 4y$

11. $x^2 = -8y$

12. $x^2 = -10y$

13. $4x^2 = 10y$

14. $3x^2 = -12y$

15. $3y^2 = 18x$

16. $2x^2 + 5y = 0$

17. $5y^2 + 15x = 0$

18. $3y^2 - 15x = 0$

19. $(y - 1)^2 = 2(x + 2)$

20. $(y + 3)^2 = 3(x - 1)$

21. $(x + 3)^2 = -3(y - 1)$

22. $(x + 2)^2 = 2(y - 1)$

23. $(x - 1)^2 = 3(y + 3)$

24. $(x - 1)^2 = -2(y + 2)$

B

Find the equation of each curve in Problems 25–28.

25. Directrix $x = -5$; focus at $(5, 0)$

26. Directrix $y = 4$; focus at $(0, -4)$

27. Vertex at $(-3, 2)$ and focus at $(-3, -1)$

28. Vertex at $(4, 2)$ and focus at $(-3, 2)$

29. *Engineering.* A radar antenna is constructed so that a cross section along its axis is a parabola with the receiver at the focus. Find the focus if the antenna is 12 m across and its depth is 4 m. See Figure 8.10 on page 312.

30. *Engineering.* If the diameter of a parabolic reflector in Problem 29 is 16 cm and the depth is 8 cm, find the focus. (See Figure 8.10.)

Courtesy of Wide World Photos, Inc.

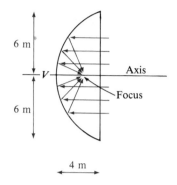

A three-dimensional model of a parabola is a **parabolic reflector** (or **parabolic mirror**). A radar antenna serves as an example. If a source of light is placed at the focus of the mirror, the light rays will reflect from the mirror as rays parallel to the axis (as in an automobile headlamp). The radar antenna works in reverse—parallel incoming rays are focused at a single location.

FIGURE 8.10   Parabolic Reflectors

Dimensions for radar antenna in Problem 29

C

31. Derive the equation of a parabola with $F(0, -c)$, where c is a positive number and the directrix is the line $y = c$.

32. Derive the equation of a parabola with $F(c, 0)$, where c is a positive number and the directrix is the line $x = -c$.

33. Derive the equation of a parabola with $F(-c, 0)$, where c is a positive number and the directrix is the line $x = c$.

34. Show that the length of the focal chord for the parabola $y^2 = 4cx$ is $4c$ $(c > 0)$.

8.2   ELLIPSES

The second conic section considered in this chapter is called an *ellipse*.

Definition and Terminology

ELLIPSE

> An **ellipse** is the set of all points in the plane such that, for each point on the ellipse, the sum of its distances from two fixed points is a constant.

The fixed points are called the **foci** (plural for **focus**). To see what an ellipse looks like, we will use the special type of graph paper shown in Figure 8.11a, where F_1 and F_2 are the foci.

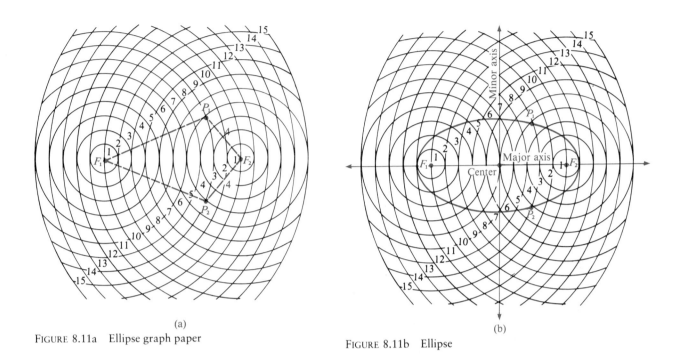

(a)

FIGURE 8.11a   Ellipse graph paper

(b)

FIGURE 8.11b   Ellipse

Let the given constant be 12. Plot all the points in the plane so that the sum of their distances from the foci is 12. If a point is 8 units from F_1, for example, then it is 4 units from F_2 and you can plot the points P_1 and P_2. The completed graph of this ellipse is shown in Figure 8.11b.

The line passing through F_1 and F_2 is called the **major axis**. The **center** is the midpoint of the segment $\overline{F_1F_2}$. The line passing through the center perpendicular to the major axis is called the **minor axis**. The ellipse is symmetric with respect to both the major and minor axes.

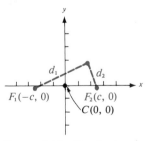

FIGURE 8.12   Developing the equation of an ellipse by using the definition

Standard-Form Ellipses

To find the equation of an ellipse, first consider a special case where the center is at the origin. Let the distance from the center to a focus be the positive number c; that is, let $F_1(-c, 0)$ and $F_2(c, 0)$ be the foci and let the constant distance be $2a$, as shown in Figure 8.12.

Notice that the center of the ellipse in Figure 8.12 is $(0, 0)$. Let $P(x, y)$ be any point on the ellipse, and use the distance formula and the definition of an ellipse to derive the equation of this ellipse:

$$d_1 + d_2 = 2a$$

or

$$\sqrt{(x+c)^2 + (y-0)^2} + \sqrt{(x-c)^2 + (y-0)^2} = 2a$$

Simplifying,

$$\sqrt{(x+c)^2 + y^2} = 2a - \sqrt{(x-c)^2 + y^2} \qquad \textit{Isolate one radical.}$$

$$(x+c)^2 + y^2 = 4a^2 - 4a\sqrt{(x-c)^2 + y^2} \qquad \textit{Square both sides.}$$
$$+ (x-c)^2 + y^2$$

$$x^2 + 2cx + c^2 + y^2 = 4a^2 - 4a\sqrt{(x-c)^2 + y^2}$$
$$+ x^2 - 2cx + c^2 + y^2$$

$$4a\sqrt{(x-c)^2 + y^2} = 4a^2 - 4cx$$

$$\sqrt{(x-c)^2 + y^2} = a - \frac{c}{a}x \qquad \textit{Since } a \neq 0, \textit{ divide by } 4a.$$

$$(x-c)^2 + y^2 = \left(a - \frac{c}{a}x\right)^2 \qquad \textit{Square both sides again.}$$

$$x^2 - 2cx + c^2 + y^2 = a^2 - 2cx + \frac{c^2}{a^2}x^2$$

$$x^2 + y^2 = a^2 - c^2 + \frac{c^2}{a^2}x^2$$

$$x^2 - \frac{c^2}{a^2}x^2 + y^2 = a^2 - c^2$$

$$\left(1 - \frac{c^2}{a^2}\right)x^2 + y^2 = a^2 - c^2$$

$$\frac{a^2 - c^2}{a^2}x^2 + y^2 = a^2 - c^2 \qquad \textit{Divide both sides by } a^2 - c^2.$$

$$\frac{x^2}{a^2} + \frac{y^2}{a^2 - c^2} = 1$$

Let $b^2 = a^2 - c^2$; then $\dfrac{x^2}{a^2} + \dfrac{y^2}{b^2} = 1$. If $x = 0$, the y-intercepts are obtained:

$$\frac{y^2}{b^2} = 1$$

$$y = \pm b$$

If $y = 0$, the x-intercepts are obtained: $x = \pm a$. The intercepts on the major axis are called the **vertices** of the ellipse, and those intercepts on the minor axis are called the **covertices**.

The equation of the ellipse with major axis vertical, $F_1(0, c)$, $F_2(0, -c)$, and constant distance $2a$ is found in a similar fashion. Simplifying the equation as before,

$$\frac{y^2}{a^2} + \frac{x^2}{b^2} = 1$$

where $b^2 = a^2 - c^2$.

Notice that in both cases a^2 must be larger than both c^2 and b^2. If it were not, a square number would be equal to a negative number, which is a contradiction in the set of real numbers.

STANDARD-FORM
EQUATIONS FOR ELLIPSES
WITH CENTER $(0, 0)$

ELLIPSE	FOCI	CONSTANT DISTANCE	CENTER	EQUATION
Horizontal	$(-c, 0)$, $(c, 0)$	$2a$	$(0, 0)$	$\dfrac{x^2}{a^2} + \dfrac{y^2}{b^2} = 1$
Vertical	$(0, c)$, $(0, -c)$	$2a$	$(0, 0)$	$\dfrac{y^2}{a^2} + \dfrac{x^2}{b^2} = 1$

where $b^2 = a^2 - c^2$ or $c^2 = a^2 - b^2$

Given equation, find graph

Example 1

Sketch $\dfrac{x^2}{9} + \dfrac{y^2}{4} = 1$.

Solution

The center of the ellipse is $(0, 0)$. The x-intercepts are ± 3 (these are the vertices) and the y-intercepts are ± 2. Sketch the ellipse as shown in Figure 8.13. The foci can also be found, since

$$c^2 = a^2 - b^2$$
$$c^2 = 9 - 4$$
$$c = \sqrt{5}$$

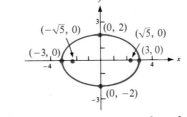

FIGURE 8.13   Graph of $\dfrac{x^2}{9} + \dfrac{y^2}{4} = 1$

Example 2

Sketch $\dfrac{x^2}{4} + \dfrac{y^2}{9} = 1$.

Solution

Here $a^2 = 9$ and $b^2 = 4$, which is an ellipse with major axis vertical. The x-intercepts are ± 2 and the y-intercepts are ± 3 (these are the vertices). The sketch is shown in Figure 8.14.

The foci are found by

$$c^2 = a^2 - b^2$$
$$c^2 = 9 - 4$$
$$c = \pm\sqrt{5}$$

Remember: The foci are always on the major axis, so plot $(0, \sqrt{5})$ and $(0, -\sqrt{5})$ for the foci.

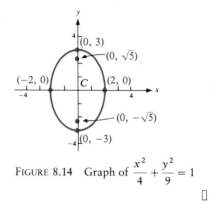

FIGURE 8.14   Graph of $\dfrac{x^2}{4} + \dfrac{y^2}{9} = 1$

Translated Ellipses

If the center of the ellipse is at (h, k), the equations can be written in terms of a translation, as shown by the following box.

STANDARD-FORM
EQUATIONS FOR
TRANSLATED ELLIPSES

ELLIPSE	EQUATION
Horizontal	$\dfrac{(x - h)^2}{a^2} + \dfrac{(y - k)^2}{b^2} = 1$
Vertical	$\dfrac{(y - k)^2}{a^2} + \dfrac{(x - h)^2}{b^2} = 1$

The segment from the center to a vertex on the major axis is called a **semimajor axis** and has length a; the segment from the center to an intercept on the minor axis is called a **semiminor axis** and has length b.

Example 3

Graph $\dfrac{(x - 3)^2}{25} + \dfrac{(y - 1)^2}{16} = 1$.

Solution

Step 1: Plot the center (h, k). By inspection, the center of this ellipse is $(3, 1)$. This becomes the center of a new translated coordinate system. The vertices and foci are now measured with reference to the new origin at $(3, 1)$.

Step 2: Plot the x' and y' intercepts. These are ± 5 and ± 4, respectively. Remember to measure these distances from $(3, 1)$, as shown in Figure 8.15.

The foci are found by

$$\begin{aligned}
c^2 &= a^2 - b^2 \\
&= 25 - 16 \\
&= 9
\end{aligned}$$

The distance from the center to either focus is 3, so the coordinates of the foci are $(6, 1)$ and $(0, 1)$.

FIGURE 8.15 Graph of $\dfrac{(x-3)^2}{25} + \dfrac{(y-1)^2}{16} = 1$

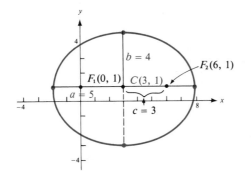

Circles

We have seen that some ellipses are more circular and some more flat than others. A measure of the amount of flatness of an ellipse is called its **eccentricity**, which is defined as*

$$\varepsilon = \frac{c}{a}$$

Notice that

$$\varepsilon = \frac{c}{a} = \frac{\sqrt{a^2 - b^2}}{a} = \sqrt{\frac{a^2 - b^2}{a^2}} = \sqrt{1 - \left(\frac{b}{a}\right)^2}$$

Since $c < a$, ε is between 0 and 1. If $a = b$, then $\varepsilon = 0$ and the conic section is a **circle**. Recall that a circle is the set of all points a given distance from a given point. If the ratio b/a is small, then the ellipse is very flat. Thus, for an ellipse,

$$0 \le \varepsilon < 1$$

and ε measures the amount of roundness of the ellipse.

Consider a circle; that is, suppose $a = b$. In this case, let $r = a = b$ and call this distance the **radius**. You can see that a circle is a special case of an ellipse:

$$\frac{(x-h)^2}{r^2} + \frac{(y-k)^2}{r^2} = 1$$

STANDARD-FORM EQUATION OF A CIRCLE

> The equation of a circle with center at (h, k) and radius r is
>
> $$(x-h)^2 + (y-k)^2 = r^2$$

---

* Many books use e for eccentricity, but we reserve e to represent an irrational number.

Example 4

Graph $(x - 1)^2 + (y - 1)^2 = 25$.

Solution

This is a circle with center at $(1, 1)$ and radius 5, as shown in Figure 8.16.

FIGURE 8.16

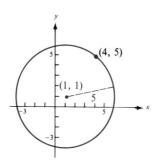

□

Given graph, find equation

Example 5

Find the equation of the ellipse with vertices at $(3, 2)$ and $(3, -4)$ and foci at $(3, \sqrt{5} - 1)$ and $(3, -\sqrt{5} - 1)$.

Solution

By inspection, the ellipse is vertical, and it is centered at $(3, -1)$, where $a = 3$ and $c = \sqrt{5}$ (see Figure 8.17). Thus,

$$c^2 = a^2 - b^2$$
$$5 = 9 - b^2$$
$$b^2 = 4$$

The equation is

$$\frac{(y + 1)^2}{9} + \frac{(x - 3)^2}{4} = 1$$

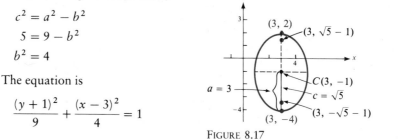

FIGURE 8.17   □

Example 6

Find the equation of the set of all points with the sum of distances from $(-3, 2)$ and $(5, 2)$ equal to 16.

Solution

By inspection, the ellipse is horizontal and is centered at $(1, 2)$. You are given

$$2a = 16 \quad \textit{This is the sum of the distances.}$$
$$a = 8$$

and $c = 4$—the distance from the center $(1, 2)$ to a focus $(5, 2)$. Thus,

$$c^2 = a^2 - b^2$$
$$16 = 64 - b^2$$
$$b^2 = 48$$

The equation is

$$\frac{(x - 1)^2}{64} + \frac{(y - 2)^2}{48} = 1$$ ▢

Example 7   Find the equation of the ellipse with foci at $(-3, 6)$ and $(-3, 2)$ with $\varepsilon = \frac{1}{5}$.

Solution   By inspection, the ellipse is vertical and is centered at $(-3, 4)$ with $c = 2$. Since

$$\varepsilon = \frac{c}{a} = \frac{1}{5}$$

and $c = 2$, then

$$\frac{2}{a} = \frac{1}{5}$$

which implies $a = 10$. Just because

$$\frac{c}{a} = \frac{1}{5}$$

you cannot assume that $c = 1$ and $a = 5$; all you know is that the reduced *ratio* of c to a is $\frac{1}{5}$. Also,

$$c^2 = a^2 - b^2$$
$$4 = 100 - b^2$$
$$b^2 = 96$$

Thus, the equation is

$$\frac{(y - 4)^2}{100} + \frac{(x + 3)^2}{96} = 1$$ ▢

Problem Set 8.2

A   *Sketch the curves in Problems 1–15.*

1. $\dfrac{x^2}{4} + \dfrac{y^2}{9} = 1$

2. $\dfrac{x^2}{25} + \dfrac{y^2}{36} = 1$

3. $x^2 + \dfrac{y^2}{9} = 1$

4. $4x^2 + 9y^2 = 36$

5. $25x^2 + 16y^2 = 400$

6. $36x^2 + 25y^2 = 900$

7. $3x^2 + 2y^2 = 6$

8. $4x^2 + 3y^2 = 12$

9. $5x^2 + 10y^2 = 7$

10. $(x - 2)^2 + (y + 3)^2 = 25$

11. $(x + 4)^2 + (y - 2)^2 = 49$

12. $(x - 1)^2 + (y - 1)^2 = \frac{1}{4}$

13. $\dfrac{(x + 3)^2}{81} + \dfrac{(y - 1)^2}{49} = 1$

14. $\dfrac{(x - 3)^2}{16} + \dfrac{(y - 2)^2}{9} = 1$

15. $\dfrac{(x + 2)^2}{25} + \dfrac{(y + 4)^2}{9} = 1$

Use the definitions of an ellipse and a circle to sketch the curves given in Problems 16–22.

16. The set of points 6 units from the point $(4, 5)$

17. The set of points 3 units from the point $(-2, 3)$

18. The set of points 6 units from $(-1, -4)$

19. The set of points such that the sum of the distances from $(-6, 0)$ and $(6, 0)$ is 20

20. The set of points such that the sum of the distances from $(0, 4)$ and $(0, -4)$ is 10

21. The ellipse with vertices at $(0, 7)$ and $(0, -7)$ and foci at $(0, 5)$ and $(0, -5)$

22. The ellipse with vertices at $(4, 3)$ and $(4, -5)$ and foci at $(4, 2)$ and $(4, -4)$

B   *Find the equations of the curves in Problems 23–30.*

23. The set of points 3 units from the point $(-2, 3)$

24. The set of points 5 units from $(-1, -4)$

25. The set of points such that the sum of the distances from $(-6, 0)$ and $(6, 0)$ is 20

26. The set of points such that the sum of the distances from $(0, 4)$ and $(0, -4)$ is 10

27. The ellipse with vertices at $(0, 7)$ and $(0, -7)$ and foci at $(0, 5)$ and $(0, -5)$

28. The ellipse with vertices at $(4, 3)$ and $(4, -5)$ and foci at $(4, 2)$ and $(4, -4)$

29. The ellipse with foci $(2, 0)$ and $(-2, 0)$ with $\varepsilon = \frac{1}{5}$

30. The ellipse with foci $(0, 3)$ and $(0, -3)$ with $\varepsilon = \frac{4}{5}$

*The following information is needed for Problems 31–33.

The orbit of a plane (see Figure 8.18) can be described by

$$\frac{x^2}{a^2} + \frac{y^2}{b^2} = 1$$

with the sun at one focus. The orbit is commonly identified by its major axis, $2a$, and its eccentricity, ε:

FIGURE 8.18   Planetary orbits: The orbits of planets or satellites serve as an example of ellipses. See the Application for Further Study at the end of this chapter for a discussion of this application.

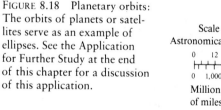

Scale
Astronomical units

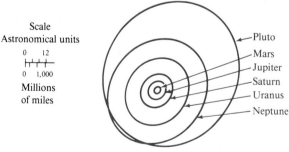

* You might also see The Application for Further Study on page 351.

$$\varepsilon = \frac{c}{a} = \sqrt{1 - \frac{b^2}{a^2}} \qquad a \geq b > 0$$

where c is the distance between the center and a focus.

31. *Astronomy.* The major axis of the earth's orbit is approximately 186,000,000 miles and its eccentricity is $\frac{1}{62}$. Find the greatest distance, called the *aphelion*, and the smallest distance, called the *perihelion*, of the sun from the earth.

32. *Astronomy.* The *perihelion* (the greatest distance) of the planet Mercury with the sun is approximately 28 million miles, and the eccentricity of the orbit is $\frac{1}{5}$. Find the major axis of Mercury's orbit. (See Problem 31).

33. *Astronomy.* The moon's orbit is elliptical with the earth at one focus. If the major axis of the orbit is approximately 378,000 miles and the *apogee* (the greatest distance) is approximately 199,000 miles, then what is the eccentricity of the moon's orbit? (See Problem 31).

C   34. Derive the equation of the ellipse with foci at $(0, c)$ and $(0, -c)$ and constant distance $2a$. Let $b^2 = a^2 - c^2$. Show all your work.

35. If we are given an ellipse with foci at $(-c, 0)$ and $(c, 0)$ and vertices at $(-a, 0)$ and $(a, 0)$, we define the *directrices* of the ellipse as the lines $x = a/\varepsilon$ and $x = -a/\varepsilon$. Show that an ellipse is the set of all points with distances from $F(c, 0)$ equal to ε times their distances from the line $x = a/\varepsilon$ $(a > 0, c > 0)$.

36. A line segment through a focus parallel to a directrix (see Problem 35) and cut off by the ellipse is called the *focal chord.* Show that the length of the focal chord of the following ellipse is $2b^2/a$:

$$\frac{x^2}{a^2} + \frac{y^2}{b^2} = 1$$

8.3   HYPERBOLAS

The last of the conic sections to be considered has a definition similar to that of the ellipse.

Definition and Terminology

HYPERBOLA

A **hyperbola** is the set of all points in the plane such that, for each point on the hyperbola, the difference of its distances from two fixed points is a constant.

The fixed points are called the **foci.** A hyperbola with foci at F_1 and F_2, where the given constant is 8, is shown in Figure 8.19.

FIGURE 8.19   Graph of a
hyperbola from the definition

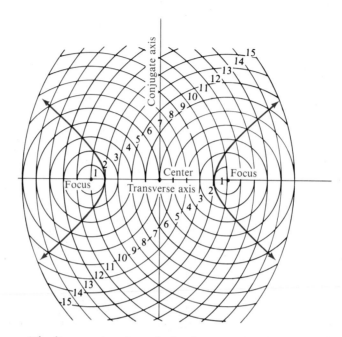

The line passing through the foci is called the **transverse axis**. The **center** is the midpoint of the segment connecting the foci. The line passing through the center perpendicular to the transverse axis is called the **conjugate axis**. The hyperbola is symmetric with respect to both the transverse and the conjugate axes.

Standard-Form Hyperbolas

If you use the definition, you can derive the equation for a hyperbola with foci at $(-c, 0)$ and $(c, 0)$ with constant distance $2a$. If (x, y) is any point on the curve, then

$$\left|\sqrt{(x + c)^2 + (y - 0)^2} - \sqrt{(x - c)^2 + (y - 0)^2}\right| = 2a$$

The procedure for simplifying this expression is the same as that shown for the ellipse, so the details are left as a problem (see Problems 29 and 30). After several steps, you should obtain

$$\frac{x^2}{a^2} - \frac{y^2}{c^2 - a^2} = 1$$

If $b^2 = c^2 - a^2$, then

$$\frac{x^2}{a^2} - \frac{y^2}{b^2} = 1$$

which is the standard-form equation. Notice that $c^2 = a^2 - b^2$ for the ellipse and that $c^2 = a^2 + b^2$ for the hyperbola. For the ellipse it is necessary that $a^2 > b^2$, but for the hyperbola there is no restriction on the relative sizes for a and b.

Repeat the argument for a hyperbola with foci $(0, c)$ and $(0, -c)$, and you will obtain the other standard-form equation for a hyperbola with a vertical transverse axis.

STANDARD-FORM
EQUATIONS FOR THE
HYPERBOLA WITH
CENTER $(0, 0)$

HYPERBOLA	FOCI	CONSTANT DISTANCE	CENTER	EQUATION
Horizontal	$(-c, 0), (c, 0)$	$2a$	$(0, 0)$	$\dfrac{x^2}{a^2} - \dfrac{y^2}{b^2} = 1$
Vertical	$(0, c), (0, -c)$	$2a$	$(0, 0)$	$\dfrac{y^2}{a^2} - \dfrac{x^2}{b^2} = 1$
where $b^2 = c^2 - a^2$ or $c^2 = a^2 + b^2$				

As with the other conic sections, we will sketch a hyperbola by determining some information about the curve directly from the equation by inspection. The points of intersection of the hyperbola with the transverse axis are called the **vertices**. For

$$\frac{x^2}{a^2} - \frac{y^2}{b^2} = 1 \quad \text{and} \quad \frac{y^2}{a^2} - \frac{x^2}{b^2} = 1$$

notice that the vertices occur at $(a, 0)$, $(-a, 0)$ and $(0, a)$, $(0, -a)$, respectively. The number $2a$ is the **length of the transverse axis**. The hyperbola does not intersect the conjugate axis, but if you plot the points $(0, b)$, $(0, -b)$ and $(-b, 0)$, $(b, 0)$, respectively, you determine a segment on the conjugate axis called the **length of the conjugate axis**.

Given Equation, Find Graph

Example 1

Sketch $\dfrac{x^2}{4} - \dfrac{y^2}{9} = 1$.

Solution

The center of the hyperbola is $(0, 0)$, $a = 2$, and $b = 3$. Plot the vertices at ± 2, as shown in Figure 8.20. The transverse axis is along the x-axis and the conjugate axis is

FIGURE 8.20

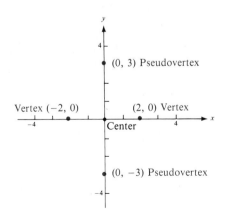

along the y-axis. Plot the length of the conjugate axis at ± 3. We call the points the **pseudovertices**, since the curve does not actually pass through these points.

Next draw lines through the vertices and pseudovertices parallel to the axes of the hyperbola. These lines form what we will call the **central rectangle**. The diagonal lines passing through the corners of the central rectangle are **slant asymptotes** for the hyperbola, as shown in Figure 8.21; they aid in sketching the hyperbola.

FIGURE 8.21   Graph of $\dfrac{x^2}{4} - \dfrac{y^2}{9} = 1$

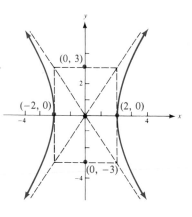

Translated Hyperbolas

If the center of the hyperbola is (h, k), the following equations for a hyperbola are obtained:

STANDARD-FORM
EQUATIONS FOR
TRANSLATED
HYPERBOLAS

HYPERBOLA	EQUATIONS
Horizontal	$\dfrac{(x-h)^2}{a^2} - \dfrac{(y-k)^2}{b^2} = 1$
Vertical	$\dfrac{(y-k)^2}{a^2} - \dfrac{(x-h)^2}{b^2} = 1$

Example 2

Sketch $16(x-4)^2 - 9(y+1)^2 = 144$.

Solution

First, put the equation into standard form:

$$\frac{(x-4)^2}{9} - \frac{(y+1)^2}{16} = 1 \quad \text{\textit{Divide both sides by}} \ 144.$$

The graph is shown in Figure 8.22. You can also find the foci, since

$$c^2 = a^2 + b^2$$

Thus,

$$c^2 = 9 + 16 \qquad c = \pm 5$$

FIGURE 8.22 Sketch of
$16(x - 4)^2 - 9(y + 1)^2 = 144$

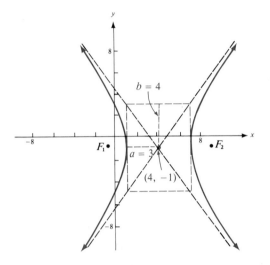

Given Graph, Find Equation

Example 3

Find the equation of the hyperbola with vertices at $(2, 4)$ and $(2, -2)$ and foci at $(2, 6)$ and $(2, -4)$.

Solution

Plot the given points as shown in Figure 8.23. Notice that the center of the hyperbola is $(2, 1)$ since it is the midpoint of the segment connecting the foci. Also, $c = 5$ and $a = 3$. Since

$$c^2 = a^2 + b^2$$

you have

$$25 = 9 + b^2$$
$$b^2 = 16$$

and the equation is

$$\frac{(y - 1)^2}{9} - \frac{(x - 2)^2}{16} = 1$$

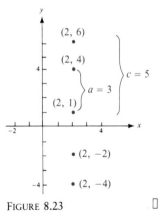

FIGURE 8.23

The eccentricity of the hyperbola and parabola is defined by the same equation that was used for the ellipse, namely

$$\varepsilon = \frac{c}{a}$$

Remember that for the ellipse, $0 \le \varepsilon < 1$; however, for the hyperbola $c > a$ so $\varepsilon > 1$, and for the parabola $c = a$ so $\varepsilon = 1$.

Example 4

Find the equation of the hyperbola with foci at $(-3, 2)$ and $(5, 2)$ and with eccentricity $\frac{3}{2}$.

Solution

This is a horizontal hyperbola with center at $(1, 2)$ and $c = 4$. Also, since

$$\varepsilon = \frac{c}{a} = \frac{3}{2}$$

you have

$$\frac{4}{a} = \frac{3}{2}$$
$$a = \frac{8}{3}$$

And since $c^2 = a^2 + b^2$,

$$16 = \frac{64}{9} + b^2$$
$$b^2 = \frac{80}{9}$$

Thus the equation is

$$\frac{(x-1)^2}{\frac{64}{9}} - \frac{(y-2)^2}{\frac{80}{9}} = 1 \quad \text{or} \quad \frac{9(x-1)^2}{64} - \frac{9(y-2)^2}{80} = 1 \qquad \square$$

Example 5

Find the set of points such that the difference of their distances from $(6, 2)$ and $(6, -5)$ is always 3.

Solution

This is a vertical hyperbola with center $(6, -\frac{3}{2})$ and $c = \frac{7}{2}$. Also $2a = 3$, so $a = \frac{3}{2}$. Since

$$c^2 = a^2 + b^2$$

you have

$$\frac{49}{4} = \frac{9}{4} + b^2$$
$$b^2 = 10$$

The equation is

$$\frac{(y + \frac{3}{2})^2}{\frac{9}{4}} - \frac{(x-6)^2}{10} = 1 \quad \text{or} \quad \frac{4(y + \frac{3}{2})^2}{9} - \frac{(x-6)^2}{10} = 1 \qquad \square$$

Problem Set 8.3

A   *Sketch the curves in Problems 1–22.*

1. $x^2 - y^2 = 1$

2. $x^2 - y^2 = 4$

3. $y^2 - x^2 = 1$

4. $\dfrac{x^2}{4} - \dfrac{y^2}{9} = 1$

5. $\dfrac{x^2}{9} - \dfrac{y^2}{4} = 1$

6. $\dfrac{y^2}{9} - \dfrac{x^2}{4} = 1$

7. $\dfrac{x^2}{16} - \dfrac{y^2}{25} = 1$

8. $\dfrac{y^2}{16} - \dfrac{x^2}{25} = 1$

9. $\dfrac{x^2}{36} - \dfrac{y^2}{9} = 1$

10. $36y^2 - 25x^2 = 900$

11. $3x^2 - 4y^2 = 12$

12. $3y^2 = 4x^2 + 12$

13. $3x^2 - 4y^2 = 5$

14. $4y^2 - 4x^2 = 5$

15. $4y^2 - x^2 = 9$

16. $\dfrac{(x-2)^2}{4} - \dfrac{(y+3)^2}{16} = 1$

17. $\dfrac{(x+3)^2}{8} - \dfrac{(y-1)^2}{5} = 1$

18. $\dfrac{(y-1)^2}{6} - \dfrac{(x+2)^2}{8} = 1$

19. $\dfrac{(x-2)^2}{16} - \dfrac{(y+1)^2}{9} = 1$

20. $\dfrac{(y+2)^2}{25} - \dfrac{(x+1)^2}{16} = 1$

21. $5(x-2)^2 - 2(y+3)^2 = 10$

22. $4(x+4)^2 - 3(y+3)^2 = -12$

B   *Find the equations of the curves in Problems 23–28.*

23. The hyperbola with vertices at $(0,5)$ and $(0,-5)$ and foci at $(0,7)$ and $(0,-7)$

24. The set of points such that the difference of their distances from $(-6,0)$ and $(6,0)$ is 10

25. The hyperbola with foci at $(5,0)$ and $(-5,0)$ and eccentricity 5

26. The hyperbola with vertices at $(4,4)$ and $(4,8)$ and foci at $(4,3)$ and $(4,9)$

27. The set of points such that the difference of their distances from $(4,-3)$ and $(-4,-3)$ is 6

28. The hyperbola with vertices at $(-2,0)$ and $(6,0)$ passing through $(10,3)$

C   29. Derive the equation of the hyperbola with foci at $(-c,0)$ and $(c,0)$ and constant distance $2a$. Let $b^2 = c^2 - a^2$. Show all your work.

30. Derive the equation of the hyperbola with foci at $(0,c)$ and $(0,-c)$ and constant distance $2a$. Let $b^2 = c^2 - a^2$. Show all your work.

31. *Physics.* Consider a person A who fires a rifle at a distant gong B. Assuming that the ground is flat, where must you stand to hear the sound of the gun and the sound of the gong simultaneously? [*Hint:* To answer this question, let x be the distance that sound travels in the length of time it takes the bullet to travel from the gun to the gong. Show that the person who hears the sounds simultaneously must stand on a branch of a hyperbola (the one nearest the target) so that the difference of the distances from A to B is x.]

8.4   GENERAL CONIC SECTIONS

Degenerate Conics

We have considered graphs of standard-form conic sections with centers at $(0,0)$ or (h,k). We will now consider the general conics of the form

$$Ax^2 + Bxy + Cy^2 + Dx + Ey + F = 0$$

Geometrically they represent the intersection of a plane and a cone, usually resulting in a line, parabola, ellipse, or hyperbola. However, there are certain positions of the plane that result in what are called **degenerate conics**. To visualize some of these degenerate conics, consider Figure 8.24.

FIGURE 8.24   Conic sections

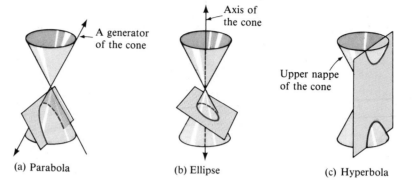

(a) Parabola   (b) Ellipse   (c) Hyperbola

For a **degenerate parabola**, visualize the plane (see Figure 8.24a) intersecting one of the generators of the cone; a line results. For a **degenerate ellipse**, visualize the plane intersecting at the vertex of the upper and lower nappes (see Figure 8.24b); a point results. And finally, for a **degenerate hyperbola**, visualize the plane situated so that the axis of the cone lies in the plane (see Figure 8.24c); a pair of intersecting lines results.

Example 1

Sketch $\dfrac{(x-2)^2}{4} + \dfrac{(y+3)^2}{9} = 0$.

Solution

There is only one point that satisfies this equation—namely $(2, -3)$. This is an example of a *degenerate ellipse*. Notice that except for the zero the equation has the "form of an ellipse." See Figure 8.25.

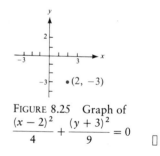

FIGURE 8.25   Graph of
$\dfrac{(x-2)^2}{4} + \dfrac{(y+3)^2}{9} = 0$ ☐

Example 2

Sketch $\dfrac{x^2}{4} - \dfrac{y^2}{9} = 0$.

Solution

This equation has the "form of a hyperbola," but because of the zero it cannot be put into standard form. You can, however, treat this as a factored form:

$$\left(\frac{x}{2} - \frac{y}{3}\right)\left(\frac{x}{2} + \frac{y}{3}\right) = 0$$

$$\frac{x}{2} - \frac{y}{3} = 0 \quad \text{or} \quad \frac{x}{2} + \frac{y}{3} = 0$$

The graph (Figure 8.26) is a *degenerate hyperbola*.

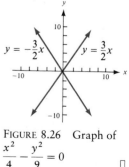

FIGURE 8.26   Graph of
$\dfrac{x^2}{4} - \dfrac{y^2}{9} = 0$ ☐

Recognition of Conics by Inspection

It is important to be able to recognize the curve by inspection of the equation before you begin. The first thing to notice is whether there is an xy term.

If $B = 0$ (that is, no xy term), then:

TYPE OF CURVE	DEGREE OF EQUATION	DEGREE IN x	DEGREE IN y	RELATIONSHIP TO GENERAL EQUATION $Ax^2 + Bxy + Cy^2 + Dx + Ey + F = 0$
1. Line	First	First	First	$A = C = 0$
2. Parabola	Second	First	Second	$A = 0$ and $C \neq 0$
		Second	First	$A \neq 0$ and $C = 0$
3. Ellipse	Second	Second	Second	A and C have same sign
4. Circle	Second	Second	Second	$A = C$
5. Hyperbola	Second	Second	Second	A and C have opposite signs

If $B \neq 0$ (that is, there is an xy term), then:

TYPE OF CURVE	RELATIONSHIP TO GENERAL EQUATION
1. Ellipse	$B^2 - 4AC < 0$
2. Parabola	$B^2 - 4AC = 0$
3. Hyperbola	$B^2 - 4AC > 0$

These tests do not distinguish degenerate cases. This means that the test may tell you the curve is an ellipse that may turn out to be a single point. Remember, too, that a circle is a special case of the ellipse. The expression $B^2 - 4AC$ is called the **discriminant**.

Example 3

Solution

Identify each curve.

a. $x^2 + 4xy + 4y^2 = 9$
   $B^2 - 4AC = 16 - 4(1)(4) = 0$; parabola

b. $2x^2 + 3xy + y^2 = 25$
   $B^2 - 4AC = 9 - 4(2)(1) > 0$; hyperbola

c. $x^2 + xy + y^2 - 8x + 8y = 0$
   $B^2 - 4AC = 1 - 4(1)(1) < 0$; ellipse

d. $xy = 5$
   $B^2 - 4AC = 1 - 4(0)(0) > 0$; hyperbola

TABLE 8.1 Summary of
Standard-Position Conic
Sections

	Parabola	Ellipse	Hyperbola
Definition	All points equidistant from a given point and a given line	All points with the sum of distances from two fixed points constant	All points with the difference of distances from two fixed points constant
Equations	Up: $x^2 = 4cy$ Down: $x^2 = -4cy$ Right: $y^2 = 4cx$ Left: $y^2 = -4cx$	$c^2 = a^2 - b^2$ Horizontal axis: $$\frac{x^2}{a^2} + \frac{y^2}{b^2} = 1$$ Vertical axis: $$\frac{y^2}{a^2} + \frac{x^2}{b^2} = 1$$	$c^2 = a^2 + b^2$ Horizontal axis: $$\frac{x^2}{a^2} - \frac{y^2}{b^2} = 1$$ Vertical axis: $$\frac{y^2}{a^2} - \frac{x^2}{b^2} = 1$$
Recognition	Second-degree equation; linear in one variable, quadratic in the other variable	Second-degree equation; coefficients of x^2 and y^2 have same sign	Second-degree equation; coefficients of x^2 and y^2 have different signs
Eccentricity	$\varepsilon = 1$	$0 \leq \varepsilon < 1$	$\varepsilon > 1$
Directrix	Perpendicular to axis c units from the vertex (one directrix)	Perpendicular to major axis $\pm a/\varepsilon$ units from center (two directrices)	Perpendicular to transverse axis $\pm a/\varepsilon$ units from center (two directrices)

Translations to the point (h, k):

$$x' = x - h \quad \text{and} \quad y' = y - k$$

All conics can be reduced to a standard-form equation by means of a translation or (as we will see in Section 8.5) a rotation. It is important to remember these standard-position equations and some basic information about these curves. The important ideas are summarized in Table 8.1. Notice that the directrices of the ellipse and hyperbola are defined in this table to be consistent with the earlier definition we gave for the directrix of a parabola.

Translating Conics; Completing the Square

Now suppose $B = 0$ and that the rest of the equation is *not* in factored form. The procedure will be to rewrite the equation in factored form by carrying out a process that in algebra was called **completing the square**. This process allows you to determine (h, k) and is best understood by considering some examples.

Example 4

Sketch $x^2 + 4y + 8x + 4 = 0$.

Solution

First, notice this is a parabola (second-degree in x and first-degree in y). Now, complete the square in x.

Step 1:   Associate together the terms involving the variable that is squared:

$$x^2 + 8x = -4y - 4$$

Step 2:   Complete the square for the variable that is squared.

⌐ *Coefficient is 1; if it is not, divide both sides by this coefficient.*

$$x^2 + 8x + \left(\frac{1}{2} \cdot 8\right)^2 = -4y - 4 + \left(\frac{1}{2} \cdot 8\right)^2$$

Take one-half of this coefficient, square it, and add it to both sides.

$$x^2 + 8x + 16 = -4y - 4 + 16$$
$$(x + 4)^2 = -4y + 12$$

Step 3:   Factor out the coefficient of the first-degree term:

$$(x + 4)^2 = -4(y - 3)$$

Step 4:   Determine the vertex by inspection. Plot (h, k); in this example, the vertex is $(-4, 3)$. (See Figure 8.27.)

Step 5:   Determine the focus. By inspection, $4c = 4$, $c = 1$, and the parabola opens downward from the vertex, as shown in Figure 8.27.

Step 6:   Plot the endpoints of the focal chord; $4c = 4$. Draw the parabola as shown in Figure 8.27.

FIGURE 8.27   Graph of $x^2 + 4y + 8x + 4 = 0$

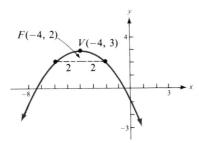

Example 5

Graph $3x^2 + 4y^2 + 24x - 16y + 52 = 0$.

Solution

First, recognize this as an ellipse (second-degree in x and y and coefficients are both positive). Next, complete the square for both squared terms.

Step 1:   Associate together the x and the y terms:

$$(3x^2 + 24x) + (4y^2 - 16y) = -52$$

Step 2: Complete the squares in both x and y. This requires that the coefficients of the squared terms be 1. You can accomplish this by factoring:

$$3(x^2 + 8x \qquad) + 4(y^2 - 4y \qquad) = -52$$

Next complete the square for both x and y, being sure to add the same number to both sides:

added 16 to both sides

$$3(x^2 + 8x + 16) + 4(y^2 - 4y + 4) = -52 + 48 + 16$$

added 48 to both sides

Step 3: Factor: $3(x + 4)^2 + 4(y - 2)^2 = 12$

Step 4: Divide both sides by 12:

$$\frac{(x + 4)^2}{4} + \frac{(y - 2)^2}{3} = 1$$

Step 5: Plot the center (h, k). By inspection, you can see the center is $(-4, 2)$. The vertices are at ± 2, and the length of the semiminor axis is $\sqrt{3}$, as shown in Figure 8.28.

FIGURE 8.28 Graph of $3x^2 + 4y^2 + 24x - 16y + 52 = 0$

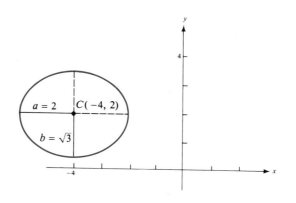

Problem Set 8.4

A *In Problems 1–12 identify and sketch the curve.*

1. $2x - y - 8 = 0$

2. $2x + y - 10 = 0$

3. $4x^2 - 16y = 0$

4. $\dfrac{(x - 3)^2}{4} - \dfrac{(y + 2)}{6} = 1$

5. $\dfrac{(x - 3)^2}{9} - \dfrac{(y + 2)^2}{25} = 1$

6. $\dfrac{x - 3}{9} + \dfrac{y - 2}{25} = 1$

7. $(x + 3)^2 + (y - 2)^2 = 0$

8. $9(x + 3)^2 + 4(y - 2) = 0$

9. $9(x + 3)^2 - 4(y - 2)^2 = 0$

10. $x^2 + 8(y - 12)^2 = 16$

11. $x^2 + 64(y + 4)^2 = 16$

12. $x^2 + y^2 - 3y = 0$

B   In Problems 13–30 identify and sketch the curve.

13. $4x^2 - 3y^2 - 24y - 112 = 0$

14. $y^2 - 4x + 2y + 21 = 0$

15. $9x^2 + 2y^2 - 48y + 270 = 0$

16. $x^2 + 4x + 12y + 64 = 0$

17. $y^2 - 6y - 4x + 5 = 0$

18. $100x^2 - 7y^2 + 98y - 368 = 0$

19. $x^2 + y^2 + 2x - 4y - 20 = 0$

20. $4x^2 + 12x + 4y^2 + 4y + 1 = 0$

21. $x^2 - 4y^2 - 6x - 8y - 11 = 0$

22. $9x^2 + 25y^2 - 54x - 200y + 256 = 0$

23. $2y^2 + 8y - 20x + 148 = 0$

24. $y^2 - 4x + 10y + 13 = 0$

25. $9x^2 + 6x + 18y - 23 = 0$

26. $x^2 + 9y - 6x + 18 = 0$

27. $x^2 + y^2 - 4x + 10y + 15 = 0$

28. $y^2 + 4x^2 + 2y - 8x + 1 = 0$

29. $x^2 + 9y^2 - 4x - 18y - 14 = 0$

30. $4x^2 + y^2 + 24x + 4y + 16 = 0$

C   31. *Engineering.* A stone tunnel is to be constructed such that the opening is a semielliptic arch as shown in Figure 8.29. It is necessary to know the height at 4-ft intervals from the center. That is, how high is the tunnel at 4, 8, 12, 16, and 20 ft from the center? (Answer to the nearest tenth foot.)

FIGURE 8.29   Semielliptic arch

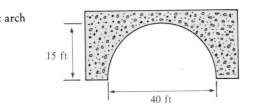

15 ft

40 ft

32. *Engineering.* A parabolic archway has the dimensions shown in Figure 8.30. Find the equation of the parabolic portion.

FIGURE 8.30   A parabolic archway

8.5 ROTATIONS

Rotation of Axes

In Section 7.2 we rotated polar-form curves. In this section we consider the idea of **rotation** of the conic sections by writing the equation in terms of a rotated coordinate system. Suppose the coordinate axes are rotated through an angle θ $(0 < \theta < 90°)$. The relationship between the old coordinates (x, y) and the new coordinates (x', y') can be found by considering Figure 8.31.

FIGURE 8.31   Rotation of axes

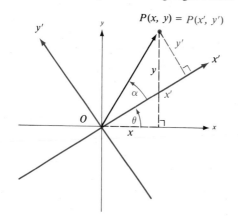

Let O be the origin and P be a point with coordinates (x, y) relative to the old coordinate system and (x', y') relative to the new rotated coordinate system. Let θ be the amount of rotation and let α be the angle between the x'-axis and $|OP|$. Then, using the definition of sine and cosine,

$$x = |OP| \cos(\theta + \alpha) \qquad x' = |OP| \cos \alpha$$
$$y = |OP| \sin(\theta + \alpha) \qquad y' = |OP| \sin \alpha$$

Now

$$
\begin{aligned}
x = |OP| \cos(\theta + \alpha) &= |OP|[\cos \theta \cos \alpha - \sin \theta \sin \alpha] \\
&= |OP| \cos \theta \cos \alpha - |OP| \sin \theta \sin \alpha \\
&= (|OP| \cos \alpha) \cos \theta - (|OP| \sin \alpha) \sin \theta \\
&= x' \cos \theta - y' \sin \theta
\end{aligned}
$$

Also,

$$
\begin{aligned}
y = |OP| \sin(\theta + \alpha) &= |OP| \sin \theta \cos \alpha + |OP| \cos \theta \sin \alpha \\
&= x' \sin \theta + y' \cos \theta
\end{aligned}
$$

Therefore,

ROTATION OF AXES FORMULAS

$$x = x' \cos \theta - y' \sin \theta \qquad y = x' \sin \theta + y' \cos \theta$$

All the curves considered in this chapter can be characterized by the general second-degree equation

$$Ax^2 + Bxy + Cy^2 + Dx + Ey + F = 0$$

Notice that the xy term has not appeared before. The presence of this term indicates that the conic has been rotated. Thus, in this section we assume that $B \neq 0$ and now need to determine the amount this conic has been rotated from standard position. That is, the new axes should be rotated the same amount as the given conic so that it will be in standard position after the rotation. To find out how much to rotate the axes, substitute

$$x = x' \cos\theta - y' \sin\theta \qquad y = x' \sin\theta + y' \cos\theta$$

into

$$Ax^2 + Bxy + Cy^2 + Dx + Ey + F = 0 \qquad (B \neq 0)$$

After a lot of simplifying you will obtain

$$\begin{aligned}
(A\cos^2\theta + B\cos\theta\sin\theta &+ C\sin^2\theta)x'^2 \\
&+ [B(\cos^2\theta - \sin^2\theta) + 2(C - A)\sin\theta\cos\theta]x'y' \\
&+ (A\sin^2\theta - B\sin\theta\cos\theta + C\cos^2\theta)y'^2 + (D\cos\theta + E\sin\theta)x' \\
&+ (-D\sin\theta + E\cos\theta)y' + F = 0
\end{aligned}$$

This looks terrible, but you want to choose θ so that the coefficient of $x'y'$ is zero. This will give you a standard position relative to the new coordinate axes. That is,

$$B(\cos^2\theta - \sin^2\theta) + 2(C - A)\sin\theta\cos\theta = 0 \qquad \textit{Using double-angle identities,}$$
$$\textit{B} \neq 0, \theta \neq 0$$

$$B\cos 2\theta + (C - A)\sin 2\theta = 0$$
$$B\cos 2\theta = (A - C)\sin 2\theta$$
$$\frac{\cos 2\theta}{\sin 2\theta} = \frac{A - C}{B}$$

Simplifying, you obtain the following result:

AMOUNT OF ROTATION
FORMULA

$$\cos 2\theta = \frac{A - C}{B}$$

Notice that we required $0 < \theta < 90°$, so 2θ is in Quadrant I or Quadrant II. This means that if $\cot 2\theta$ is positive, then 2θ must be in Quadrant I; if $\cot 2\theta$ is negative, then 2θ is in Quadrant II.

In Examples 1–4, find the appropriate rotation so that the given curve will be in standard position relative to the rotated axes. Also find the x and y values in the new coordinate system.

Example 1 $xy = 6$

Solution

$$\cot 2\theta = \frac{A - C}{B} \qquad x = x' \cos \theta - y' \sin \theta \qquad y = x' \sin \theta + y' \cos \theta$$

$$= \frac{0 - 0}{1} \qquad x = x' \cos 45° - y' \sin 45° \qquad y = x' \sin 45° + y' \cos 45°$$

$$= 0 \qquad = x'\left(\frac{1}{\sqrt{2}}\right) - y'\left(\frac{1}{\sqrt{2}}\right) \qquad = x'\left(\frac{1}{\sqrt{2}}\right) + y'\left(\frac{1}{\sqrt{2}}\right)$$

Thus $2\theta = 90°$
and $\theta = 45°$.
$$= \frac{1}{\sqrt{2}}(x' - y') \qquad\qquad = \frac{1}{\sqrt{2}}(x' + y') \qquad \square$$

Example 2 $7x^2 - 6\sqrt{3}\,xy + 13y^2 - 16 = 0$

Solution

$$\cot 2\theta = \frac{A - C}{B}$$

$$= \frac{7 - 13}{-6\sqrt{3}} \qquad x = x' \cos \theta - y' \sin \theta. \qquad y = x' \sin \theta + y' \cos \theta.$$

$$= \frac{1}{\sqrt{3}} \qquad\qquad = x'\left(\frac{\sqrt{3}}{2}\right) - y'\left(\frac{1}{2}\right) \qquad = \frac{1}{2}(x' + \sqrt{3}\,y')$$

Thus $2\theta = 60°$
and $\theta = 30°$.
$$= \frac{1}{2}(\sqrt{3}\,x' - y') \qquad\qquad\qquad \square$$

Example 3 $x^2 - 4xy + 4y^2 + 5\sqrt{5}\,y - 10 = 0$

Solution

$$\cot 2\theta = \frac{1 - 4}{-4} = \frac{3}{4}$$

Since this is not an exact value for θ (as it was in Examples 1 and 2), you will need to use some trigonometric identities to find $\cos \theta$ and $\sin \theta$. If $\cot 2\theta = \frac{3}{4}$, then $\tan 2\theta = \frac{4}{3}$ and $\sec 2\theta = \sqrt{1 + (\frac{4}{3})^2} = \frac{5}{3}$. Then $\cos 2\theta = \frac{3}{5}$ and you can now apply the half-angle identities:

$$\cos \theta = \pm \sqrt{\frac{1 + \cos 2\theta}{2}} \qquad \sin \theta = \pm \sqrt{\frac{1 - \cos 2\theta}{2}}$$

$$\cos \theta = \sqrt{\frac{1 + (\frac{3}{5})}{2}} \qquad \sin \theta = \sqrt{\frac{1 - (\frac{3}{5})}{2}} \qquad \textit{Positive because } \theta \textit{ is in}$$
$$\qquad\qquad\qquad\qquad\qquad\qquad\qquad\qquad\qquad\qquad \textit{Quadrant } \mathrm{I}$$

$$= \frac{2}{\sqrt{5}} \qquad\qquad\qquad = \frac{1}{\sqrt{5}}$$

To find the amount of rotation, use a calculator and one of the preceding equations to find $\theta \approx 26.6°$. Finally, the rotation of axes formulas provide

$$x = x' \cos \theta - y' \sin \theta \qquad\qquad y = x' \sin \theta + y' \cos \theta$$

$$= x'\left(\frac{2}{\sqrt{5}}\right) - y'\left(\frac{1}{\sqrt{5}}\right) \qquad = x'\left(\frac{1}{\sqrt{5}}\right) - y'\left(\frac{2}{\sqrt{5}}\right)$$

$$= \frac{1}{\sqrt{5}}(2x' - y') \qquad\qquad = \frac{1}{\sqrt{5}}(x' + 2y')$$

$\square$

Example 4

$$10x^2 + 24xy + 17y^2 - 9 = 0$$

Solution

$$\cot 2\theta = \frac{10 - 17}{24}$$

$$= -\frac{7}{24}$$

Since this is negative, 2θ must be in Quadrant II; this means that $\sec 2\theta$ is negative in the following sequence of identities: Since $\cot 2\theta = -\frac{7}{24}$, then $\tan 2\theta = -\frac{24}{7}$ and $\sec 2\theta = -\sqrt{1 + (-\frac{24}{7})^2} = -\frac{25}{7}$. Thus $\cos 2\theta = -\frac{7}{25}$, which gives

$$\cos \theta = \sqrt{\frac{1 + (-\frac{7}{25})}{2}} \qquad \sin \theta = \sqrt{\frac{1 - (-\frac{7}{25})}{2}}$$

$$= \frac{3}{5} \qquad\qquad = \frac{4}{5}$$

Using either of these equations and a calculator, you find that the rotation is $\theta \approx 53.1°$. The rotation of axes formulas provide

$$x = x' \cos \theta - y' \sin \theta \qquad y = x' \sin \theta + y' \cos \theta$$

$$= x'(\tfrac{3}{5}) - y'(\tfrac{4}{5}) \qquad\qquad = x'(\tfrac{4}{5}) + y'(\tfrac{3}{5})$$

$$= \tfrac{1}{5}(3x' - 4y') \qquad\qquad = \tfrac{1}{5}(4x' + 3y')$$

$\square$

Sketching Rotated Conics

PROCEDURE FOR
SKETCHING A ROTATED
CONIC

1. Find the angle of rotation.
2. Find x and y in the new coordinate system.
3. Substitute the values found in Step 2 into the given equation and simplify.
4. Sketch the resulting equation relative to the new x'- and y'-axes. You may have to complete the square if it is not centered at the origin.

Example 5

Sketch $xy = 6$.

Solution

From Example 1, the rotation is $\theta = 45°$ and

$$x = \frac{1}{\sqrt{2}}(x' - y') \qquad y = \frac{1}{\sqrt{2}}(x' + y')$$

Substitute these values into the original equation $xy = 6$:

$$\left[\frac{1}{\sqrt{2}}(x' - y')\right]\left[\frac{1}{\sqrt{2}}(x' + y')\right] = 6$$

Simplify (see Problem 1 for the details) to obtain

$$x'^2 - y'^2 = 12$$

$$\frac{x'^2}{12} - \frac{y'^2}{12} = 1$$

This curve is a hyperbola that has been rotated 45°. Draw the rotated axis and sketch this equation relative to the rotated axes. The result is shown in Figure 8.32.

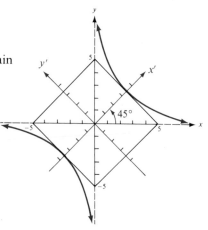

FIGURE 8.32   Graph of $xy = 6$

Example 6

Sketch $7x^2 - 6\sqrt{3}\,xy + 13y^2 - 16 = 0$.

Solution

From Example 2, $\theta = 30°$ and

$$x = \tfrac{1}{2}(\sqrt{3}\,x' - y') \qquad y = \tfrac{1}{2}(x' + \sqrt{3}\,y')$$

Substitute into the original equation:

$$7(\tfrac{1}{2})^2(\sqrt{3}\,x' - y')^2 - 6\sqrt{3}\,(\tfrac{1}{2})(\sqrt{3}\,x' - y')(\tfrac{1}{2})(x' + \sqrt{3}\,y')$$
$$+ 13(\tfrac{1}{2})^2(x' + \sqrt{3}\,y')^2 - 16 = 0$$

Simplify (see Problem 2 for the details) to obtain

$$\frac{x'^2}{4} + \frac{y'^2}{1} = 1$$

This curve is a rotated ellipse with a 30° rotation. The sketch is shown in Figure 8.33.

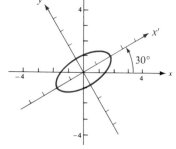

FIGURE 8.33   Graph of
$7x^2 - 6\sqrt{3}\,xy + 13y^2 - 16 = 0$

Example 7

Sketch $x^2 - 4xy + 4y^2 + 5\sqrt{5}\,y - 10 = 0$.

Solution

From Example 3, the rotation is $\theta \approx 26.6°$ and

$$x = \tfrac{1}{\sqrt{5}}(2x' - y')$$
$$y = \tfrac{1}{\sqrt{5}}(x' + 2y')$$

Substitute

$$\tfrac{1}{5}(2x' - y')^2 - 4\left(\tfrac{1}{5}\right)(2x' - y')(x' + 2y') + 4\left(\tfrac{1}{5}\right)(x' + 2y')^2$$
$$+ 5\sqrt{5}\left(\tfrac{1}{\sqrt{5}}\right)(x' + 2y') - 10 = 0$$

Simplify (see Problem 3 for the details) to obtain

$$y'^2 + 2y' = -x' + 2$$

This curve is a parabola with a rotation of about 26.6°. Next complete the square to obtain

$$(y' + 1)^2 = -(x' - 3)$$

This sketch is shown in Figure 8.34.

FIGURE 8.34   Graph of
$x^2 - 4xy + 4y^2 + 5\sqrt{5}y - 10 = 0$

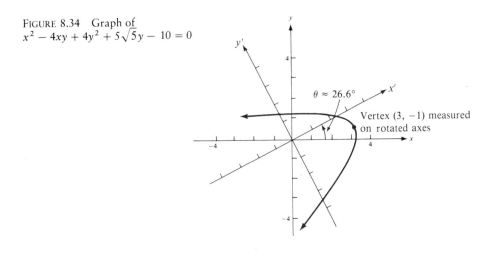

$\theta \approx 26.6°$

Vertex $(3, -1)$ measured
on rotated axes

Conic Sections Graphing Summary

To graph the general second-degree equation

$$Ax^2 + Bxy + Cy^2 + Dx + Ey + F = 0$$

follow the steps shown in Figure 8.35.

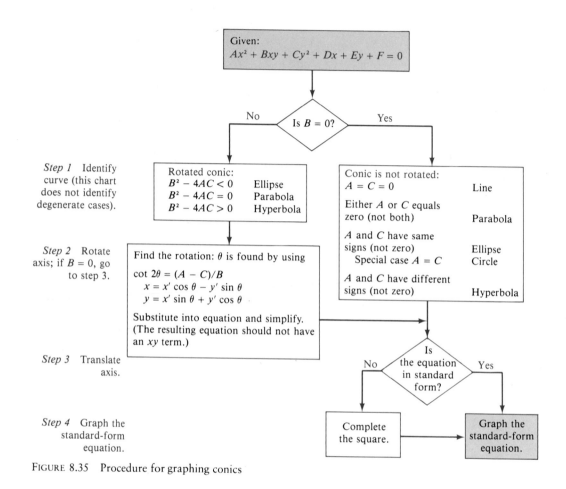

FIGURE 8.35   Procedure for graphing conics

Problem Set 8.5

A   *Because of the amount of arithmetic and algebra involved, many algebraic steps were left out of the examples of this section. In Problems 1–3 fill in the details left out of the indicated example.*

1. Example 5   2. Example 6   3. Example 7

Identify the curves whose equations are given in Problems 4–17.

4. $xy = 10$

5. $xy = -1$

6. $xy = -4$

7. $13x^2 - 10xy + 13y^2 - 72 = 0$

8. $5x^2 - 26xy + 5y^2 + 72 = 0$

9. $x^2 + 4xy + 4y^2 + 10\sqrt{5}x = 9$

10. $5x^2 - 4xy + 8y^2 = 36$

11. $23x^2 + 26\sqrt{3}xy - 3y^2 - 144 = 0$

12. $3x^2 + 2\sqrt{3}xy + y^2 + 16x - 16\sqrt{3}y = 0$

13. $24x^2 + 16\sqrt{3}xy + 8y^2 - x + \sqrt{3}y - 8 = 0$

14. $3x^2 - 2\sqrt{3}xy + y^2 + 24x + 24\sqrt{3}y = 0$

15. $13x^2 - 6\sqrt{3}xy + 7y^2 + (16\sqrt{3} - 8)x + (-16 - 8\sqrt{3})y + 16 = 0$

16. $3x^2 - 10xy + 3y^2 - 32 = 0$

17. $5x^2 - 3xy + y^2 + 65x - 25y + 203 = 0$

B *Find the appropriate rotation in Problems 18–33 so that the given curve will be in standard position relative to the rotated axes. Also find the x and y values in the new coordinate system by using the rotation of axes formulas. Finally, sketch the curve.*

18. $xy = 10$

19. $xy = -1$

20. $xy = -4$

21. $xy = 8$

C 22. $8x^2 - 4xy + 5y^2 = 36$

23. $13x^2 - 10xy + 13y^2 - 72 = 0$

24. $5x^2 - 26xy + 5y^2 + 72 = 0$

25. $x^2 + 4xy + 4y^2 + 10\sqrt{5}x = 9$

26. $5x^2 - 4xy + 8y^2 = 36$

27. $23x^2 + 26\sqrt{3}xy - 3y^2 - 144 = 0$

28. $3x^2 + 2\sqrt{3}xy + y^2 + 16x - 16\sqrt{3}y = 0$

29. $24x^2 + 16\sqrt{3}xy + 8y^2 - x + \sqrt{3}y - 8 = 0$

30. $3x^2 - 2\sqrt{3}xy + y^2 + 24x + 24\sqrt{3}y = 0$

31. $3x^2 - 10xy + 3y^2 - 32 = 0$

32. $x^2 + 2xy + y^2 + 12\sqrt{2}x - 6 = 0$

33. $10x^2 + 24xy + 17y^2 - 9 = 0$

8.6   CONIC SECTIONS IN POLAR FORM

In Table 8.1 of Section 8.4 it was noted that the conic sections could be distinguished by looking at the eccentricity, ε. Up to now in this chapter, we let the origin lie at the center of the conic, or else we translated the origin to point (h, k). However, there are many important applications of conics which put a focus at the origin (or pole). In this section, we give an alternate definition of the conic sections which puts a focus at the pole and also uses ε to distinguish one conic from another.

ALTERNATE DEFINITION OF A CONIC

The set of points in a plane such that the distance from a fixed point (called the **focus**) is in constant ratio to its distance from a fixed line (called the **directrix**) is called a **conic**. The constant ratio is called the **eccentricity** of the conic and is denoted by ε. Moreover, the conic is:

an **ellipse** if $0 < \varepsilon < 1$;

a **parabola** if $\varepsilon = 1$;

a **hyperbola** if $\varepsilon > 1$.

(See Figure 8.36.) The single equation for all of these conic sections is

$$r = \frac{\varepsilon p}{1 - \varepsilon \cos \theta}$$

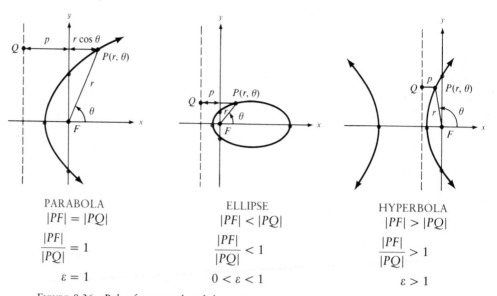

PARABOLA
$|PF| = |PQ|$

$$\frac{|PF|}{|PQ|} = 1$$

$\varepsilon = 1$

ELLIPSE
$|PF| < |PQ|$

$$\frac{|PF|}{|PQ|} < 1$$

$0 < \varepsilon < 1$

HYPERBOLA
$|PF| > |PQ|$

$$\frac{|PF|}{|PQ|} > 1$$

$\varepsilon > 1$

FIGURE 8.36   Polar-form graphs of the conic sections

We can show that this definition is equivalent for our previous definitions for the conics. For example, if $\varepsilon = 1$, then

$$r = \frac{p}{1 - \cos \theta}$$

$$r(1 - \cos \theta) = p$$

$$r - r \cos \theta = p$$

In order to transform this equation into rectangular form, we need to remember the conversion formulas: $r = \sqrt{x^2 + y^2}$ and $x = r \cos \theta$:

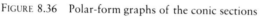

$$\sqrt{x^2 + y^2} - x = p$$

$$\sqrt{x^2 + y^2} = x + p$$

$$x^2 + y^2 = (x + p)^2$$

$$x^2 + y^2 = x^2 + 2px + p^2$$

$$y^2 = 2px + p^2$$

$$y^2 = 2p\left(x + \tfrac{p}{2}\right)$$

This is the equation of a parabola with vertex (in rectangular coordinates) $(h, k) = \left(-\tfrac{p}{2}, 0\right)$ and focal chord length $2p$. The steps for verifying the polar-form equations for ellipses and hyperbolas are the same as illustrated for the parabola.

Example 1

Describe and sketch the graph of the equation

$$r = \frac{4}{3 - 3 \cos \theta}$$

Solution

The first thing to look at is the first term in the denominator; multiply the fraction by 1 written as the reciprocal of this first term divided by that same number. In this problem, the first term in the denominator is 3 so the fraction is to be multiplied by $\frac{1}{3}/\frac{1}{3}$:

$$r = \frac{4}{3 - 3\cos\theta} \cdot \frac{\frac{1}{3}}{\frac{1}{3}} = \frac{\frac{4}{3}}{1 - \cos\theta}$$

By inspection you can now see that $\varepsilon = 1$ so this is a parabola. Furthermore, $\varepsilon p = p = \frac{4}{3}$. Compare with Figure 8.36 to see that the directrix is $\frac{4}{3}$ to the left of the focus, and thus the vertex is $\frac{1}{2}(\frac{4}{3}) = \frac{2}{3}$ from the focus. Plot a few points; those shown in color in Figure 8.36 are generally the easiest to plot. If $\theta = \frac{\pi}{2}$, then

$$r = \frac{4}{3 - 3\cos\frac{\pi}{2}} = \frac{4}{3}$$

You can plot other points, say $\theta = \pi$, or $\theta = \frac{3\pi}{2}$, but it is not necessary because you know where the vertex is located and you also know the point where the parabola crosses the negative y-axis by using symmetry. The graph is shown in Figure 8.37. □

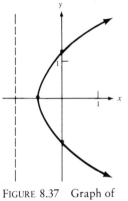

FIGURE 8.37   Graph of
$$r = \frac{4}{3 - 3\cos\theta}$$

Example 2

Describe and sketch the graph of the equation

$$r = \frac{4}{4 - 3\cos\theta}$$

Solution

Notice the leading coefficient in the denominator:

$$r = \frac{4}{4 - 3\cos\theta} \cdot \frac{\frac{1}{4}}{\frac{1}{4}} = \frac{1}{1 - \frac{3}{4}\cos\theta}$$

We see that $\varepsilon = \frac{3}{4}$, which means this is an ellipse. Plot some values:

If $\theta = 0$, $r = \dfrac{4}{4 - 3(1)} = 4$; plot this point.

If $\theta = \frac{\pi}{2}$, $r = \dfrac{4}{4 - 3(0)} = 1$; plot this point and the one at $\theta = \frac{3\pi}{2}$ (by symmetry).

If $\theta = \pi$, $r = \dfrac{4}{4 - 3(-1)} = \frac{4}{7}$; plot this point.

It looks like we could use another point or two:

If $\theta = \frac{\pi}{4}$, $r = \dfrac{4}{4 - 3\cos\frac{3}{4}} \approx 2.129$.

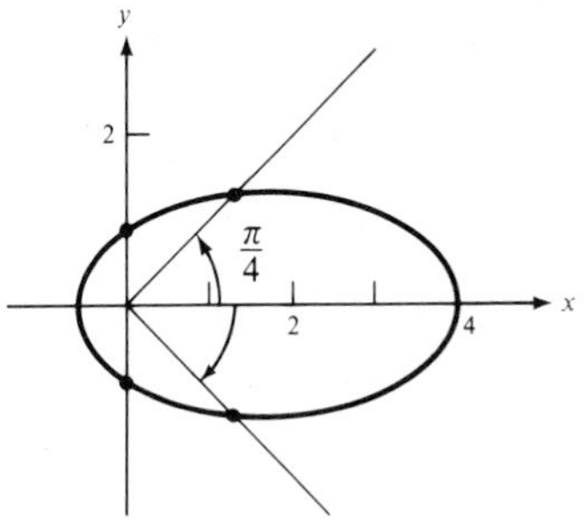

By symmetry, plot the point for $\theta = -\frac{\pi}{4}$ and connect the points as shown in Figure 8.38. □

FIGURE 8.38   Graph of
$$r = \frac{4}{4 - 3\cos\theta}$$

Example 3

Describe and sketch the graph of $r = \dfrac{4}{3 - 4\cos\theta}$.

Solution

Look at the leading term in the denominator to decide the form of 1 for the multiplication step:

$$r = \frac{4}{3 - 4\cos\theta} \cdot \frac{\frac{1}{3}}{\frac{1}{3}} = \frac{\frac{4}{3}}{1 - \frac{4}{3}\cos\theta}$$

This is called the standard form. You see that $\varepsilon = \frac{4}{3}$ so this is a hyperbola. First find the intercepts:

If $\theta = 0$, then $r = \dfrac{4}{3 - 4(1)} = -4$.   *This is $(-4, 0)$ or $(4, \pi)$.*

If $\theta = \pi$, then $r = \dfrac{4}{3 - 4(-1)} = \frac{4}{7}$.   *This is $(\frac{4}{7}, \pi)$.*

The center of the hyperbola will be midway between these points: $\dfrac{4 + \frac{4}{7}}{2} = \frac{16}{7}$. The y-intercepts are $(\frac{\pi}{2}, \frac{4}{3})$ and $(\frac{3\pi}{2}, \frac{4}{3})$. Even with these points you still do not have a clear picture of the hyperbola. Hyperbolas are the most difficult conic to sketch because the domain of θ determines on which branch of the hyperbola you are plotting points. It is helpful to find the asymptotes. These are the values of θ that cause the denominator to be 0. Solve this equation:

$$3 - 4\cos\theta = 0$$
$$\cos\theta = \tfrac{3}{4}$$
$$\theta = \cos^{-1} .75 \approx .72, 5.56$$

If you draw the lines through the center with these values of θ you will have the asymptotes and can finish the sketch of the hyperbola as shown in Figure 8.39.

FIGURE 8.39   Graph of $r = \dfrac{4}{3 - 4\cos\theta}$

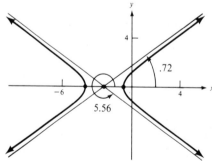

Rotated Conics

It is easy (compared with what you needed to do in Section 8.5) to rotate the conic sections when the equation is given in polar form. Remember that if the argument of a trigonometric function is $(\theta - \alpha)$ then the curve is rotated through an angle α.

Example 4

Graph $r = \dfrac{16}{5 - 3\cos\left(\theta - \frac{\pi}{6}\right)}$.

Solution

First determine the type of conic by writing the equation in standard form:

$$r = \frac{16}{5 - 3\cos\left(\theta - \frac{\pi}{6}\right)} \cdot \frac{\frac{1}{5}}{\frac{1}{5}} = \frac{\frac{16}{5}}{1 - \frac{3}{5}\cos\left(\theta - \frac{\pi}{6}\right)}$$

This is an ellipse $\left(\varepsilon = \frac{3}{5}\right)$ that has been rotated $\frac{\pi}{6}$. First rotate the conic and then plot points on the rotated axis using the equation

$$r' = \frac{\frac{16}{5}}{1 - \frac{3}{5}\cos\theta'}$$

If $\theta' = 0$, then $r' = 8$; if $\theta' = \frac{\pi}{2}$, then $r' = \frac{16}{5}$; if $\theta' = \pi$, then $r' = 2$. Plot these points as shown in Figure 8.40.

Figure 8.40   Graph of $r = \dfrac{16}{5 - 3\cos\left(\theta - \frac{\pi}{6}\right)}$

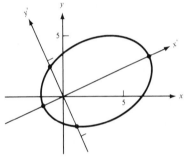

As we observed in the last chapter, a rotation of $90°$ changes the cosine function to a sine function; a rotation of $180°$ will change the sign of the cosine; and a rotation of $270°$ will change the sign and the function.

ALTERNATE EQUATIONS
FOR POLAR-FORM CONICS

The conic sections can be written as one of the following polar-form equations:

$$r = \frac{\varepsilon p}{1 \pm \varepsilon\cos\theta} \quad \text{or} \quad r = \frac{\varepsilon p}{1 \pm \varepsilon\sin\theta}$$

Example 5

Describe and graph $r = \dfrac{8}{4 + 4\sin\theta}$.

Solution

Write in standard form:

$$r = \frac{8}{4 + 4\sin\theta} \cdot \frac{\frac{1}{4}}{\frac{1}{4}} = \frac{2}{1 + \sin\theta}$$

Recognize this as a parabola since $\varepsilon = 1$; it is also rotated $270°$ so instead of opening to the right it opens down. The graph is shown in Figure 8.41.

FIGURE 8.41   Graph of
$$r = \frac{8}{4 + 4\sin\theta}$$

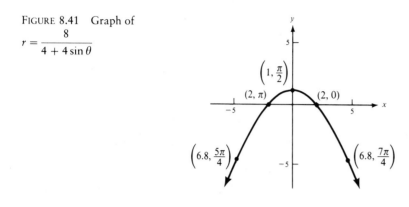

Problem Set 8.6

A   *Identify the conic in each of Problems 1–16. In each case write the standard form of the equation and give ε.*

1. $r = \dfrac{3}{1 - \cos\theta}$

2. $r = \dfrac{5}{1 + \sin\theta}$

3. $r = \dfrac{5}{3 + 5\cos\theta}$

4. $r = \dfrac{6}{2 - 5\sin\theta}$

5. $r = \dfrac{5}{5 - 4\cos\theta}$

6. $r = \dfrac{16}{8 + \cos\theta}$

7. $r(1 - 2\cos\theta) = 4$

8. $r(1 + 3\sin\theta) = 5$

9. $r(2 - 2\cos\theta) = 8$

10. $r(4 - 4\sin\theta) = 32$

11. $r = \dfrac{3}{1 - \cos(\theta - \frac{4}{\pi})}$

12. $r = \dfrac{4}{2 - 2\cos(\theta - \frac{\pi}{3})}$

13. $r = \dfrac{5}{4 - 2\cos(\theta - \frac{\pi}{6})}$

14. $r = \dfrac{2}{2 - \cos(\theta + 1)}$

15. $r = \dfrac{\sqrt{2}}{1 + \cos(\theta + .5)}$

16. $r = \dfrac{\sqrt{3}}{1 - \sin(\theta - 1)}$

B   *Graph each of the curves given in Problems 17–32. Notice that these are the equations you analyzed in Problems 1–16.*

17. $r = \dfrac{3}{1 - \cos\theta}$

18. $r = \dfrac{5}{1 + \sin\theta}$

19. $r = \dfrac{5}{3 + 5\cos\theta}$

20. $r = \dfrac{6}{2 - 5\sin\theta}$

21. $r = \dfrac{5}{5 - 4\cos\theta}$

22. $r = \dfrac{16}{8 + \cos\theta}$

23. $r(1 - 2\cos\theta) = 4$

24. $r(1 + 3\sin\theta) = 5$

25. $r(2 - 2\cos\theta) = 8$

26. $r(4 - 4\sin\theta) = 32$

27. $r = \dfrac{3}{1 - \cos(\theta - \frac{\pi}{4})}$

28. $r = \dfrac{4}{2 - 2\cos(\theta - \frac{\pi}{3})}$

29. $r = \dfrac{5}{4 - 2\cos\left(\theta - \frac{\pi}{6}\right)}$ **30.** $r = \dfrac{2}{2 - \cos(\theta + 1)}$ **31.** $r = \dfrac{\sqrt{2}}{1 + \cos(\theta + .5)}$

32. $r = \dfrac{\sqrt{3}}{1 - \sin(\theta - 1)}$

C   **33.** Assume that $\varepsilon = \frac{1}{2}$ in the definition. Convert the polar-form equation to rectangular form and show it is an ellipse.

   34. Assume that $\varepsilon = \frac{3}{2}$ in the definition. Convert the polar-form equation to rectangular form and show it is a hyperbola.

8.7 SUMMARY AND REVIEW

	OBJECTIVES	PAGES/EXAMPLES
8.1 PARABOLAS	1. Graph standard-position parabolas centered at $(0, 0)$.	p. 307; Examples 1–3; Problems 7–18.
	2. Graph standard-position parabolas centered at (h, k).	p. 310; Example 5; Problems 19–24.
	3. Find the equations of parabolas given certain information about the graph.	p. 309; Examples 4, 6: Problems 25–30.
8.2 ELLIPSES	4. Graph standard-position ellipses centered at $(0, 0)$.	p. 313; Examples 1–2; Problems 1–9.
	5. Graph standard-position circles with radius r centered at (h, k).	p. 317; Example 4; Problems 10–12.
	6. Graph standard-position ellipses centered at (h, k).	p. 316; Example 3; Problems 13–15.
	7. Find the equations of ellipses and circles given certain information about the graph.	p. 318; Examples 5–7; Problems 16–30.
8.3 HYPERBOLAS	8. Graph standard-position hyperbolas centered at $(0, 0)$.	p. 323; Example 1; Problems 1–15.
	9. Graph standard-position hyperbolas centered at (h, k).	p. 324; Example 2; Problems 16–22.
	10. Find the equations of hyperbolas given certain information about the graph.	p. 325; Examples 3–5; Problems 23–28.
8.4 GENERAL CONIC SECTIONS	11. Graph degenerate conics.	p. 328; Examples 1–2; Problems 7, 9.

Terms

Axis [8.1]	Directrix [8.1, 8.6]	Pseudovertices [8.3]
Center [8.2, 8.3]	Discriminant [8.4]	Radius [8.2]
Central rectangle [8.3]	Eccentricity [8.2]	Rotation [8.5, 8.6]
Circle [8.2]	Ellipse [8.2, 8.4, 8.6]	Second-degree equation
Completing the square [8.4]	First-degree equation [8.1]	[8.1]
Conic sections [8.1, 8.4, 8.6]	Focal chord [8.1]	Semimajor axis [8.2]
	Focus [8.1, 8.2, 8.3, 8.6]	Semiminor axis [8.2]
Conjugate axis [8.3]	Hyperbola [8.3, 8.4, 8.6]	Slant asymptote [8.3]
Covertices [8.2]	Major axis [8.2]	Translation [8.4]
Degenerate conic [8.4]	Minor axis [8.2]	Transverse axis [8.3]
	Parabola [8.1, 8.4, 8.6]	Vertex [8.1, 8.2]

Chapter 8 Review of Objectives

The problem numbers correspond to the objectives listed in Section 8.7.

1. Graph the given curves.
   a. $x^2 = y$ b. $2y^2 = x$
2. Graph the given curves.
   a. $(y - 1)^2 = 8(x + 2)$ b. $8(y - 2)^2 = x + 1$
3. Find the equations for the given parabolas.
   a. Vertex $(6, 3)$; directrix $x = 1$
   b. Directrix $y - 3 = 0$; focus $(-3, -2)$
   c. Vertex $(-3, 5)$; focus $(-3, -1)$

4. Graph the given curves.

   a. $25x^2 + 16y^2 = 400$ b. $\dfrac{x^2}{25} + \dfrac{y^2}{4} = 1$

5. Graph the given curves.
   a. $4x^2 + 4y^2 = 9$ b. $(x - 3)^2 + (y + 1)^2 = 16$

6. Graph the given curves.

   a. $9(x + 3)^2 + 5(y - 2)^2 = 45$ b. $\dfrac{(x + 1)^2}{5} + \dfrac{(y - 2)^2}{9} = 1$

7. Find the equations of the given ellipses.
   a. The ellipse with center $(4, 1)$, a focus $(5, 1)$, and a semimajor axis 2
   b. The set of points with the sum of distances from $(-3, 4)$ and $(-7, 4)$ equal to 12
   c. The set of points 8 units from the point $(-1, -2)$

8. Graph the given curves.

   a. $y^2 - x^2 = 1$ b. $\dfrac{x^2}{4} - \dfrac{y^2}{6} = 1$

9. Graph the given curves.

   a. $5(x + 2)^2 - 3(y + 4)^2 = 60$ b. $\dfrac{(x + \frac{1}{2})^2}{3} - \dfrac{(y + \frac{1}{2})^2}{3} = 1$

10. Find the equations of the given hyperbolas.
    a. The hyperbola with vertices at $(-3, 0)$ and $(3, 0)$ and foci at $(5, 0)$ and $(-5, 0)$
    b. Foci at $(0, -5)$ and $(0, 5)$; eccentricity $\frac{5}{3}$
    c. Vertices at $(-3, 1)$ and $(-5, 1)$; foci at $(-4 - \sqrt{6}, 1)$ and $(-4 + \sqrt{6}, 1)$

11. Graph:

    a. $(x - 2)^2 + (y + 3)^2 = 0$ b. $\dfrac{x^2}{4} - \dfrac{y^2}{9} = 0$

12. State the appropriate standard-form equation.
    a. Horizontal ellipse   b. Vertical hyperbola
    c. Parabola opening right   d. Circle

13. Graph the given curves.
    a. $y^2 + 4x + 4y = 0$ b. $x^2 + y^2 = 4x + 2y - 3$
    c. $16x^2 + 9y^2 - 160x - 18y + 265 = 0$ d. $3x^2 - 4y^2 + 6x - 8y - 16 = 0$

14. Name the curve (by inspection).
    a. $3x - 2y^2 - 4y + 7 = 0$ b. $\dfrac{x}{16} + \dfrac{y}{4} = 1$

    c. $25x^2 + 9y^2 = 225$ d. $25(x - 2)^2 + 25(y + 1)^2 = 400$
    e. $xy = 5$ f. $5x^2 + 4xy + 5y^2 + 3x = 2$
    g. $xy + x^2 - 3x = 5$ h. $x^2 + y^2 + 2xy + 3x - y = 3$
    i. $x^2 + 2xy + y^2 = 10$ j. $(x - 1)(y + 1) = 7$

15. What is the rotation for each of the following conic sections?
    a. $5x^2 + 4xy + 5y^2 + 3x - 2y + 5 = 0$ b. $4x^2 + 4xy + y^2 + 3x - 2y + 7 = 0$

16. Graph $10x^2 + 17y^2 = 9 - 24xy$.

17. Identify and graph $r = \dfrac{10}{5 - 5\cos\theta}$.   18. Identify and graph $r = \dfrac{6}{3 - 2\sin\theta}$.

Miscellaneous Problems

Graph each of the curves in Problems 1–30.

1. $y = 10x^2$

2. $y = \frac{1}{5}x^2$

3. $x = 3y^2$

4. $x = .01y^2$

5. $y - 3 = (x + 1)^2$

6. $y = 2x^2 + 12x + 20$

7. $9x^2 + 25y^2 = 225$

8. $4x^2 + 9y^2 = 36$

9. $3x - 4y = 0$

10. $\dfrac{x}{16} + \dfrac{y}{4} = 1$

11. $\dfrac{x^2}{16} + \dfrac{y}{4} = 1$

12. $\dfrac{x^2}{16} + \dfrac{y^2}{4} = 1$

13. $\dfrac{x^2}{16} - \dfrac{y^2}{4} = 1$

14. $3y - 2x^2 - 4x + 7 = 0$

15. $x^2 + y^2 + 2x - 4y = 0$

16. $5x^2 + 4y^2 + 10x - 8y + 4 = 0$

17. $3x - 2y^2 - 4y + 7 = 0$

18. $25(x - 1)^2 + 25(y + 2)^2 = 400$

19. $3x^2 - y^2 + 6y - 4 = 0$

20. $2x^2 + 8x - 3y + 2 = 0$

21. $xy = 10$

22. $25x^2 + 14xy + 25y^2 + 4\sqrt{2}\,x - 20\sqrt{2}\,y + 2 = 2$

23. $r = \dfrac{\sqrt{3}}{2 - 2\cos\theta}$

24. $r = \dfrac{1}{4 - 5\cos\theta}$

25. $r = \dfrac{2}{6 - 5\cos\theta}$

26. $r = \dfrac{\sqrt{2}}{3 - 3\sin\theta}$

27. $r = \dfrac{3}{2 + 3\cos\theta}$

28. $r = \dfrac{4}{3 + 2\sin\theta}$

29. $r = \dfrac{5}{2 + 2\cos(\theta - \frac{\pi}{6})}$

30. $r = \dfrac{3}{2 + \cos(\theta - .2)}$

A 'Planet X' Way Out There?

LIVERMORE, Calif. (UPI)—A scientist at the Lawrence Livermore Laboratory suggested Friday the existence of a "Planet X" three times as massive as Saturn and nearly six billion miles from earth.

The planet, far beyond Pluto which is currently the outermost of the nine known planets of the solar system, was predicted on sophisticated mathematical computations of the movements of Halley's Comet.

Joseph L. Brady, a Lawrence mathematician and an authority on the comet, reported the calculations in the Journal of the Astronomical Society of the Pacific.

Brady said he and his colleagues, Edna M. Carpenter and Francis H. McMahon, used a computer to process mathematical observations of the strange deviations in Halley's Comet going back to before Christ.

Lawrence officials said the existence of a 10th planet has been predicted before, but Brady is the first to predict its orbit, mass and position.

Brady said the planet was about 65 times as far from the sun as earth, which is about 93 million miles from the sun. From earth "Planet X" would be located in the constellation Casseiopeia on the border of the Milky Way.

The size and location of "Planet X" were proposed to account for mysterious deviations in the orbit of Halley's Comet. But the calculations subsequently were found to account for deviations in the orbits of two other reappearing comets, Olbers and Pons-Brooks, Brady said.

No contradiction between the proposed planet and the known orbits of comets and other planets has been found.

The prediction of unseen planets is not new. The location of Neptune was predicted in 1846 on the basis of deviations in the orbit of Uranus. Deviations in Neptune's orbit led to a prediction of Pluto's location in 1915.

Although no such deviations of Pluto have been found, Brady pointed out that since its discovery in 1930, Pluto has been observed through less than one-fourth of its revolution around the sun and a complete picture of its orbit is not available.

Brady said "Planet X" may be as elusive as Pluto was a half century ago. It took 15 years to find it from the time of its prediction.

"The proposed planet is located in the densely populated Milky Way where even a tiny area encompasses thousands of stars, many of which are brighter than we expect this planet to be," he said. "If it exists, it will be extremely difficult to find."

"Planet X," in its huge orbit, takes 600 years to complete a revolution around the sun, he said.

After the planet Uranus was discovered in 1781, its motion revealed gravitational perturbations caused by an unknown planet. Independent mathematical calculations by Urbain Leverrier and John Couch Adams predicted the position of this unknown planet—the discovery of Neptune in 1846 is one of the greatest triumphs of celestial mechanics and mathematics in the history of astronomy. A similar search led to the discovery of Pluto in 1930. The article reproduced here is dated April 30, 1972, but at the time of this printing no planet has yet been found. Remember, though, that it took 15 years to find Pluto after the time of its prediction.

The orbits of the planets are elliptical in shape. If the sun is placed at one of the foci of a giant ellipse, the orbit of the earth is elliptical. The *perihelion* is the point where the planet comes closest to the sun; the *aphelion* is the farthest distance the planet travels from the sun. The eccentricity of a planet tells us the amount of roundness of that planet's orbit. The eccentricity of a circle is 0 and that of a parabola is 1. The eccentricity for each planet in our solar system is given here:

PLANET	ECCENTRICITY
Mercury	.194
Venus	.007
Earth	.017
Mars	.093
Jupiter	.048
Saturn	.056
Uranus	.047
Neptune	.009
Pluto	.249

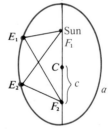

Example 1

The orbit of the earth around the sun is elliptical with the sun at one focus. If the semimajor axis of this orbit is 9.3×10^7 mi and the eccentricity is about .017, determine the greatest and least distance of the earth from the sun (correct to two significant digits).

Solution

We are given $a = 9.3 \times 10^7$ and $\varepsilon = .017$. Now

$$\varepsilon = \frac{c}{a}$$

$$.017 = \frac{c}{9.3 \times 10^7}$$

$$c \approx 1.581 \times 10^6$$

The greatest distance is

$$a + c \approx 9.5 \times 10^7$$

The least distance is

$$a - c \approx 9.1 \times 10^7$$

The orbit of a satellite can be calculated from Kepler's third law:

$$\frac{\text{Mass of (planet + satellite)}}{\text{Mass of (sun + planet + satellite)}} = \frac{(\text{semimajor axis of satellite orbit})^3}{(\text{semimajor axis of planet orbit})^3} \times \frac{(\text{period of planet})^2}{(\text{period of satellite})^2}$$

Example 2

If the mass of the earth is 6.58×10^{21} tons, the sun 2.2×10^{27} tons, and the moon 8.1×10^{19} tons, calculate the orbit of the moon. Assume that the semimajor axis of the earth is 9.3×10^7 mi, the period of the earth 365.25 days, and the period of the moon 27.3 days.

352

Solution   First solve the equation given as Kepler's third law for the unknown—the semimajor axis of the satellite orbit:

(Semimajor axis of satellite orbit)^3

$$= \frac{\text{mass of (planet + satellite)}}{\text{mass of (sun + planet + satellite)}} \times \frac{(\text{period of satellite})^2}{(\text{period of planet})^2} \times (\text{semimajor axis of planet orbit})^3$$

Thus,

(Semimajor axis of satellite orbit)^3

$$= \frac{(6.58 \cdot 10^{21}) + (8.1 \cdot 10^{19})}{(2.2 \cdot 10^{27}) + (6.58 \cdot 10^{21}) + (8.1 \cdot 10^{19})} \times \frac{(27.3)^2}{(365.25)^2} \times (9.3 \cdot 10^7)^3$$

$$\approx \frac{3.993131141 \cdot 10^{48}}{2.934975261 \cdot 10^{32}}$$

$$\approx 1.360533151 \cdot 10^{16}$$

Semimajor axis of satellite orbit $\approx 2.387278258 \cdot 10^5$

The moon's orbit has a semimajor axis of about 239,000 mi.   ☐

Problems for Further Study—Planetary Orbits

1. *Astronomy.* The orbit of Mars about the sun is elliptical with the sun at one focus. If the semimajor axis of this orbit is 1.4×10^8 mi and the eccentricity is about .093, determine the greatest and least distance of Mars from the sun, correct to two significant digits.

2. *Astronomy.* The orbit of Venus about the sun is elliptical with the sun at one focus. If the semimajor axis of this orbit is 6.7×10^7 mi and the eccentricity is about .007, determine the greatest and least distance of Venus from the sun, correct to two significant digits.

3. *Astronomy.* The orbit of Neptune about the sun is elliptical with the sun at one focus. If the semimajor axis of this orbit is 3.66×10^9 mi and the eccentricity is about .009, determine the greatest and least distance of Neptune from the sun, correct to two significant digits.

4. *Astronomy.* If a Planet X has an elliptical orbit about the sun with a semimajor axis of 6.89×10^{10} mi and eccentricity of about .35, what is the least distance between Planet X and the sun?

5. *Astronomy.* If the mass of Mars is 7.05×10^{20} tons, its satellite Phobos 4.16×10^{12} tons, and the sun 2.2×10^{27} tons, calculate the distance of the semimajor axis of Phobos. Assume that the semimajor axis of Mars is 1.4×10^8 mi, the period of Mars 693.5 days, and the period of Phobos .3 day.

6. *Astronomy.* Use the information in Problem 5 to find the approximate mass of the Martian satellite Deimos if its elliptical orbit has a semimajor axis of 15,000 mi and a period of 1.26 days.

7. *Astronomy.* If the perihelion distance of Mercury is 2.9×10^7 mi and the aphelion distance is 4.3×10^7 mi, write the equation for the orbit of Mercury.
8. *Astronomy.* If the perihelion distance of Venus is 6.7234×10^7 mi and the aphelion distance is 6.8174×10^7 mi, write the equation for the orbit of Venus.
9. *Astronomy.* If the perihelion distance of Earth is 9.225×10^7 mi and the aphelion distance is 9.542×10^7 mi, write the equation for the orbit of Earth.
10. *Astronomy.* If the perihelion distance of Mars is 1.29×10^8 mi and the aphelion distance is 1.55×10^8 mi, write the equation for the orbit of Mars.

CALCULATORS

Appendix A provides a brief introduction to calculators and calculator usage. In the last few years, pocket calculators have been one of the fastest growing items in the United States. There are probably two reasons for this increase in popularity. Most people (including mathematicians!) do not like to do arithmetic, and a good calculator can be purchased for under $20.

This book was written with the assumption that students have access to a calculator. Throughout the text, references concerning the use of calculators are made. Trigonometry is the first course most students take that *requires* the use of a calculator, and it is important that the text instruct the students in how to use a calculator. This appendix is included to help the student choose a calculator and understand the calculator comments in this book.

Calculators are classified by the types of problems they are equipped to handle, as well as by the type of logic for which they are programmed. The task of selecting a calculator is compounded by the multiplicity of brands from which to choose.

The different types of calculators are distinguished primarily by their price.

1. *Four-function calculators* (*under* $10).
   These calculators have a keyboard consisting of the numerals and the four arithmetic operations, or functions: addition $\boxed{+}$, subtraction $\boxed{-}$, multiplication $\boxed{\times}$, and division $\boxed{\div}$.

2. *Four-function calculators with memory* ($10–$20).
   Usually no more expensive than four-function calculators, these offer a memory register: $\boxed{M}$, $\boxed{STO}$, or $\boxed{M^+}$. The more expensive models may have more than one memory register. Memory registers allow you to store partial calculations for later recall. Some models will even remember the total when they are turned off.

3. *Scientific calculators* ($20–$50).
   These calculators add additional mathematical functions, such as square root $\boxed{\sqrt{}}$, trigonometric $\boxed{\sin}$, $\boxed{\cos}$, and $\boxed{\tan}$, and logarithmic $\boxed{\log}$ and $\boxed{\exp}$. Depending on the particular brand, a scientific model may have other keys as well.

4. *Special-purpose calculators* ($40–$400).
   Special-use calculators for business, statistics, surveying, medicine, or even gambling and chess are available.

5. *Programmable calculators* ($50–$600).
   With these calculators you can enter a *sequence* of steps for the calculator to repeat on your command. Some of these calculators allow the insertion of different cards that "remember" the sequence of steps for complex calculations.

6. *Graphics calculators* ($60–$300).

These calculators (which include the Casio fx-7000g, the Sharp EL-5200, and the HP-35) not only can do arithmetic operations but also can graph and do algebraic manipulations.

For most nonscientific purposes, a four-function calculator with memory will be sufficient for everyday usage. **For this book you will need a scientific calculator.** Three types of logic are used by scientific calculators: arithmetic, algebraic, and RPN. You will need to know the type of logic used by your calculator. To determine the type of logic used by a particular calculator, try this test problem:

If the answer shown is 20, it is an arithmetic-logic calculator. If the answer is 14 (the correct answer), then it is an algebraic-logic calculator. If the calculator has no equal key $\boxed{=}$ but has an $\boxed{\text{ENTER}}$ or $\boxed{\text{SAVE}}$ key, then it is an RPN-logic calculator. An RPN-logic calculator will give the answer as 14. In algebra you learn to perform multiplication before addition, so that the correct value for $2 + 3 \times 4$ is 14 (multiply first). An algebraic calculator will "know" this fact and will give the correct answer, whereas an arithmetic calculator will simply work from left to right to obtain the incorrect answer, 20. Therefore, if you have an arithmetic-logic calculator, you will need to be careful about the order of operations. Some arithmetic-logic calculators provide parentheses $\boxed{(}$ $\boxed{)}$ so that operations can be grouped as in

$$\boxed{2}\ \boxed{+}\ \boxed{(}\ \boxed{3}\ \boxed{\times}\ \boxed{4}\ \boxed{)}\ \boxed{=}$$

but then you must remember to insert the parentheses.

With an RPN calculator, the operation symbol is entered after the numbers have been entered. These three types of logic can be illustrated by the problem $2 + 3 \times 4$.

Arithmetic Logic	Algebraic Logic	RPN Logic
$\boxed{3}$	$\boxed{2}$	$\boxed{2}$
$\boxed{\times}$	$\boxed{+}$	$\boxed{\text{ENTER}}$
$\boxed{4}$	$\boxed{3}$	$\boxed{3}$
$\boxed{=}$	$\boxed{\times}$	$\boxed{\text{ENTER}}$
$\boxed{+}$	$\boxed{4}$	$\boxed{4}$
$\boxed{2}$	$\boxed{=}$	$\boxed{\times}$
$\boxed{=}$		$\boxed{+}$

In this book we illustrate the examples using algebraic- and RPN-logic calculators. The keys to be pressed are indicated by boxes drawn around the numbers and operational signs as shown in the preceding table. Numerals for calculator display are illustrated as

0 1 2 3 4 5 6 7 8 9

This Sharp calculator is an example of a calculator with arithmetic logic. Courtesy of Sharp Electronics Corporation.

The Texas Instruments SR-51-II is an example of a calculator with algebraic logic. Courtesy of Texas Instruments, Inc.

The Hewlett-Packard calculator uses RPN logic. Courtesy of Hewlett-Packard.

Regardless of the type of logic your calculator uses, it is a good idea to check your owner's manual for each type of problem illustrated in the text because there are many different brands of calculators on the market, and many have slight variations in keyboards.

Even though most of you may not yet have a graphics calculator, I have included some Calculator Graphics applications in the text so you can see what you can do with these calculators. These sections are set apart in a box and are optional. In writing these sections, I used a Casio fx-7000g calculator for which I paid the nondiscounted price of $79.

There is also a limit to the accuracy of your calculator. You may have a calculator with a 6-, 8-, or 10-digit display. You can test the accuracy with the following example.

Example 1

Find $2 \div 3$.

Solution

Algebraic: $\boxed{2}$ $\boxed{\div}$ $\boxed{3}$ $\boxed{=}$

RPN: $\boxed{2}$ $\boxed{\text{ENTER}}$ $\boxed{3}$ $\boxed{\div}$

DISPLAY: 2   2.   3   0.6666666667   □

There may be some discrepancy between this answer and the one you obtain on your calculator. Some machines will not round the answer as shown here but will show the display

0.6666666666

Others will show a display such as

6.6666−01

This is a number in scientific notation and should be interpreted as

6.6666×10^{-1} or   .66666

Most calculators will also use scientific notation when the numbers become larger than allowed by their display register. Example 2 will show you how your calculator handles large numbers. Some calculators will simply show an overflow and will not accept larger numbers. You will want a calculator that will accept and handle large and small numbers.

Example 2

Find 50^6.

Solution

Check your owner's manual to find out how to use an exponent key. Most calculators will work as shown below:

Algebraic: $\boxed{50}$ $\boxed{y^x}$ $\boxed{6}$

RPN: $\boxed{50}$ $\boxed{\text{ENTER}}$ $\boxed{6}$ $\boxed{y^x}$

When the maximum size of the display has been reached, the calculator should automatically switch to scientific notation. The point at which a calculator will do this varies from one type or brand to another. The answer for this example is

$1.5625 \times 10^{10} = 15,625,000,000$ ❑

Answers on a calculator can sometimes be inverted so that a word is shown in the display.

Example 3

The answer for the calculation

$5 \times .9547 + 2 \times 353$

gives an old Arab proverb. What is the proverb?

Solution

Algebraic: $\boxed{5}$ $\boxed{\times}$ $\boxed{.9547}$ $\boxed{+}$ $\boxed{2}$ $\boxed{\times}$ $\boxed{353}$ $\boxed{=}$

RPN: $\boxed{5}$ $\boxed{\text{ENTER}}$ $\boxed{.9547}$ $\boxed{\times}$ $\boxed{2}$ $\boxed{\text{ENTER}}$ $\boxed{353}$ $\boxed{\times}$ $\boxed{+}$

The display is

710.7735

Now turn the display upside-down to read the answer:

SELL·OIL   ❑

Some problems involving calculator computations are given in Appendix B.

ACCURACY AND ROUNDING

Applications involving measurements can never be exact. Appendix B discusses some agreements concerning the accuracy of our calculations. It is particularly important that you pay attention to the accuracy of your measurements when you have access to a calculator, because calculator processes can give you a false sense of security about the accuracy in a particular problem. For example, if you measure a triangle and find the sides are approximately 1.2 and 3.4 and then find the ratio of $1.2/3.4 \approx .35294117$, it appears that the result is more accurate than the original measurements! Some discussion about the accuracy of results follows in this section.

The digits known to be correct in a number obtained by a measurement are called **significant digits**. The digits 1, 2, 3, 4, 5, 6, 7, 8, and 9 are always significant, whereas the digit 0 may or may not be significant.

1. Zeros that come between two other digits are significant, as in 203 or 10.04.

2. If the zero's only function is to place the decimal point, it is not significant, as in

$$.000023 \qquad \text{or} \qquad 23,000$$
$$\uparrow \qquad\qquad\qquad \uparrow$$
$$\textit{Placeholders} \qquad \textit{Placeholders}$$

If it does more than fix the decimal point, it is significant, as in

$$.00230 \qquad\qquad \text{or} \qquad\qquad 23,000.01$$
$$\uparrow \qquad\qquad\qquad\qquad\qquad\qquad \uparrow$$

This digit is significant.   *These are significant, since they come between two other digits.*

This second rule can, of course, result in certain ambiguities, such as in 23,000 (measured to the *exact* unit). To avoid such confusion, we use scientific notation in this case.

2.3×10^4 has two significant digits

2.3000×10^4 has five significant digits

Numbers that come about by counting are considered to be exact and are correct to any number of significant digits. Since exponents are usually counting numbers, in this book we will not consider exponents in deciding on the proper number of significant digits used in a particular calculation.

Example 1

a. Two significant digits: 46,   .00083, 4.0×10^1,   .050
b. Three significant digits: 523,   403, 4.00×10^2,   .000800
c. Four significant digits: 600.1, 4.000×10^1,   .0002345

When we are doing calculations with approximate numbers (particularly when using a calculator), it is often necessary to round off results. In this book we use the following rounding procedure.

ROUNDING PROCEDURE

> To round off numbers:
>
> 1. Increase the last retained digit by 1 if the remainder is greater than or equal to 5; or
> 2. Retain the last digit unchanged if the remainder is less than 5.
> 3. In Problems requiring rounding but involving several steps, round only once, at the end. That is, do not work with rounded results, since round-off errors can accumulate.

Elaborate rules for computation of approximate data can be developed (when it is necessary for some applications, such as in chemistry), but there are two simple rules that will work satisfactorily for the material in this text.

RULES FOR
SIGNIFICANT DIGITS

> *Addition–Subtraction:* Add or subtract in the usual fashion, and then round off the result so that the last digit retained is in the column farthest to the right in which both given numbers have significant digits.
>
> *Multiplication–Division:* Multiply or divide in the usual fashion, and then round off the results to the smaller number of significant digits found in either of the given numbers.
>
> *Counting numbers:* Numbers used to count or whole numbers used as exponents are considered to be correct to any number of significant digits.
>
> *Multiple operations:* Apply the rounding rules given above for significant digits only when stating your final answer, in order to avoid errors due to repeated rounding.

Example 2

Calculate

$$b = \frac{50}{\tan 35°}$$

by division, by multiplication, and by calculator. Compare the results and then give the answer to the correct number of significant digits.

Solution

By division, use Appendix D Table: $\tan 35° = .7002$. (When we say $\tan 35° = .7002$, we mean, of course, $\tan 35° \approx .7002$. We often write the former for convenience, but keep in mind that table values are approximate. Appendix D Table is correct to four significant digits.) Thus,

$$b \approx \frac{50}{.7002} = 71.408169090\ldots$$

By multiplication:

$$b = \frac{50}{\tan 35°}$$
$$= 50 \cot 35°$$
$$\approx 50(1.428)$$
$$= 71.400$$

By calculator with algebraic logic (set to degrees):

| 50 | | ÷ | | 35 | | tan | | = | DISPLAY: 71.407401

By calculator with RPN logic (set to degrees):

| 50 | | ENTER | | 35 | | tan | | ÷ | DISPLAY: 71.40740034

Notice that all the answers above differ. We now use the multiplication–division rule.

50:   This number has one or two significant digits; there is no ambiguity if we write 5×10^1 or 5.0×10^1. *In this book*, if the given data include a number whose degree of accuracy is doubtful, *we assume the maximum degree of accuracy.* Thus, 50 has two significant digits.

tan 35°:   From Appendix D Table we find this number to four significant digits; on a calculator you may have 8, 10, or 12 significant digits (depending on the calculator).

The result of this division is correct to two significant digits—namely,

$$b = \frac{50}{\tan 35°} = 71$$

which agrees with all the above methods of solution.   ▯

In working with triangles in this text, we assume a certain relationship in the accuracy of the measurement between the sides and the angles.

Accuracy in Sides	Equivalent Accuracy in Angles
Two significant digits	Nearest degree
Three significant digits	Nearest tenth of a degree
Four significant digits	Nearest hundredth of a degree

This chart means that, if the data include one side given with two significant digits and another with three significant digits, the angle would be computed to the nearest degree. If one side is given to four significant digits and an angle to the nearest tenth of a degree, then the other sides would be given to three significant digits and the angles computed to the nearest tenth of a degree. In general, results computed from the above table should not be more accurate than the least accurate item of the given data.

WARNING! Note this agreement about accuracy.

If you have access only to a four-function calculator, you can use Appendix D Table in conjunction with your calculator. For example, to find b, you first find $\tan 35° = .7002$ and then calculate

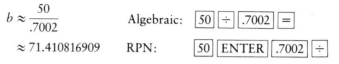

$$b \approx \frac{50}{.7002}$$

Algebraic: 50 ÷ .7002 =

$$\approx 71.410816909$$

RPN: 50 ENTER .7002 ÷

or, to two significant digits, $b = 71$.

The following problems require a calculator. If you do not have a calculator, you may either skip these problems or work them using logarithms.

Problem Set B

A   *Use a calculator to evaluate each of the expressions in Problems 1–12. Be sure to round off each answer to the appropriate number of significant digits. If you have a calculator with ten decimal places, you can use the hints to check your numerical answers before rounding. To answer the question in each hint with words, perform the numerical calculation, turn your calculator over, and read the word formed. Then round your answer. For example, if an answer is*

     53 1907018

and you turn the calculator over, it will spell

     8I OLObIES

1. (14)(351)   (*Hint:* What is the opposite of low?)

2. (218)(263)   (*Hint:* You step on them.)

3. $(2.00)^4(1,245)(277)$ (*Hint:* How do you like your calculator?)

4. $(3.00)^3(182)$ (*Hint:* How are you feeling?)

5. $\dfrac{(1,979)(1,356)}{452}$ (*Hint:* What is above your feet?)

6. $\dfrac{(515)(20,600)}{200}$ (*Hint:* What is below your feet?)

7. $(990)(1,117)(342) - 89$ (*Hint:* This problem is illegible.)

8. $[.14 + (197)(25.08)](19)$ (*Hint:* How will you raise the money?)

9. $\dfrac{1.00}{.005 + .020}$ (*Hint:* This is fun!)

10. $\dfrac{1.500 \times 10^4 + (7.000)(67.00)}{20,000}$ (*Hint:* Which credit card will you use to purchase these tires?)

11. $3,478.06 + 2,256.028 + 1,979.919 + .00091$ (*Hint:* Where do you live on the pipeline?)

12. $57,300 + .094 + 32.3 + 2.09 + .0074$ (*Hint:* How can you look taller?)

B   *Use a calculator to evaluate each of the expressions in Problems 13–30. Be sure to round off each answer to the appropriate number of significant digits.*

13. $(6.28)^{1/2}(4.85)$

14. $(8.23)^{1/2}(6.14)$

15. $\dfrac{1.00}{\sqrt{4.83} + \sqrt{2.51}}$

16. $\dfrac{1.00}{\sqrt{8.48} - \sqrt{21.3}}$

17. $[(4.083)^2(4.283)^3]^{-2/3}$

18. $[(6.128)^4(3.412)^2]^{-1/2}$

19. $\dfrac{(2.51)^2 + (6.48)^2 - (2.51)(6.48)(.3462)}{(2.51)(6.48)}$

20. $\dfrac{241^2 + 568^2 - (241)(568)(.5213)}{(241)(568)}$

21. $\dfrac{23.7 \sin 36.2°}{45.1}$

22. $\dfrac{461 \sin 43.8°}{1,215}$

23. $\dfrac{62.8 \sin 81.5°}{\sin 42.3°}$

24. $\dfrac{4.381 \sin 49.86°}{\sin 71.32°}$

25. $\dfrac{16 \sin 22°}{25}$

26. $\dfrac{42 \sin 52°}{61}$

27. $\dfrac{16^2 + 25^2 - 9.0^2}{(2.0)(16)(25)}$

28. $\dfrac{216^2 + 418^2 - 315^2}{(2.00)(216)(418)}$

29. $\dfrac{4.82^2 + 6.14^2 - 9.13^2}{(2.00)(4.82)(6.14)}$

30. $\dfrac{18.361^2 + 15.215^2 - 13.815^2}{(2.0000)(18.361)(15.215)}$

REVIEW OF GEOMETRY

This appendix will review those ideas from plane geometry that are necessary for the study of trigonometry.

Mensuration Formulas

Let A = area, P = perimeter, C = circumference, S = surface area, and V = volume.

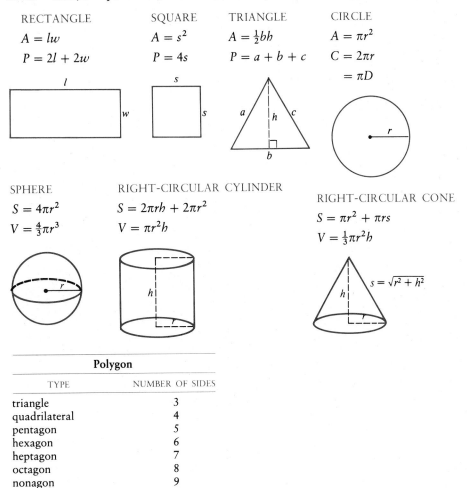

RECTANGLE

$A = lw$

$P = 2l + 2w$

SQUARE

$A = s^2$

$P = 4s$

TRIANGLE

$A = \frac{1}{2}bh$

$P = a + b + c$

CIRCLE

$A = \pi r^2$

$C = 2\pi r$

$\quad = \pi D$

SPHERE

$S = 4\pi r^2$

$V = \frac{4}{3}\pi r^3$

RIGHT-CIRCULAR CYLINDER

$S = 2\pi rh + 2\pi r^2$

$V = \pi r^2 h$

RIGHT-CIRCULAR CONE

$S = \pi r^2 + \pi rs$

$V = \frac{1}{3}\pi r^2 h$

$s = \sqrt{r^2 + h^2}$

Polygon	
TYPE	NUMBER OF SIDES
triangle	3
quadrilateral	4
pentagon	5
hexagon	6
heptagon	7
octagon	8
nonagon	9
decagon	10
undecagon	11
dodecagon	12

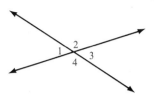

FIGURE C.1   Vertical angles. ∠1 and ∠3 are vertical, and ∠2 and ∠4 are also vertical.

Classification of Angles

Complementary angles are two angles whose measures add up to $\frac{\pi}{2}$ or $90°$.

Supplementary angles are two angles whose measures add up to π or $180°$.

Vertical angles are two nonadjacent angles formed by two intersecting lines.

Vertical angles are equal. The measures of angles 1 and 3 are equal and the measures of angles 2 and 4 are equal. (See Figure C.1.)

Adjacent angles are two angles that have the same vertex and a common side between them. In Figure C.1, ∠1 and ∠2 are adjacent, and ∠2 and ∠3 are adjacent. Can you name other adjacent pairs?

Parallel Lines

Definition: **Parallel lines** are straight lines in the same plane that do not intersect. A **transversal** of two or more lines is a line that cuts across these lines (see Figure C.2).
  Properties:

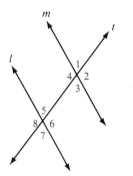

FIGURE C.2   Parallel lines l and m cut by a transversal t

1. **Interior angles** formed by two lines cut by a transversal are the angles between the two lines.
   a. ∠3, ∠4, ∠5, and ∠6 in Figure C.2 are interior angles.
   b. Adjacent interior angles are supplementary.
   c. If the lines are parallel, then each pair of interior angles on the same side of the transversal is supplementary.

2. **Exterior angles** formed by two lines cut by a transversal are the angles outside the two lines.
   a. ∠1, ∠2, ∠7, and ∠8 in Figure C.2 are exterior angles.
   b. Adjacent exterior angles are supplementary.
   c. If the lines are parallel, then each pair of exterior angles on the same side of the transversal is supplementary.

3. **Corresponding angles** formed by two lines cut by a transversal are the angles on the same side of the transversal and on the same side of the lines.
   a. ∠1 and ∠5, ∠2 and ∠6, ∠3 and ∠7, and ∠4 and ∠8 in Figure C.2 are corresponding angles.
   b. If two lines are parallel, then each pair of corresponding angles is equal.

4. **Alternate interior angles** of two lines cut by a transversal are nonadjacent angles between the two lines and on opposite sides of the transversal.
   a. ∠4 and ∠6, and ∠3 and ∠5 in Figure C.2 are alternate interior angles.
   b. If two lines are parallel, then each pair of alternate interior angles is equal.

Triangles

Definition: A **triangle** is a figure formed by three line segments (called **sides**) joining three noncollinear points (called **vertices**—plural of **vertex**). The **height** or **altitude** of a triangle is the length of a segment drawn from any vertex perpendicular to the opposite side, called the **base**.

Classification of triangles:

BY SIDES:	BY ANGLES:
scalene—no equal sides	**acute**—three acute angles
isosceles—at least two equal sides	**right**—one right angle
equilateral—three equal sides	**obtuse**—one obtuse angle

Congruent triangles: **Congruent triangles** are triangles that have the same size and shape (symbolized by ≅). (Remember, in the text it was agreed that when we say "angles are equal," we mean "the measures of the angles are equal," and when we say "line segments are equal," we mean "the measures of the line segments are equal.")

1. Corresponding parts of congruent triangles are equal.
2. (SAS) If two sides and the included angle of one triangle are equal to the corresponding parts of another, then the triangles are congruent.
3. (ASA) If two angles and the included side of one triangle are equal to the corresponding parts of another, then the triangles are congruent.
4. (SSS) If three sides of a triangle are equal to three sides of another, then the triangles are congruent.

Isosceles triangles:

1. If two sides of a triangle are equal, then the angles opposite these sides are equal.
2. If two angles of a triangle are equal, then the sides opposite these angles are equal.
3. An altitude of an isosceles triangle bisects its base.

Equilateral triangles:

1. If a triangle is equilateral, then it is equiangular.
2. If a triangle is equiangular, then it is equilateral.

Angles of a triangle:

1. The sum of the angles of a triangle is π or $180°$.
2. If two angles of one triangle are equal to two angles of another triangle, respectively, then the remaining angles are equal.
3. The acute angles of a right triangle are complementary.
4. A triangle can have no more than one right angle.
5. A triangle can have no more than one obtuse angle.

Right triangles:

1. *Pythagorean theorem.*   In a right triangle, the square of the hypotenuse equals the sum of the squares of the legs. (See Figure C.3.)
2. $30°$–$60°$–$90°$ *triangle:* The leg opposite the $30°$ angle equals one-half the hypotenuse, and the leg opposite the $60°$ angle equals one-half the hypotenuse

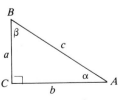

FIGURE C.3 Pythagorean theorem. If C is a right angle in $\triangle ABC$, then $a^2 + b^2 = c^2$.

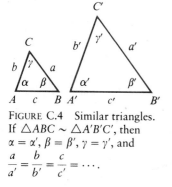

FIGURE C.4   Similar triangles. If $\triangle ABC \sim \triangle A'B'C'$, then $\alpha = \alpha'$, $\beta = \beta'$, $\gamma = \gamma'$, and
$$\frac{a}{a'} = \frac{b}{b'} = \frac{c}{c'} = \cdots .$$

times the square root of 3. In Figure C.3, if $\alpha = 30°$, then $\beta = 60°$ and $a = \frac{1}{2}c$ and $b = a\sqrt{3}$.

3. $45°{-}45°{-}90°$ *triangle:*   The leg opposite a 45° angle equals one-half the hypotenuse times the square root of 2. In Figure C.3, if $\alpha = 45°$, then $\beta = 45°$ and $a = \frac{1}{2}c\sqrt{2}$ and $b = \frac{1}{2}c\sqrt{2}$.

Similar triangles: Triangles whose corresponding angles are congruent and whose corresponding sides are in proportion (symbolized by $\sim$) are similar triangles.

1. The **ratio** of two quantities is the first divided by the second.
2. A **proportion** is a statement of equality of two ratios. (See Figure C.4.)
3. Two triangles are similar if two angles of one triangle are equal, respectively, to two angles of the other.
4. If two triangles are similar, then their corresponding sides are proportional:

$$\frac{a}{a'} = \frac{b}{b'}; \frac{a}{a'} = \frac{c}{c'}; \frac{b}{b'} = \frac{c}{c'}; \frac{a}{b} = \frac{a'}{b'}; \cdots$$

Parallelograms

Definition: A **parallelogram** is a quadrilateral whose opposite sides are parallel.
   Properties:

1. The opposite sides of a parallelogram are parallel.
2. A diagonal of a parallelogram divides it into two congruent triangles.
3. The opposite sides of a parallelogram are equal.
4. The opposite angles of a parallelogram are equal.
5. The consecutive angles of a parallelogram are supplementary.

Circles

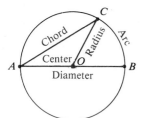

FIGURE C.5   Circle terminology.

Definition: A **circle** is the set of all points in a plane a given distance (called the **radius**) from a given point (called the **center**). (See Figure C.5.)
   Terminology:

1. **Circumference:**   distance around the circle, $2\pi r$.
2. **Chord:**   a line joining two points of the circle.
3. **Diameter:**   a chord through the center; AB in Figure C.5.
4. **Arc:**   part of a circle: $\overset{\frown}{BC}$, $\overset{\frown}{AC}$, or $\overset{\frown}{ACB}$ in Figure C.5.
5. To **intercept an arc** is to cut off the arc; in Figure C.5, $\angle COB$ intercepts $\overset{\frown}{BC}$.
6. **Tangent** of a circle is a line that intersects the circle at one and only one point.
7. **Secant** of a circle is a line that intersects the circle at two points.
8. **Inscribed polygon** is a polygon all of whose sides are chords of a circle.
9. **Inscribed circle** is a circle to which all the sides of a polygon are tangents.
10. **Circumscribed polygon** is a polygon all of whose sides are tangents to a circle.
11. **Circumscribed circle** is a circle passing through each vertex of a polygon.

TABLE OF TRIGONOMETRIC FUNCTIONS

Trigonometric Functions

RAD	DEG	COS	SIN	TAN		
.00	.0	1.0000	.0000	.0000	90.0	1.57
.00	.1	1.0000	.0017	.0017	89.9	1.57
.00	.2	1.0000	.0035	.0035	89.8	1.57
.01	.3	1.0000	.0052	.0052	89.7	1.57
.01	.4	1.0000	.0070	.0070	89.6	1.56
.01	.5	1.0000	.0087	.0087	89.5	1.56
.01	.6	.9999	.0105	.0105	89.4	1.56
.01	.7	.9999	.0122	.0122	89.3	1.56
.01	.8	.9999	.0140	.0140	89.2	1.56
.02	.9	.9999	.0157	.0157	89.1	1.56
.02	1.0	.9998	.0175	.0175	89.0	1.55
.02	1.1	.9998	.0192	.0192	88.9	1.55
.02	1.2	.9998	.0209	.0209	88.8	1.55
.02	1.3	.9997	.0227	.0227	88.7	1.55
.02	1.4	.9997	.0244	.0244	88.6	1.55
.03	1.5	.9997	.0262	.0262	88.5	1.54
.03	1.6	.9996	.0279	.0279	88.4	1.54
.03	1.7	.9996	.0297	.0297	88.3	1.54
.03	1.8	.9995	.0314	.0314	88.2	1.54
.03	1.9	.9995	.0332	.0332	88.1	1.54
.03	2.0	.9994	.0349	.0349	88.0	1.54
.04	2.1	.9993	.0366	.0367	87.9	1.53
.04	2.2	.9993	.0384	.0384	87.8	1.53
.04	2.3	.9992	.0401	.0402	87.7	1.53
.04	2.4	.9991	.0419	.0419	87.6	1.53
.04	2.5	.9990	.0436	.0437	87.5	1.53
.05	2.6	.9990	.0454	.0454	87.4	1.53
.05	2.7	.9989	.0471	.0472	87.3	1.52
.05	2.8	.9988	.0488	.0489	87.2	1.52
.05	2.9	.9987	.0506	.0507	87.1	1.52
.05	3.0	.9986	.0523	.0524	87.0	1.52
.05	3.1	.9985	.0541	.0542	86.9	1.52
.06	3.2	.9984	.0558	.0559	86.8	1.51
.06	3.3	.9983	.0576	.0577	86.7	1.51
.06	3.4	.9982	.0593	.0594	86.6	1.51
.06	3.5	.9981	.0610	.0612	86.5	1.51
.06	3.6	.9980	.0628	.0629	86.4	1.51
.06	3.7	.9979	.0645	.0647	86.3	1.51
.07	3.8	.9978	.0663	.0664	86.2	1.50
.07	3.9	.9977	.0680	.0682	86.1	1.50
.07	4.0	.9976	.0698	.0699	86.0	1.50
		SIN	COS	COT	DEG	RAD

RAD	DEG	COS	SIN	TAN		
.07	4.0	.9976	.0698	.0699	86.0	1.50
.07	4.1	.9974	.0715	.0717	85.9	1.50
.07	4.2	.9973	.0732	.0734	85.8	1.50
.08	4.3	.9972	.0750	.0752	85.7	1.50
.08	4.4	.9971	.0767	.0769	85.6	1.49
.08	4.5	.9969	.0785	.0787	85.5	1.49
.08	4.6	.9968	.0802	.0805	85.4	1.49
.08	4.7	.9966	.0819	.0822	85.3	1.49
.08	4.8	.9965	.0837	.0840	85.2	1.49
.09	4.9	.9963	.0854	.0857	85.1	1.49
.09	5.0	.9962	.0872	.0875	85.0	1.48
.09	5.1	.9960	.0889	.0892	84.9	1.48
.09	5.2	.9959	.0906	.0910	84.8	1.48
.09	5.3	.9957	.0924	.0928	84.7	1.48
.09	5.4	.9956	.0941	.0945	84.6	1.48
.10	5.5	.9954	.0958	.0963	84.5	1.47
.10	5.6	.9952	.0976	.0981	84.4	1.47
.10	5.7	.9951	.0993	.0998	84.3	1.47
.10	5.8	.9949	.1011	.1016	84.2	1.47
.10	5.9	.9947	.1028	.1033	84.1	1.47
.10	6.0	.9945	.1045	.1051	84.0	1.47
.11	6.1	.9943	.1063	.1069	83.9	1.46
.11	6.2	.9942	.1080	.1086	83.8	1.46
.11	6.3	.9940	.1097	.1104	83.7	1.46
.11	6.4	.9938	.1115	.1122	83.6	1.46
.11	6.5	.9936	.1132	.1139	83.5	1.46
.12	6.6	.9934	.1149	.1157	83.4	1.46
.12	6.7	.9932	.1167	.1175	83.3	1.45
.12	6.8	.9930	.1184	.1192	83.2	1.45
.12	6.9	.9928	.1201	.1210	83.1	1.45
.12	7.0	.9925	.1219	.1228	83.0	1.45
.12	7.1	.9923	.1236	.1246	82.9	1.45
.13	7.2	.9921	.1253	.1263	82.8	1.45
.13	7.3	.9919	.1271	.1281	82.7	1.44
.13	7.4	.9917	.1288	.1299	82.6	1.44
.13	7.5	.9914	.1305	.1317	82.5	1.44
.13	7.6	.9912	.1323	.1334	82.4	1.44
.13	7.7	.9910	.1340	.1352	82.3	1.44
.14	7.8	.9907	.1357	.1370	82.2	1.43
.14	7.9	.9905	.1374	.1388	82.1	1.43
.14	8.0	.9903	.1392	.1405	82.0	1.43
		SIN	COS	COT	DEG	RAD

Trigonometric Functions (*continued*)

RAD	DEG	COS	SIN	TAN		
.14	**8.0**	.9903	.1392	.1405	**82.0**	1.43
.14	8.1	.9900	.1409	.1423	81.9	1.43
.14	8.2	.9898	.1426	.1441	81.8	1.43
.14	8.3	.9895	.1444	.1459	81.7	1.43
.15	8.4	.9893	.1461	.1477	81.6	1.42
.15	**8.5**	.9890	.1478	.1495	**81.5**	1.42
.15	8.6	.9888	.1495	.1512	81.4	1.42
.15	8.7	.9885	.1513	.1530	81.3	1.42
.15	8.8	.9882	.1530	.1548	81.2	1.42
.16	8.9	.9880	.1547	.1566	81.1	1.42
.16	**9.0**	.9877	.1564	.1584	**81.0**	1.41
.16	9.1	.9874	.1582	.1602	80.9	1.41
.16	9.2	.9871	.1599	.1620	80.8	1.41
.16	9.3	.9869	.1616	.1638	80.7	1.41
.16	9.4	.9866	.1633	.1655	80.6	1.41
.17	**9.5**	.9863	.1650	.1673	**80.5**	1.40
.17	9.6	.9860	.1668	.1691	80.4	1.40
.17	9.7	.9857	.1685	.1709	80.3	1.40
.17	9.8	.9854	.1702	.1727	80.2	1.40
.17	9.9	.9851	.1719	.1745	80.1	1.40
.17	**10.0**	.9848	.1736	.1763	**80.0**	1.40
.18	10.1	.9845	.1754	.1781	79.9	1.39
.18	10.2	.9842	.1771	.1799	79.8	1.39
.18	10.3	.9839	.1788	.1817	79.7	1.39
.18	10.4	.9836	.1805	.1835	79.6	1.39
.18	**10.5**	.9833	.1822	.1853	**79.5**	1.39
.19	10.6	.9829	.1840	.1871	79.4	1.39
.19	10.7	.9826	.1857	.1890	79.3	1.38
.19	10.8	.9823	.1874	.1908	79.2	1.38
.19	10.9	.9820	.1891	.1926	79.1	1.38
.19	**11.0**	.9816	.1908	.1944	**79.0**	1.38
.19	11.1	.9813	.1925	.1962	78.9	1.38
.20	11.2	.9810	.1942	.1980	78.8	1.38
.20	11.3	.9806	.1959	.1998	78.7	1.37
.20	11.4	.9803	.1977	.2016	78.6	1.37
.20	**11.5**	.9799	.1994	.2035	**78.5**	1.37
.20	11.6	.9796	.2011	.2053	78.4	1.37
.20	11.7	.9792	.2028	.2071	78.3	1.37
.21	11.8	.9789	.2045	.2089	78.2	1.36
.21	11.9	.9785	.2062	.2107	78.1	1.36
.21	**12.0**	.9871	.2079	.2126	**78.0**	1.36
.21	12.1	.9778	.2096	.2144	77.9	1.36
.21	12.2	.9774	.2113	.2162	77.8	1.36
.21	12.3	.9770	.2130	.2180	77.7	1.36
.22	12.4	.9767	.2147	.2199	77.6	1.35
.22	**12.5**	.9763	.2164	.2217	**77.5**	1.35
.22	12.6	.9759	.2181	.2235	77.4	1.35
.22	12.7	.9755	.2198	.2254	77.3	1.35
.22	12.8	.9751	.2215	.2272	77.2	1.35
.23	12.9	.9748	.2233	.2290	77.1	1.35
.23	**13.0**	.9744	.2250	.2309	**77.0**	1.34
		SIN	COS	COT	DEG	RAD

RAD	DEG	COS	SIN	TAN		
.23	**13.0**	.9744	.2250	.2309	**77.0**	1.34
.23	13.1	.9740	.2267	.2327	76.9	1.34
.23	13.2	.9736	.2284	.2345	76.8	1.34
.23	13.3	.9732	.2300	.2364	76.7	1.34
.23	13.4	.9728	.2317	.2382	76.6	1.34
.24	**13.5**	.9724	.2334	.2401	**76.5**	1.34
.24	13.6	.9720	.2351	.2419	76.4	1.33
.24	13.7	.9715	.2368	.2438	76.3	1.33
.24	13.8	.9711	.2385	.2456	76.2	1.33
.24	13.9	.9707	.2402	.2475	76.1	1.33
.24	**14.0**	.9703	.2419	.2493	**76.0**	1.33
.25	14.1	.9699	.2436	.2512	75.9	1.32
.25	14.2	.9694	.2453	.2530	75.8	1.32
.25	14.3	.9690	.2470	.2549	75.7	1.32
.25	14.4	.9686	.2487	.2568	75.6	1.32
.25	**14.5**	.9681	.2504	.2586	**75.5**	1.32
.25	14.6	.9677	.2521	.2605	75.4	1.32
.26	14.7	.9673	.2538	.2623	75.3	1.31
.26	14.8	.9668	.2554	.2642	75.2	1.31
.26	14.9	.9664	.2571	.2661	75.1	1.31
.26	**15.0**	.9659	.2588	.2679	**75.0**	1.31
.26	15.1	.9655	.2605	.2698	74.9	1.31
.27	15.2	.9650	.2622	.2717	74.8	1.31
.27	15.3	.9646	.2639	.2736	74.7	1.30
.27	15.4	.9641	.2656	.2754	74.6	1.30
.27	**15.5**	.9636	.2672	.2773	**74.5**	1.30
.27	15.6	.9632	.2689	.2792	74.4	1.30
.27	15.7	.9627	.2706	.2811	74.3	1.30
.28	15.8	.9622	.2723	.2830	74.2	1.30
.28	15.9	.9617	.2740	.2849	74.1	1.29
.28	**16.0**	.9613	.2756	.2867	**74.0**	1.29
.28	16.1	.9608	.2773	.2886	73.9	1.29
.28	16.2	.9603	.2790	.2905	73.8	1.29
.28	16.3	.9598	.2807	.2924	73.7	1.29
.29	16.4	.9593	.2823	.2943	73.6	1.28
.29	**16.5**	.9588	.2840	.2962	**73.5**	1.28
.29	16.6	.9583	.2857	.2981	73.4	1.28
.29	16.7	.9578	.2874	.3000	73.3	1.28
.29	16.8	.9573	.2890	.3019	73.2	1.28
.29	16.9	.9568	.2907	.3038	73.1	1.28
.30	**17.0**	.9563	.2924	.3057	**73.0**	1.27
.30	17.1	.9558	.2940	.3076	72.9	1.27
.30	17.2	.9553	.2957	.3096	72.8	1.27
.30	17.3	.9548	.2974	.3115	72.7	1.27
.30	17.4	.9542	.2990	.3134	72.6	1.27
.31	**17.5**	.9537	.3007	.3153	**72.5**	1.27
.31	17.6	.9532	.3024	.3172	72.4	1.26
.31	17.7	.9527	.3040	.3191	72.3	1.26
.31	17.8	.9521	.3057	.3211	72.2	1.26
.31	17.9	.9516	.3074	.3230	72.1	1.26
.31	**18.0**	.9511	.3090	.3249	**72.0**	1.26
		SIN	COS	COT	DEG	RAD

Trigonometric Functions (*continued*)

RAD	DEG	COS	SIN	TAN		
.31	**18.0**	.9511	.3090	.3249	**72.0**	1.26
.32	18.1	.9505	.3107	.3269	71.9	1.25
.32	18.2	.9500	.3123	.3288	71.8	1.25
.32	18.3	.9494	.3140	.3307	71.7	1.25
.32	18.4	.9489	.3156	.3327	71.6	1.25
.32	**18.5**	.9483	.3173	.3346	**71.5**	1.25
.32	18.6	.9478	.3190	.3365	71.4	1.25
.33	18.7	.9472	.3206	.3385	71.3	1.24
.33	18.8	.9466	.3223	.3404	71.2	1.24
.33	18.9	.9461	.3239	.3424	71.1	1.24
.33	**19.0**	.9455	.3256	.3443	**71.0**	1.24
.33	19.1	.9449	.3272	.3463	70.9	1.24
.34	19.2	.9444	.3289	.3482	70.8	1.24
.34	19.3	.9438	.3305	.3502	70.7	1.23
.34	19.4	.9432	.3322	.3522	70.6	1.23
.34	**19.5**	.9426	.3338	.3541	**70.5**	1.23
.34	19.6	.9421	.3355	.3561	70.4	1.23
.34	19.7	.9415	.3371	.3581	70.3	1.23
.35	19.8	.9409	.3387	.3600	70.2	1.23
.35	19.9	.9403	.3404	.3620	70.1	1.22
.35	**20.0**	.9397	.3420	.3640	**70.0**	1.22
.35	20.1	.9391	.3437	.3659	69.9	1.22
.35	20.2	.9385	.3453	.3679	69.8	1.22
.35	20.3	.9379	.3469	.3699	69.7	1.22
.36	20.4	.9373	.3486	.3719	69.6	1.21
.36	**20.5**	.9367	.3502	.3739	**69.5**	1.21
.36	20.6	.9361	.3518	.3759	69.4	1.21
.36	20.7	.9354	.3535	.3779	69.3	1.21
.36	20.8	.9348	.3551	.3799	69.2	1.21
.36	20.9	.9342	.3567	.3819	69.1	1.21
.37	**21.0**	.9336	.3584	.3839	**69.0**	1.20
.37	21.1	.9330	.3600	.3859	68.9	1.20
.37	21.2	.9323	.3616	.3879	68.8	1.20
.37	21.3	.9317	.3633	.3899	68.7	1.20
.37	21.4	.9311	.3649	.3919	68.6	1.20
.38	**21.5**	.9304	.3665	.3939	**68.5**	1.20
.38	21.6	.9298	.3681	.3959	68.4	1.19
.38	21.7	.9291	.3697	.3979	68.3	1.19
.38	21.8	.9285	.3714	.4000	68.2	1.19
.38	21.9	.9278	.3730	.4020	68.1	1.19
.38	**22.0**	.9272	.3746	.4040	**68.0**	1.19
.39	22.1	.9265	.3762	.4061	67.9	1.19
.39	22.2	.9259	.3778	.4081	67.8	1.18
.39	22.3	.9252	.3795	.4101	67.7	1.18
.39	22.4	.9245	.3811	.4122	67.6	1.18
.39	**22.5**	.9239	.3827	.4142	**67.5**	1.18
.39	22.6	.9232	.3843	.4163	67.4	1.18
.40	22.7	.9225	.3859	.4183	67.3	1.17
.40	22.8	.9219	.3875	.4204	67.2	1.17
.40	22.9	.9212	.3891	.4224	67.1	1.17
.40	**23.0**	.9205	.3907	.4245	**67.0**	1.17
		SIN	COS	COT	DEG	RAD

RAD	DEG	COS	SIN	TAN		
.40	**23.0**	.9205	.3907	.4245	**67.0**	1.17
.40	23.1	.9198	.3923	.4265	66.9	1.17
.40	23.2	.9191	.3939	.4286	66.8	1.17
.41	23.3	.9184	.3955	.4307	66.7	1.16
.41	23.4	.9178	.3971	.4327	66.6	1.16
.41	**23.5**	.9171	.3987	.4348	**66.5**	1.16
.41	23.6	.9164	.4003	.4369	66.4	1.16
.41	23.7	.9157	.4019	.4390	66.3	1.16
.42	23.8	.9150	.4035	.4411	66.2	1.16
.42	23.9	.9143	.4051	.4431	66.1	1.15
.42	**24.0**	.9135	.4067	.4452	**66.0**	1.15
.42	24.1	.9128	.4083	.4473	65.9	1.15
.42	24.2	.9121	.4099	.4494	65.8	1.15
.42	24.3	.9114	.4115	.4515	65.7	1.15
.43	24.4	.9107	.4131	.4536	65.6	1.14
.43	**24.5**	.9100	.4147	.4557	**65.5**	1.14
.43	24.6	.9092	.4163	.4578	65.4	1.14
.43	24.7	.9085	.4179	.4599	65.3	1.14
.43	24.8	.9078	.4195	.4621	65.2	1.14
.43	24.9	.9070	.4210	.4642	65.1	1.14
.44	**25.0**	.9063	.4226	.4663	**65.0**	1.13
.44	25.1	.9056	.4242	.4684	64.9	1.13
.44	25.2	.9048	.4258	.4706	64.8	1.13
.45	25.3	.9041	.4274	.4727	64.7	1.13
.45	25.4	.9033	.4289	.4748	64.6	1.13
.45	**25.5**	.9026	.4305	.4770	**64.5**	1.13
.45	25.6	.9018	.4321	.4791	64.4	1.12
.45	25.7	.9011	.4337	.4813	64.3	1.12
.45	25.8	.9003	.4352	.4834	64.2	1.12
.45	25.9	.8996	.4368	.4856	64.1	1.12
.45	**26.0**	.8988	.4384	.4877	**64.0**	1.12
.46	26.1	.8980	.4399	.4899	63.9	1.12
.46	26.2	.8973	.4415	.4921	63.8	1.11
.46	26.3	.8965	.4431	.4942	63.7	1.11
.46	26.4	.8957	.4446	.4964	63.6	1.11
.46	**26.5**	.8949	.4462	.4986	**63.5**	1.11
.46	26.6	.8942	.4478	.5008	63.4	1.11
.47	26.7	.8934	.4493	.5029	63.3	1.10
.47	26.8	.8926	.4509	.5051	63.2	1.10
.47	26.9	.8918	.4524	.5073	63.1	1.10
.47	**27.0**	.8910	.4540	.5095	**63.0**	1.10
.47	27.1	.8902	.4555	.5117	62.9	1.10
.47	27.2	.8894	.4571	.5139	62.8	1.10
.48	27.3	.8886	.4586	.5161	62.7	1.09
.48	27.4	.8878	.4602	.5184	62.6	1.09
.48	**27.5**	.8870	.4617	.5206	**62.5**	1.09
.48	27.6	.8862	.4633	.5228	62.4	1.09
.48	27.7	.8854	.4648	.5250	62.3	1.09
.49	27.8	.8846	.4664	.5272	62.2	1.09
.49	27.9	.8838	.4679	.5295	62.1	1.08
.49	**28.0**	.8829	.4695	.5317	**62.0**	1.08
		SIN	COS	COT	DEG	RAD

Trigonometric Functions (*continued*)

RAD	DEG	COS	SIN	TAN		
.49	28.0	.8829	.4695	.5317	62.0	1.08
.49	28.1	.8821	.4710	.5340	61.9	1.08
.49	28.2	.8813	.4726	.5362	61.8	1.08
.49	28.3	.8805	.4741	.5384	61.7	1.08
.50	28.4	.8796	.4756	.5407	61.6	1.08
.50	28.5	.8788	.4772	.5430	61.5	1.07
.50	28.6	.8780	.4787	.5452	61.4	1.07
.50	28.7	.8771	.4802	.5475	61.3	1.07
.50	28.8	.8763	.4818	.5498	61.2	1.07
.50	28.9	.8755	.4833	.5520	61.1	1.07
.51	29.0	.8746	.4848	.5543	61.0	1.06
.51	29.1	.8738	.4863	.5566	60.9	1.06
.51	29.2	.8729	.4879	.5589	60.8	1.06
.51	29.3	.8721	.4894	.5612	60.7	1.06
.51	29.4	.8712	.4909	.5635	60.6	1.06
.51	29.5	.8704	.4924	.5658	60.5	1.06
.52	29.6	.8695	.4939	.5681	60.4	1.05
.52	29.7	.8686	.4955	.5704	60.3	1.05
.52	29.8	.8678	.4970	.5727	60.2	1.05
.52	29.9	.8669	.4985	.5750	60.1	1.05
.52	30.0	.8660	.5000	.5774	60.0	1.05
.53	30.1	.8652	.5015	.5797	59.9	1.05
.53	30.2	.8643	.5030	.5820	59.8	1.04
.53	30.3	.8634	.5045	.5844	59.7	1.04
.53	30.4	.8625	.5060	.5867	59.6	1.04
.53	30.5	.8616	.5075	.5890	59.5	1.04
.53	30.6	.8607	.5090	.5914	59.4	1.04
.54	30.7	.8599	.5105	.5938	59.3	1.03
.54	30.8	.8590	.5120	.5961	59.2	1.03
.54	30.9	.8581	.5135	.5985	59.1	1.03
.54	31.0	.8572	.5150	.6009	59.0	1.03
.54	31.1	.8563	.5165	.6032	58.9	1.03
.54	31.2	.8554	.5180	.6056	58.8	1.03
.55	31.3	.8545	.5195	.6080	58.7	1.02
.55	31.4	.8536	.5210	.6104	58.6	1.02
.55	31.5	.8526	.5225	.6128	58.5	1.02
.55	31.6	.8517	.5240	.6152	58.4	1.02
.55	31.7	.8508	.5255	.6176	58.3	1.02
.56	31.8	.8499	.5270	.6200	58.2	1.02
.56	31.9	.8490	.5284	.6224	58.1	1.01
.56	32.0	.8480	.5299	.6249	58.0	1.01
.56	32.1	.8471	.5314	.6273	57.9	1.01
.56	32.2	.8462	.5329	.6297	57.8	1.01
.56	32.3	.8453	.5344	.6322	57.7	1.01
.57	32.4	.8443	.5358	.6346	57.6	1.01
.57	32.5	.8434	.5373	.6371	57.5	1.00
		SIN	COS	COT	DEG	RAD

RAD	DEG	COS	SIN	TAN		
.57	32.5	.8434	.5373	.6371	57.5	1.00
.57	32.6	.8425	.5388	.6395	57.4	1.00
.57	32.7	.8415	.5402	.6420	57.3	1.00
.57	32.8	.8406	.5417	.6445	57.2	1.00
.57	32.9	.8396	.5432	.6469	57.1	1.00
.58	33.0	.8387	.5446	.6494	57.0	.99
.58	33.1	.8377	.5461	.6519	56.9	.99
.58	33.2	.8368	.5476	.6544	56.8	.99
.58	33.3	.8358	.5490	.6569	56.7	.99
.58	33.4	.8348	.5505	.6594	56.6	.99
.58	33.5	.8339	.5519	.6619	56.5	.99
.59	33.6	.8329	.5534	.6644	56.4	.98
.59	33.7	.8320	.5548	.6669	56.3	.98
.59	33.8	.8310	.5563	.6694	56.2	.98
.59	33.9	.8300	.5577	.6720	56.1	.98
.59	34.0	.8290	.5592	.6745	56.0	.98
.60	34.1	.8281	.5606	.6771	55.9	.98
.60	34.2	.8271	.5621	.6796	55.8	.97
.60	34.3	.8261	.5635	.6822	55.7	.97
.60	34.4	.8251	.5650	.6847	55.6	.97
.60	34.5	.8241	.5664	.6873	55.5	.97
.60	34.6	.8231	.5678	.6899	55.4	.97
.61	34.7	.8221	.5693	.6924	55.3	.97
.61	34.8	.8211	.5707	.6950	55.2	.96
.61	34.9	.8202	.5721	.6976	55.1	.96
.61	35.0	.8192	.5736	.7002	55.0	.96
.61	35.1	.8181	.5750	.7028	54.9	.96
.61	35.2	.8171	.5764	.7054	54.8	.96
.62	35.3	.8161	.5779	.7080	54.7	.95
.62	35.4	.8151	.5793	.7107	54.6	.95
.62	35.5	.8141	.5807	.7133	54.5	.95
.62	35.6	.8131	.5821	.7159	54.4	.95
.62	35.7	.8121	.5835	.7186	54.3	.95
.62	35.8	.8111	.5850	.7212	54.2	.95
.63	35.9	.8100	.5864	.7239	54.1	.94
.63	36.0	.8090	.5878	.7265	54.0	.94
.63	36.1	.8080	.5892	.7292	53.9	.94
.63	36.2	.8070	.5906	.7319	53.8	.94
.63	36.3	.8059	.5920	.7346	53.7	.94
.64	36.4	.8049	.5934	.7373	53.6	.94
.64	36.5	.8039	.5948	.7400	53.5	.93
.64	36.6	.8028	.5962	.7427	53.4	.93
.64	36.7	.8018	.5976	.7454	53.3	.93
.64	36.8	.8007	.5990	.7481	53.2	.93
.64	36.9	.7997	.6004	.7508	53.1	.93
.65	37.0	.7986	.6018	.7536	53.0	.93
		SIN	COS	COT	DEG	RAD

Trigonometric Functions (*continued*)

RAD	DEG	COS	SIN	TAN		
.65	37.0	.7986	.6018	.7536	53.0	.93
.65	37.1	.7976	.6032	.7563	52.9	.92
.65	37.2	.7965	.6046	.7590	52.8	.92
.65	37.3	.7955	.6060	.7618	52.7	.92
.65	37.4	.7944	.6074	.7646	52.6	.92
.65	37.5	.7934	.6088	.7673	52.5	.92
.66	37.6	.7923	.6101	.7701	52.4	.91
.66	37.7	.7912	.6115	.7729	52.3	.91
.66	37.8	.7902	.6129	.7757	52.2	.91
.66	37.9	7891	.6143	.7785	52.1	.91
.66	38.0	.7880	.6157	.7813	52.0	.91
.66	38.1	.7869	.6170	.7841	51.9	.91
.67	38.2	.7859	.6184	.7869	51.8	.90
.67	38.3	.7848	.6198	.7898	51.7	.90
.67	38.4	.7837	.6211	.7926	51.6	.90
.67	38.5	.7826	.6225	.7954	51.5	.90
.67	38.6	.7815	.6239	.7983	51.4	.90
.68	38.7	.7804	.6252	.8012	51.3	.90
.68	38.8	.7793	.6266	.8040	51.2	.89
.68	38.9	.7782	.6280	.8069	51.1	.89
.68	39.0	.7771	.6293	.8098	51.0	.89
.68	39.1	.7760	.6307	.8127	50.9	.89
.68	39.2	.7749	.6320	.8156	50.8	.89
.69	39.3	.7738	.6334	.8185	50.7	.88
.69	39.4	.7727	.6347	.8214	50.6	.88
.69	39.5	.7716	.6361	.8243	50.5	.88
.69	39.6	.7705	.6374	.8273	50.4	.88
.69	39.7	.7694	.6388	.8302	50.3	.88
.69	39.8	.7683	.6401	.8332	50.2	.88
.70	39.9	.7672	.6414	.8361	50.1	.87
.70	40.0	.7660	.6428	.8391	50.0	.87
.70	40.1	.7649	.6441	.8421	49.9	.87
.70	40.2	.7638	.6455	.8451	49.8	.87
.70	40.3	.7627	.6468	.8481	49.7	.87
.71	40.4	.7615	.6481	.8511	49.6	.87
.71	40.5	.7604	.6494	.8541	49.5	.86
.71	40.6	.7593	.6508	.8571	49.4	.86
.71	40.7	.7581	.6521	.8601	49.3	.86
.71	40.8	.7570	.6534	.8632	49.2	.86
.71	40.9	.7559	.6547	.8662	49.1	.86
.72	41.0	.7547	.6561	.8693	49.0	.86
.72	41.1	.7536	.6574	.8724	48.9	.85
.72	41.2	.7524	.6587	.8754	48.8	.85
.72	41.3	.7513	.6600	.8785	48.7	.85
.72	41.4	.7501	.6613	.8816	48.6	.85
.72	41.5	.7490	.6626	.8847	48.5	.85
		SIN	COS	COT	DEG	RAD

RAD	DEG	COS	SIN	TAN		
.72	41.5	.7490	.6626	.8847	48.5	.85
.73	41.6	.7478	.6639	.8878	48.4	.84
.73	41.7	.7466	.6652	.8910	48.3	.84
.73	41.8	.7455	.6665	.8941	48.2	.84
.73	41.9	.7443	.6678	.8972	48.1	.84
.73	42.0	.7431	.6691	.9004	48.0	.84
.73	42.1	.7420	.6704	.9036	47.9	.84
.74	42.2	.7408	.6717	.9067	47.8	.83
.74	42.3	.7396	.6730	.9099	47.7	.83
.74	42.4	.7385	.6743	.9131	47.6	.83
.74	42.5	.7373	.6756	.9163	47.5	.83
.74	42.6	.7361	.6769	.9195	47.4	.83
.75	42.7	.7349	.6782	.9228	47.3	.83
.75	42.8	.7337	.6794	.9260	47.2	.82
.75	42.9	.7325	.6807	.9293	47.1	.82
.75	43.0	.7314	.6820	.9325	47.0	.82
.75	43.1	.7302	.6833	.9358	46.9	.82
.75	43.2	.7290	.6845	.9391	46.8	.82
.76	43.3	.7278	.6858	.9424	46.7	.82
.76	43.4	.7266	.6871	.9457	46.6	.81
.76	43.5	.7254	.6884	.9490	46.5	.81
.76	43.6	.7242	.6896	.9523	46.4	.81
.76	43.7	.7230	.6909	.9556	46.3	.81
.76	43.8	.7218	.6921	.9590	46.2	.81
.77	43.9	.7206	.6934	.9623	46.1	.80
.77	44.0	.7193	.6947	.9657	46.0	.80
.77	44.1	.7181	.6959	.9691	45.9	.80
.77	44.2	.7169	.6972	.9725	45.8	.80
.77	44.3	.7157	.6984	.9759	45.7	.80
.77	44.4	.7145	.6997	.9793	45.6	.80
.78	44.5	.7133	.7009	.9827	45.5	.79
.78	44.6	.7120	.7022	.9861	45.4	.79
.78	44.7	.7108	.7034	.9896	45.3	.79
.78	44.8	.7096	.7046	.9930	45.2	.79
.78	44.9	.7083	.7059	.9965	45.1	.79
.79	45.0	.7071	.7071	1.0000	45.0	.79
		SIN	COS	COT	DEG	RAD

APPENDIX E

SELECTED ANSWERS

Problem Set 1.1, pages 6–7

1. a. theta b. alpha c. phi d. omega e. mu 3. a. δ b. ϕ c. θ d. ω e. μ

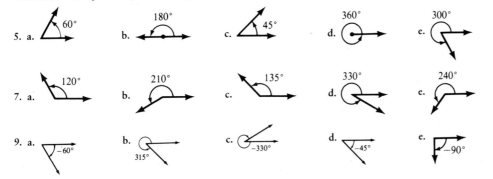

11. a. $40°$ b. $180°$ c. $80°$ d. $30°$ e. $120°$ 13. a. $240°$ b. $10°$ c. $220°$ d. $280°$ e. $40°$ 15. a. $305°$ b. $40°$ c. $90°$ d. $320°$
e. $190°$ 17. $65.67°$ 19. $85.33°$ 21. $240.5°$ 23. $16.70°$ 25. $143.38°$ 27. $128.178°$ 29. $48.469°$ 31. $210.671°$ 33. $.645°$ 35. $.102°$
37. Answer varies

Problem Set 1.2, pages 15–17

1. a. $\frac{\pi}{6}$ b. $\frac{\pi}{2}$ c. $\frac{3\pi}{2}$ d. $\frac{\pi}{4}$ e. 2π f. $\frac{\pi}{3}$ g. π 3. a.

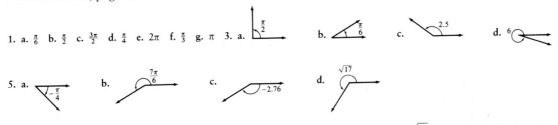

7. a. π b. $\frac{\pi}{6}$ c. π d. $7 - 2\pi \approx .7168$ 9. a. $9 - 2\pi \approx 2.7168$ b. $2\pi - 5 \approx 1.2832$ c. $\sqrt{50} - 2\pi \approx .7879$ d. $2\pi - 6 \approx .2832$
11. a. $\frac{2\pi}{9}$ b. $\frac{\pi}{9}$ 13. a. $\frac{127\pi}{90}$ b. $\frac{17\pi}{36}$ 15. a. -1.10 b. 6.11 17. a. $40.00°$ b. $18.00°$ 19. a. $-171.89°$ b. $-14.32°$ 21. 14.04 cm
23. 70.69 cm 25. 12.57 dm 27. 87 cm 29. $2,900$ mi 31. 10 33. about $6,366$ km 35. about 890 km

Problem Set 1.3, pages 28–30

1. Let θ be an angle in standard position with the point (a, b) the intersection of the terminal side of θ and the unit circle. Then the six trigonometric functions are defined as follows: $\cos\theta = a$; $\sin\theta = b$; $\tan\theta = \frac{b}{a}, a \neq 0$; $\sec\theta = \frac{1}{a}, a \neq 0$; $\csc\theta = \frac{1}{b}, b \neq 0$; and $\cot\theta = \frac{a}{b}, b \neq 0$.
3. $\cos\theta = \pm\sqrt{1 - \sin^2\theta}$ 5. $\cot\theta = \pm\sqrt{\csc^2\theta - 1}$ 7. a. $-.3$ b. $.4$ c. $-.4$ 9. a. $-.4$ b. -7.0 c. 7.1 11. a. $.8788$ b. 1.0551
c. $.2605$ 13. a. $.8415$ b. 14.1368 c. $.7936$ 15. a. $.0998$ b. 1.0468 c. 1.3154 17. a. $\sec 3$ b. $\cos 1$ c. $\tan 2$ d. $\dfrac{1}{\csc 19°}$
e. $\dfrac{1}{\cos 181°}$ f. $\dfrac{1}{\tan 210°}$ 19. a. I, IV b. III, IV c. I, II d. II, III 21. a. II b. II c. III d. II 23. a. $\cot(A + B)$ b. $\cos 75°$

25. a. $\sin 210°$ **b.** $-\sec 210°$ **27.** 1 **29.** -1 **31.** 1 **33.** $\cot\theta = \dfrac{a}{b}$ *Definition*

$$= \frac{\cos\theta}{\sin\theta} \quad \textit{Definition and substitution}$$

35. $a^2 + b^2 = 1$

$\dfrac{a^2}{b^2} + \dfrac{b^2}{b^2} = \dfrac{1}{b^2}$ *Divide both sides by b^2.*

$\left(\dfrac{a}{b}\right)^2 + 1 = \left(\dfrac{1}{b}\right)^2$ *Properties of exponents*

$(\cot\theta)^2 + 1 = (\csc\theta)^2$ *Definition and substitution*

$\cot^2\theta + 1 = \csc^2\theta$

37. Answers vary.

Problem Set 1.4, pages 35–36

1. a. $30°$ **b.** $60°$ **c.** $60°$ **d.** $60°$ **e.** $50°$ **3. a.** $40°$ **b.** $70°$ **c.** $20°$ **d.** $50°$ **e.** $30°$ **5. a.** $30°$ **b.** $60°$ **c.** $90°$ **d.** $60°$ **e.** $40°$
7. a. .642788 **b.** .342020 **c.** 5.671282 **d.** .529919 **e.** .087156 **9. a.** 1.220775 **b.** 1.345633 **c.** 4.346886 **d.** 3.765682 **e.** 3.077684
11. a. $-.363970$ **b.** $-.642788$ **c.** .642788 **d.** 1.051462 **e.** -1.279942 **13. a.** .6428 **b.** .3420 **c.** 5.6721 **d.** .5299 **e.** .2588
15. a. .5000 **b.** 1.2799 **c.** .9703 **d.** $-.7431$ **e.** 57.1429 **17. a.** 2.1445 **b.** 1.0353 **c.** 2.1299 **d.** -3.7327 **e.** -1.0263
19.

$x = 3,\, y = 4,\, r = 5$

$\cos\theta = \frac{3}{5}$ $\sec\theta = \frac{5}{3}$

$\sin\theta = \frac{4}{5}$ $\csc\theta = \frac{5}{4}$

$\tan\theta = \frac{4}{3}$ $\cot\theta = \frac{3}{4}$

21.

$x = -5,\, y = -12,\, r = 13$

$\cos\theta = \frac{-5}{13}$ $\sec\theta = \frac{-13}{5}$

$\sin\theta = \frac{-12}{13}$ $\csc\theta = \frac{-13}{12}$

$\tan\theta = \frac{12}{5}$ $\cot\theta = \frac{5}{12}$

23.

$x = -6,\, y = 1,\, r = \sqrt{37}$

$\cos\theta = \frac{-6}{\sqrt{37}} = \frac{-6}{37}\sqrt{37}$ $\sec\theta = -\frac{\sqrt{37}}{6}$

$\sin\theta = \frac{1}{\sqrt{37}} = \frac{1}{37}\sqrt{37}$ $\csc\theta = \sqrt{37}$

$\tan\theta = -\frac{1}{6}$ $\cot\theta = -6$

25. a. .5000 **b.** 1.7321 **c.** $-.2500$ **d.** .5000 **27. a.** 1.0000 **b.** 1.0000 **c.** 1.0000 **29. a.** .5000 **b.** .2969 **c.** .5000
31. $\theta^g = \frac{10}{9}\theta°$ or $\theta° = .9\theta^g$ **33. a.** .2181 **b.** .2639 **c.** 1.4229

Problem Set 1.5, pages 41–43

1. a. .8415 **b.** .9801 **c.** .3093 **d.** 1.1395 **e.** 1.3940 **3. a.** .9004 **b.** 4.4552 **c.** .8820 **d.** 1.2658 **e.** 1.7465 **5. a.** 2.1850 **b.** $-.9975$
c. .6600 **d.** -1.6310 **e.** -4.0420 **7. a.** 4.8097 **b.** .8855 **c.** $-.5440$ **d.** 1.5882 **e.** -3.2361 **9. a.** 1 **b.** 1 **c.** $\frac{\sqrt{3}}{2}$ **d.** 2
e. undefined **11. a.** 1 **b.** 1 **c.** 0 **d.** $\frac{2}{\sqrt{3}} = \frac{2}{3}\sqrt{3}$ **e.** $\frac{\sqrt{2}}{2}$ **13. a.** 1 **b.** -1 **c.** -1 **d.** 0 **e.** undefined **15. a.** $\frac{\sqrt{3}}{2}$ **b.** 0 **c.** 0 **d.** 2
e. undefined **17. a.** -1 **b.** $-\frac{\sqrt{2}}{2}$ **c.** $-\frac{\sqrt{2}}{2}$ **d.** 1 **e.** -1 **19. a.** -1 **b.** $\frac{1}{2}$ **c.** $\frac{\sqrt{2}}{2}$ **d.** 1 **e.** $\frac{\sqrt{2}}{2}$ **21. a.** $\frac{\sqrt{3}}{2}$ **b.** $\frac{1}{2}$ **c.** $-\sqrt{3}$ **d.** 2
e. 2 **23. a.** 1 **b.** $\frac{1}{2}$ **25. a.** $\sqrt{3}$ **b.** $\frac{2\sqrt{3}}{3}$ **27. a.** $\frac{1}{2}$ **b.** $\frac{1}{2}$ **29. a.** $\frac{\sqrt{3}}{2}$ **b.** $\frac{\sqrt{3}}{2}$ **31. a.** $\frac{\sqrt{3}}{2}$ **b.** $\frac{\sqrt{3}}{2}$
33. $\cos = 1 - 1^2/2! + 1^4/4! - 1^6/6! + \cdots$

$\approx 1 - .5 + .041667 - .001389 + \cdots$

$\approx .5403$

Chapter 1 Review, pages 43–46

Review of Objectives

1. a. alpha **b.** theta **c.** gamma **d.** lambda **e.** beta **2. a.** $30°$ **b.** $45°$ **c.** $60°$

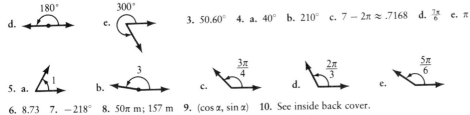

d. 180°   **e.** 300°   **3.** 50.60°   **4. a.** 40°   **b.** 210°   **c.** $7 - 2\pi \approx .7168$ **d.** $\frac{7\pi}{6}$ **e.** π

5. a.   **b.**   **c.** $\frac{3\pi}{4}$ **d.** $\frac{2\pi}{3}$ **e.** $\frac{5\pi}{6}$

6. 8.73   **7.** $-218°$ **8.** 50π m; 157 m   **9.** $(\cos\alpha, \sin\alpha)$ **10.** See inside back cover.

11. $a^2 + b^2 = 1$ *Pythagorean theorem*   **12.** -1

$\dfrac{a^2}{a^2} + \dfrac{b^2}{a^2} = \dfrac{1}{a^2}$ *Divide both sides by a^2*

$1 + \left(\dfrac{b}{a}\right)^2 = \left(\dfrac{1}{a}\right)^2$ *Property of exponents*

$1 + (\tan\theta)^2 = (\sec\theta)^2$ *Definition of tangent and secant*

$1 + \tan^2\theta = \sec^2\theta$

13. $\cos 340° \approx .9$; $\sin 340° \approx -.3$; $\tan 340° \approx -.3$; $\sec 340° \approx 1.1$; $\csc 340° \approx -3.3$; $\cot 340° \approx -3.0$ **14. a.** all   **b.** none
c. sine and cosecant   **d.** cosine, secant, tangent, and cotangent   **e.** tangent and cotangent   **f.** cosine, secant, sine, and cosecant
g. cosine and secant   **h.** sine, cosecant, tangent, and cotangent
15. $\cos 42.5° \approx .7373$; $\sin 42.5° \approx .6756$; $\tan 42.5° \approx .9163$; $\sec 42.5° \approx 1.3563$; $\csc 42.5° \approx 1.4802$; $\cot 42.5° \approx 1.0913$
16. Let θ be any angle in standard position with a point $P(x, y)$ on the terminal side a distance of r from the origin $(r \neq 0)$. Then, $\cos\theta = \frac{x}{r}$;
$\sin\theta = \frac{y}{r}$; $\tan\theta = \frac{y}{x}$, $x \neq 0$; $\sec\theta = \frac{r}{x}$, $x \neq 0$; $\csc\theta = \frac{r}{y}$, $y \neq 0$; $\cot\theta = \frac{x}{y}$, $y \neq 0$
17. $\cos\delta = \frac{-1}{3}$; $\sin\delta = \frac{2\sqrt{2}}{3}$; $\tan\delta = -2\sqrt{2}$; $\sec\delta = -3$; $\csc\delta = \frac{3\sqrt{2}}{4}$; $\cot\delta = \frac{-\sqrt{2}}{4}$ **18. a.** 40°   **b.** 60°   **c.** $\frac{\pi}{3}$ **d.** $\frac{\pi}{6}$ **e.** 82°   **19. a.** $-.9397$
b. $.3420$ **c.** $-.2126$ **d.** -2.7475 **e.** -1.1547 **20. a.** $-.4161$ **b.** $-.7568$ **c.** 1.1383 **d.** -1.0025 **e.** -6.8729 **21. a.** $\frac{1}{2}$ **b.** 0
c. $-\frac{\sqrt{3}}{2}$ **d.** $\frac{1}{2}$ **e.** $-\frac{\sqrt{3}}{2}$ **f.** $\frac{1}{2}$ **g.** -1 **h.** $-\frac{\sqrt{2}}{2}$ **i.** $\frac{\sqrt{3}}{2}$ **j.** 1   **k.** $\frac{1}{2}$ **l.** $-\frac{\sqrt{3}}{2}$ **m.** $\frac{1}{2}$ **n.** $\frac{\sqrt{3}}{2}$ **o.** 0   **p.** $-\frac{\sqrt{2}}{2}$ **q.** $\sqrt{3}$ **r.** undefined
s. $-\frac{\sqrt{3}}{3}$ **t.** $-\sqrt{3}$ **u.** $-\frac{\sqrt{3}}{3}$ **v.** $\sqrt{3}$ **w.** 0   **x.** 1

Miscellaneous Problems, pages 46–47

1. $(2\cos\alpha, 2\sin\alpha)$ **3.** $(\cos 23°, \sin 23°)$ **5.** 2.62   **7.** -1.75 **9.** 120°   **11.** $-143°$ **13.** 30 in.   **15.** 7 ft   **17.** I; 15°   **19.** II; 95°   **21.** .9965
23. $-.4161$ **25.** $-.9365$ **27.** -3.4364 **29.** $-.6873$ **31.** .4794   **33.** $\cos\theta = \frac{4}{\sqrt{41}}$; $\sin\theta = \frac{5}{\sqrt{41}}$; $\tan\theta = \frac{5}{4}$; $\sec\theta = \frac{\sqrt{41}}{4}$; $\csc\theta = \frac{\sqrt{41}}{5}$; $\cot\theta = \frac{4}{5}$
35. $\cos\theta = \frac{5}{6}$; $\sin\theta = \frac{\sqrt{11}}{6}$; $\tan\theta = \frac{\sqrt{11}}{5}$; $\sec\theta = \frac{6}{5}$; $\csc\theta = \frac{6}{\sqrt{11}}$; $\cot\theta = \frac{5}{\sqrt{11}}$ **37.** $\frac{\sqrt{3}}{2}$ **39.** $\sqrt{3}$ **41.** $\frac{-\sqrt{2}}{2}$ **43.** -1 **45.** 1   **47.** 0   **49.** $-\frac{2}{\sqrt{3}}$
51. $\sqrt{3}$

Application for Further Study: Windmills to Pistons, page 53

1. 8 radians per second   **3.** The angular velocity is $3{,}600\pi$ radians per minute or 678,584 radians per hour, and the linear velocity is 1,357 kph.
5. It is at the top of the stroke.   **7.** $\omega \approx 4.73$ radians per hour
9. The angular velocity is $4{,}800\pi$ radians per hour, and the linear velocity is about 357 mph.   **11.** 6.8 cm from the top of the stroke.

Problem Set 2.1, pages 63–65

1. a. undefined   **b.** 0   **c.** 0   **d.** $\frac{\sqrt{3}}{2}$ **e.** undefined   **f.** 0   **3. a.** undefined   **b.** -1 **c.** 0   **d.** 1   **e.** 1   **f.** $\frac{1}{2}$ **5. a.** 1   **b.** $\sqrt{2}$ **c.** $\sqrt{3}$
d. 1   **e.** 0   **f.** -1 **7.** See Figure 2.4 in text.   **9.**   **11.**   **13.**

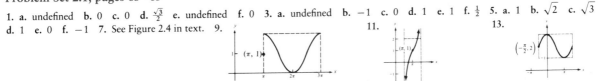

15.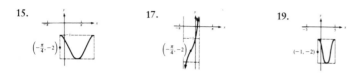

17.

19.

21. $(-1, .54), (0, 1), (1, .54), (2, -.42), (3, -.99),$ $(4, -.65), (5, .28), (6, .96), (7, .75)$

23. $(-1, -1.56), (0, 0), (1, 1.56), (2, -2.19), (3, -.14), (4, 1.16), (5, -3.38), (6, -.29), (7, .87)$

25. $(-1, 1.85), (0, 1), (1, 1.85), (2, -2.40), (3, -1.01), (4, -1.53), (5, 3.53), (6, 1.04), (7, 1.33)$

27. $(-1, -1.19),$ undefined at $x = 0, \pi; (1, 1.19), (2, 1.10), (3, 7.09), (4, -1.32), (5, -1.04), (6, -3.58), (7, 1.52)$

29. $(-1, -.64),$ undefined at $x = 0, (1, .64), (2, -.46), (3, -7.02), (4, .86), (5, -.30), (6, -3.44), (7, 1.15)$ **31.** Answers vary.

33.

35. a. Answers vary.   **b.** Cut A will produce an edge that is a line; cut B will produce an edge that is a sine or a cosine curve.

37. $\cos\theta = \dfrac{x'}{r'}$ *By definition*   **39.** See Figure 2.2 in text   **41.** See page facing inside back cover.

$\quad\quad = \dfrac{|CQ|}{|PC|}$

$\quad\quad = |CQ|$ *Since* $|PC| = 1$

$\quad\quad = |PR|$ *Since* $|CQ| = |PR|$

Problem Set 2.2, page 74

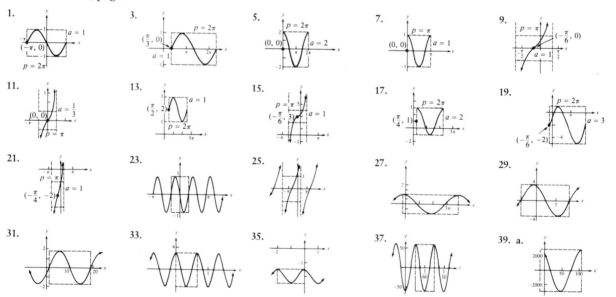

1.   **3.**   **5.**   **7.**   **9.**

11.   **13.**   **15.**   **17.**   **19.**

21.   **23.**   **25.**   **27.**   **29.**

31.   **33.**   **35.**   **37.**   **39. a.**

b. 3,000 miles   **c.** 2 hours

Problem Set 2.3, pages 80–81

1. a. 1 **b.** undefined **c.** 0 **d.** undefined **e.** -1 **f.** 0 **3. a.** $\frac{\sqrt{3}}{3}$ **b.** 2 **c.** $\frac{\sqrt{2}}{2}$ **d.** $\frac{\sqrt{2}}{2}$ **e.** $\sqrt{3}$ **f.** undefined
5. a. $\sqrt{3}$ **b.** 0 **c.** $\frac{2\sqrt{3}}{3}$ **d.** $\sqrt{2}$ **e.** -1 **f.** $\frac{\sqrt{3}}{3}$ **7.** See inside back cover.   **9.** See inside back cover.

11.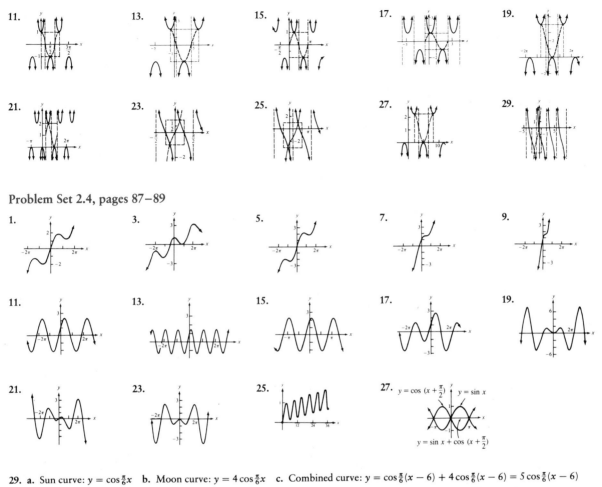

13.

15.

17.

19.

21.

23.

25.

27.

29.

Problem Set 2.4, pages 87–89

1.

3.

5.

7.

9.

11.

13.

15.

17.

19.

21.

23.

25.

27. $y = \cos\left(x + \frac{\pi}{2}\right)$ $y = \sin x$

$y = \sin x + \cos\left(x + \frac{\pi}{2}\right)$

29. a. Sun curve: $y = \cos\frac{\pi}{6}x$ **b.** Moon curve: $y = 4\cos\frac{\pi}{6}x$ **c.** Combined curve: $y = \cos\frac{\pi}{6}(x - 6) + 4\cos\frac{\pi}{6}(x - 6) = 5\cos\frac{\pi}{6}(x - 6)$

31.

33.

35.

37.

39.

41.

Chapter 2 Review, pages 89–91

Review of Objectives

1. a. b. c. 2. a. b.

c. d. 3. a. b.

4. a. (h, k) b. $2|a|$ c. $\frac{2\pi}{b}$ d. $\frac{2\pi}{b}$ e. $\frac{\pi}{b}$ f. amplitude g. period

5. a. b. c. d. e.

6. a. b. c. d. e.

7. a. b. c. 8. a. b.

c. d. 9. a. b. 10. a. b.

Miscellaneous Problems, page 91

1. 3. 5. 7. 9.

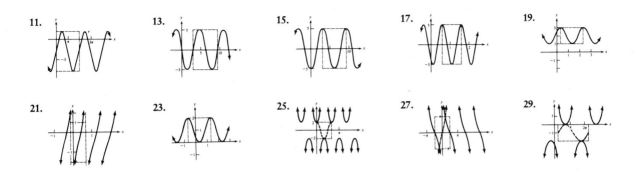

11. **13.** **15.** **17.** **19.**

21. **23.** **25.** **27.** **29.**

Application for Further Study: Computer Graphing of Trigonometric Curves, page 99

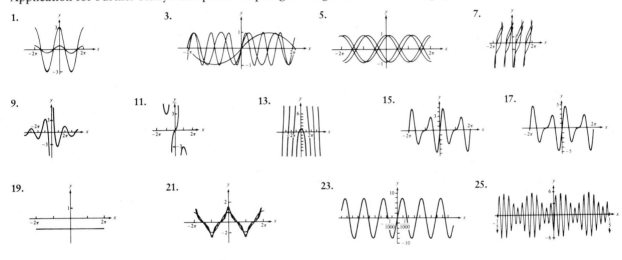

1. **3.** **5.** **7.**

9. **11.** **13.** **15.** **17.**

19. **21.** **23.** **25.**

Cumulative Review for Chapters 1 and 2, pages 100–101

Objective Test

1. See inside front cover. **2. a.** sine **b.** secant **c.** cotangent **d.** cotangent **e.** cosecant **f.** cosine **g.** secant **h.** sine **i.** cosine **j.** cosecant **3.** $\cos\alpha = \frac{\sqrt{7}}{4}$; $\sin\alpha = \frac{3}{4}$; $\tan\alpha = \frac{3\sqrt{7}}{7}$; $\sec\alpha = \frac{4\sqrt{7}}{7}$; $\csc\alpha = \frac{4}{3}$; $\cot\alpha = \frac{\sqrt{7}}{3}$ **4.** 2,800 mi **5. a.** $180°$ **b.** $-240°$ **c.** $-\frac{\pi}{4}$ **d.** $\frac{7\pi}{6}$ **e.** $\frac{\pi}{2}$ **f.** $-\frac{\sqrt{2}}{2}$ **g.** 0 **h.** $\frac{\sqrt{3}}{2}$ **i.** $-\frac{1}{2}$ **j.** 1 **k.** $\frac{\sqrt{2}}{2}$ **l.** -1 **m.** $-\frac{1}{2}$ **n.** $-\frac{\sqrt{3}}{2}$ **o.** 0 **p.** -1 **q.** 0 **r.** $-\sqrt{3}$ **s.** $\frac{\sqrt{3}}{3}$ **t.** undefined
6. a. **b.** **c.** **7.** **8.**

9. **10.**

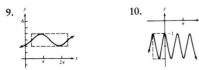

Multiple-Choice Test

1. $\frac{y}{r}$; $\frac{r}{x}$, $x \neq 0$; $\frac{x}{y}$, $y \neq 0$ **2.** II   **3.** negative   **4.** cosine (or secant); tangent (or cotangent)   **5.** 2π **6.** C   **7.** A   **8.** D   **9.** D   **10.** C   **11.** A
12. C   **13.** C   **14.** E   **15.** E   **16.** A   **17.** B   **18.** C   **19.** E   **20.** D

Problem Set 3.1, pages 107–108

1. $\dfrac{1 - \cos^2\theta}{\sin\theta} = \dfrac{\sin^2\theta}{\sin\theta}$
$= \sin\theta$

3. $\dfrac{\dfrac{\sin\theta}{\cos\theta} + \dfrac{\cos\theta}{\sin\theta}}{\dfrac{1}{\sin\theta\cos\theta}} = \dfrac{\sin^2\theta + \cos^2\theta}{\cos\theta\sin\theta} \cdot \dfrac{\sin\theta\cos\theta}{1}$
$= 1$

5. $\dfrac{\cos\theta + \dfrac{\sin^2\theta}{\cos\theta}}{\sin\theta} = \dfrac{\cos^2\theta + \sin^2\theta}{\cos\theta\sin\theta}$
$= \dfrac{1}{\cos\theta\sin\theta}$

7. $\sin\theta + \cot\theta = \sin\theta + \dfrac{\cos\theta}{\sin\theta}$
$= \dfrac{\sin^2\theta + \cos\theta}{\sin\theta}$

9. $\dfrac{\tan\theta + \cot\theta}{\sec\theta\csc\theta} = \dfrac{\dfrac{\sin\theta}{\cos\theta} + \dfrac{\cos\theta}{\sin\theta}}{\dfrac{1}{\cos\theta} \cdot \dfrac{1}{\sin\theta}}$
$= \dfrac{\sin^2\theta + \cos^2\theta}{\cos\theta\sin\theta} \cdot \dfrac{\cos\theta\sin\theta}{1}$
$= 1$

11. $\sec^2\theta + \tan^2\theta = \dfrac{1}{\cos^2\theta} + \dfrac{\sin^2\theta}{\cos^2\theta}$
$= \dfrac{1 + \sin^2\theta}{\cos^2\theta}$

13.–41. Proofs vary.

Problem Set 3.2, pages 110–111

Proofs vary.

Problem Set 3.3, page 114

1.–10. Answers vary.   (**11.–31.** Proofs vary.)   **11.** identity   **13.** identity   **15.** identity   **17.** not an identity   **19.** not an identity
21. identity   **23.** identity   **25.** identity   **27.** identity   **29.** identity   **31.** identity

Problem Set 3.4, page 121

1. $\sqrt{89}$ **3.** $2\sqrt{53}$ **5.** $\sqrt{2 - 2\sin\alpha}$ **7.** $\sin 75°$ **9.** $\tan 49°$ **11.** $-\sin\frac{\pi}{3}$ **13.** $\cos 18°$ **15.** $-\sin 41°$ **17.** $\cos 39°$ **19. a.** $\cos(\theta - \frac{\pi}{3})$
b. $-\sin(\theta - \frac{\pi}{3})$ **c.** $-\tan(\theta - \frac{\pi}{3})$ **21. a.** $\cos(\alpha - \beta)$ **b.** $-\sin(\alpha - \beta)$ **c.** $-\tan(\alpha - \beta)$

23.   **25.**   **27.**   **29.**

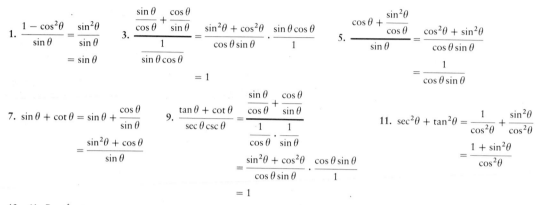

31. Proof varies.

Problem Set 3.5, pages 124–125

1. a. $\dfrac{\sqrt{3}\cos\theta - \sin\theta}{2}$ **b.** $\dfrac{\sqrt{2}\,(\sin\theta - \cos\theta)}{2}$ **3. a.** $\frac{1}{2}(\cos\theta + \sqrt{3}\sin\theta)$ **b.** $\dfrac{\sqrt{2}}{2}(\cos\theta + \sin\theta)$ **5. a.** $\cos^2\theta - \sin^2\theta$ **b.** $2\sin\theta\cos\theta$

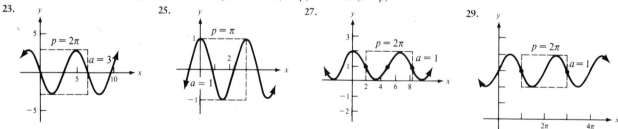

7. .9135 **9.** .2079 **11.** .6428 **13.** 1.1918 **15.** .6009 **17.** $\cos(-15°) = \dfrac{\sqrt{6}+\sqrt{2}}{4}$; $\sin(-15°) = \dfrac{\sqrt{2}-\sqrt{6}}{4}$; $\tan(-15°) = -2+\sqrt{3}$

19. $\cos 75° = \dfrac{\sqrt{6}-\sqrt{2}}{4}$; $\sin 75° = \dfrac{\sqrt{6}+\sqrt{2}}{4}$; $\tan 75° = 2+\sqrt{3}$ **21.** $\cos 345° = \dfrac{\sqrt{6}+\sqrt{2}}{4}$; $\sin 345° = \dfrac{\sqrt{2}-\sqrt{6}}{4}$; $\tan 345° = -2+\sqrt{3}$

23.

$\sin(\alpha + \beta) = \sin\alpha\cos\beta + \cos\alpha\sin\beta$ *Identity* 17

$\sin[\alpha + (-\beta)] = \sin\alpha\cos(-\beta) + \cos\alpha\sin(-\beta)$ *Replace β by $-\beta$.*

$\sin(\alpha - \beta) = \sin\alpha\cos\beta + \cos\alpha(-\sin\beta)$ *Identities* 12 *and* 13

$\therefore \sin(\alpha - \beta) = \sin\alpha\cos\beta - \cos\alpha\sin\beta$

25. $\cot(\alpha + \beta) = \dfrac{\cos(\alpha + \beta)}{\sin(\alpha + \beta)}$

$= \dfrac{\cos\alpha\cos\beta - \sin\alpha\sin\beta}{\sin\alpha\cos\beta + \cos\alpha\sin\beta} \cdot \dfrac{\dfrac{1}{\sin\alpha\sin\beta}}{\dfrac{1}{\sin\alpha\sin\beta}}$

$= \dfrac{\cot\alpha\cot\beta - 1}{\cot\beta\cot\alpha}$

27. $\dfrac{\cos 5\theta}{\sin\theta} - \dfrac{\sin 5\theta}{\cos\theta} = \dfrac{\cos 5\theta\cos\theta - \sin 5\theta\sin\theta}{\sin\theta\cos\theta}$

$= \dfrac{\cos(5\theta + \theta)}{\sin\theta\cos\theta}$

$= \dfrac{\cos 6\theta}{\sin\theta\cos\theta}$

29. $\sin(\alpha + \beta)\cos\beta - \cos(\alpha + \beta)\sin\beta$

$= (\sin\alpha\cos\beta + \cos\alpha\sin\beta)\cos\beta - (\cos\alpha\cos\beta - \sin\alpha\sin\beta)\sin\beta$

$= \sin\alpha\cos^2\beta + \cos\alpha\cos\beta\sin\beta - \cos\alpha\cos\beta\sin\beta + \sin\alpha\sin^2\beta$

$= \sin\alpha(\cos^2\beta + \sin^2\beta)$

$= \sin\alpha$

31.–47. Proofs vary.

Problem Set 3.6, pages 131–133

1. $\dfrac{\sqrt{2}}{2}$ **3.** $\dfrac{1}{2}$ **5.** -1 **7.** $\dfrac{1}{2}\sqrt{2-\sqrt{2}}$ **9.** $\sqrt{2}-1$

	$\cos 2\theta$	$\sin 2\theta$	$\tan 2\theta$		$\cos\frac{1}{2}\theta$	$\sin\frac{1}{2}\theta$	$\tan\frac{1}{2}\theta$		$\cos\theta$	$\sin\theta$	$\tan\theta$
11.	$\frac{119}{169}$	$\frac{-120}{169}$	$\frac{-120}{119}$	**17.**	$\frac{\sqrt{26}}{26}$	$\frac{5\sqrt{26}}{26}$	5	**23.**	$\frac{\sqrt{2}}{2}$	$\frac{\sqrt{2}}{2}$	1
13.	$\frac{7}{25}$	$\frac{-24}{25}$	$\frac{-24}{7}$	**19.**	$\frac{\sqrt{10}}{10}$	$\frac{3\sqrt{10}}{10}$	3	**25.**	$\frac{1}{2}$	$\frac{\sqrt{3}}{2}$	$\sqrt{3}$
15.	$\frac{-119}{169}$	$\frac{120}{169}$	$\frac{-120}{119}$	**21.**	$\frac{-2\sqrt{13}}{13}$	$\frac{3\sqrt{13}}{13}$	$\frac{-3}{2}$	**27.**	$\frac{3\sqrt{10}}{10}$	$\frac{\sqrt{10}}{10}$	$\frac{1}{3}$

29. a. $r \approx 2.6$ **b.** $r = 2\sqrt{2 - \sqrt{2}} + \sqrt{4 - 2\sqrt{2}}$ This is $\dfrac{2}{\sqrt{2-\sqrt{2}}}$ after it has been rationalized.

31. $\cos 2\theta = \cos^2\theta - \sin^2\theta$
Let $\theta = 2\theta$; then $\cos 4\theta = \cos^2 2\theta - \sin^2 2\theta$.

33. $\tan 2\theta = \dfrac{2\tan\theta}{1 - \tan^2\theta}$
Let $\theta = \dfrac{3\beta}{4}$; then $2\theta = \dfrac{3\beta}{2}$
and the result is proved.

35. $\tan\frac{1}{2}\theta = \dfrac{1 - \cos\theta}{\sin\theta}$ *(Problem 34)*

$= \dfrac{(1 - \cos\theta)(1 + \cos\theta)}{\sin\theta(1 + \cos\theta)}$

$= \dfrac{1 - \cos^2\theta}{\sin\theta(1 + \cos\theta)}$

$= \dfrac{\sin\theta}{1 + \cos\theta}$

37.–43. Proofs vary.

Problem Set 3.7, pages 136–137

1. $\cos 40° + \cos 110°$ 3. $\cos 11° - \cos 59°$ 5. $\frac{1}{2}\cos 18° - \frac{1}{2}\cos 158°$ 7. $\frac{1}{2}\sin 60° + \frac{1}{2}\sin 22°$ 9. $\frac{1}{2}\cos 75° + \frac{1}{2}\cos 165°$
11. $\frac{1}{2}\cos 2\theta + \frac{1}{2}\cos 4\theta$ 13. $2\sin 53° \cos 10°$ 15. $-2\sin 80° \sin 1°$ 17. $2\sin 257.5° \cos 42.5°$ 19. $-2\sin \frac{x}{2}\cos \frac{3x}{2}$ 21. $2\sin \frac{3x}{2}\cos \frac{x}{2}$
23. $2\cos 7y \cos 2y$ 25. $2\sin \alpha \sin \beta = \cos(\alpha - \beta) - \cos(\alpha + \beta)$ *Product formula*

$$\text{Let } x = \alpha - \beta \text{ and } y = \alpha + \beta; \text{ then } \alpha = \frac{x+y}{2} \text{ and } \beta = \frac{y-x}{2}.$$

$$2\sin\left(\frac{x+y}{2}\right)\sin\left(\frac{y-x}{2}\right) = \cos x - \cos y \quad \textit{Substitution}$$

$$-2\sin\left(\frac{x+y}{2}\right)\sin\left(\frac{x-y}{2}\right) = \cos x - \cos y \quad \textit{Since } \sin(-\theta) = -\sin\theta$$

27. $2\cos \alpha \sin \beta = \sin(\alpha + \beta) - \sin(\alpha - \beta)$ *Product formula*

$$\text{Let } x = \alpha + \beta \text{ and } y = \alpha - \beta; \text{ then } \alpha = \frac{x+y}{2} \text{ and } \beta = \frac{x-y}{2}.$$

$$2\cos\left(\frac{x+y}{2}\right)\sin\left(\frac{x-y}{2}\right) = \sin x - \sin y \quad \textit{Substitution}$$

29.–31. Answers vary.

Chapter 3 Review, pages 137–140

Review of Objectives

1.
$$\frac{\csc^2\alpha}{1 + \cot^2\alpha} = \frac{\csc^2\alpha}{\csc^2\alpha}$$
$$= 1$$

2.
$$\frac{\sec^2 x + \tan^2 x + 1}{\sec x} = \frac{\sec^2 x + (\sec^2 x - 1) + 1}{\sec x}$$
$$= \frac{2\sec^2 x}{\sec x}$$
$$= 2\sec x$$

3.
$$\frac{1 + \csc\beta}{\cos\beta \csc\beta} = \frac{1}{\cos\beta \csc\beta} + \frac{\csc\beta}{\cos\beta \csc\beta}$$
$$= \frac{1}{\frac{\cos\beta}{\sin\beta}} + \frac{1}{\cos\beta}$$
$$= \frac{\sin\beta}{\cos\beta} + \sec\beta$$
$$= \tan\beta + \sec\beta$$

4.
$$\frac{1}{\sin\theta + \cos\theta} + \frac{1}{\sin\theta - \cos\theta} = \frac{\sin\theta - \cos\theta + \sin\theta + \cos\theta}{\sin^2\theta - \cos^2\theta}$$
$$= \frac{2\sin\theta}{\sin^2\theta - \cos^2\theta} \cdot \frac{\sin^2\theta + \cos^2\theta}{\sin^2\theta + \cos^2\theta}$$
$$= \frac{2\sin\theta(\sin^2\theta + \cos^2\theta)}{\sin^4\theta - \cos^4\theta}$$
$$= \frac{2\sin\theta}{\sin^4\theta - \cos^4\theta}$$

5.
$$\frac{\sin^2\theta - \cos^2\theta}{\sin\theta + \cos\theta} = \frac{(\sin\theta + \cos\theta)(\sin\theta - \cos\theta)}{\sin\theta + \cos\theta}$$
$$= \sin\theta - \cos\theta$$

6.
$$\frac{\cos\theta}{\sec\theta} - \frac{\sin\theta}{\cot\theta} = \cos^2\theta - \frac{\sin^2\theta}{\cos\theta} = \frac{\cos^3\theta - \sin^2\theta}{\cos\theta} \quad \text{Also,}$$

$$\frac{\cos\theta\cot\theta - \tan\theta}{\csc\theta} = \frac{\cos\theta \cdot \dfrac{\cos\theta}{\sin\theta} - \dfrac{\sin\theta}{\cos\theta}}{\dfrac{1}{\sin\theta}}$$
$$= \left(\frac{\cos^2\theta}{\sin\theta} - \frac{\sin\theta}{\cos\theta}\right)\sin\theta$$
$$= \frac{(\cos^3\theta - \sin^2\theta)\sin\theta}{\sin\theta\cos\theta} = \frac{\cos^3\theta - \sin^2\theta}{\cos\theta}$$

7. $\dfrac{1 - \cos x}{1 + \cos x} = \dfrac{1 - \cos x}{1 + \cos x} \cdot \dfrac{1 - \cos x}{1 - \cos x}$

$ = \dfrac{1 - 2\cos x + \cos^2 x}{1 - \cos^2 x}$

$ = \dfrac{1 - 2\cos x + (1 - \sin^2 x)}{\sin^2 x}$

$ = \dfrac{2 - 2\cos x - \sin^2 x}{\sin^2 x}$

8. $\dfrac{1 + \tan^2\theta}{\csc\theta} = \dfrac{\sec^2\theta}{\csc\theta}$

$ = \sec\theta \cdot \dfrac{\sec\theta}{\csc\theta}$

$ = \sec\theta \cdot \dfrac{\sin\theta}{\cos\theta}$

$ = \sec\theta \tan\theta$

9. Answers vary.   10. a. False; let $t = 1$. Then $(\sin t + \cos t)^2 \approx 1.9$.   b. False; let $t = 1$. Then $\cos t \tan t \csc t \sec t \approx 1.85$.

11. $|P_\alpha P_\beta| = \sqrt{(\cos\beta - \cos\alpha)^2 + (\sin\beta - \sin\alpha)^2}$

$ = \sqrt{2 - 2(\cos\alpha \cos\beta + \sin\alpha \sin\beta)}$

12. $|AP_\theta| = \sqrt{2 - 2\cos\theta}$ 13. a. $\cos 52°$ b. $\cot\frac{3\pi}{8}$ c. $\sin(\frac{\pi}{2} - 2)$ or $-\sin(.43)$ 14. a. $\cos(\theta - \frac{\pi}{6})$ b. $-\sin(\theta - \frac{\pi}{6})$ c. $-\tan(\theta - \frac{\pi}{6})$

15.

16. $\frac{\sqrt{3}}{2}\cos\theta + \frac{1}{2}\sin\theta$ 17. $\frac{\sqrt{6}+\sqrt{2}}{4}$

18. Let P_α and P_β be defined as in Problem 11. Then,

$|P_\alpha P_\beta| = \sqrt{2 - 2(\cos\alpha \cos\beta + \sin\alpha \sin\beta)}$

The angle between rays through P_α and P_β is $\alpha - \beta$, so

$|P_\alpha P_\beta| = \sqrt{2 - 2\cos(\alpha - \beta)}$ from Problem 12.

Then, $\sqrt{2 - 2\cos(\alpha - \beta)} = \sqrt{2 - 2(\cos\alpha \cos\beta + \sin\alpha \sin\beta)}$

$ 2 - 2\cos(\alpha - \beta) = 2 - 2(\cos\alpha \cos\beta + \sin\alpha \sin\beta)$

$ \cos(\alpha - \beta) = \cos\alpha \cos\beta + \sin\alpha \sin\beta$

19. $\dfrac{2\tan\frac{\pi}{6}}{1 - \tan^2\frac{\pi}{6}} = \tan(2 \cdot \frac{\pi}{6})$ 20. $\cos 2\theta = \frac{7}{25}$; $\sin 2\theta = -\frac{24}{25}$; $\tan 2\theta = -\frac{24}{7}$

$ = \tan\frac{\pi}{6}$

$ = \sqrt{3}$

21. $-\sqrt{\dfrac{1 + \cos 240°}{2}} = \cos 120°$ 22. $\cos\theta = \frac{1}{10}\sqrt{10}$; $\sin\theta = \frac{3}{10}\sqrt{10}$; $\tan\theta = 3$

$ = -\frac{1}{2}$

23. $\cos 2\alpha = 2\cos^2\alpha - 1$

$ 2\cos^2\alpha = \cos 2\alpha + 1$

$ \cos^2\alpha = \dfrac{\cos 2\alpha + 1}{2}$

$ \cos\alpha = \pm\sqrt{\dfrac{1 + \cos 2\alpha}{2}}$

Let $\alpha = \frac{1}{2}\theta$; then

$\cos\frac{1}{2}\theta = \pm\sqrt{\dfrac{1 + \cos\theta}{2}}$

24. $\frac{1}{2}\sin 4\theta + \frac{1}{2}\sin 2\theta$

25. $2\sin\left(\dfrac{h}{2}\right)\cos\left(\dfrac{2x + h}{2}\right)$

26. $\dfrac{\sin 5\theta + \sin 3\theta}{\cos 5\theta - \cos 3\theta} = \dfrac{2\sin 4\theta \cos\theta}{-2\sin 4\theta \sin\theta}$

$ = -\cot\theta$

Miscellaneous Problems, pages 141–142

1.–37. Proofs vary. **39.** $\frac{\sqrt{6}+\sqrt{2}}{4}$ **41.** $\frac{1}{2}$ **43.** 1 **45.** $\sqrt{3}$ **47.** $\frac{\sqrt{2}}{2}$ **49.** $\frac{-7}{25}$

Application for Further Study: Sunrise, Sunset, and Linear Interpolation, page 146

1. 16 hr 50 min or 4:50 P.M. **3.** 19 hr 1 min or 7:01 P.M. **5.** 6 hr 18 min or 6:18 A.M.
7.

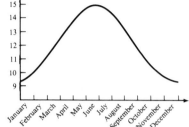

11. Answers vary.

Problem Set 4.1, pages 156–158

1. a. $30° + 360°k, 150° + 360°k$ **b.** $60° + 360°k, 300° + 360°k$ **3. a.** $23° + 360°k, 157° + 360°k$ **b.** $73° + 360°k, 287° + 360°k$
5. a. $60° + 360°k, 120° + 360°k$ **b.** $45° + 360°k, 225° + 360°k$ or $45° + 180°k$ **7. a.** $\frac{\pi}{6} + 2\pi k, \frac{11\pi}{6} + 2\pi k$ **b.** $\frac{\pi}{4} + 2\pi k, \frac{3\pi}{4} + 2\pi k$
9. a. $\frac{3\pi}{4} + 2\pi k, \frac{7\pi}{4} + 2\pi k$ or $\frac{3\pi}{4} + \pi k$ **b.** $\frac{5\pi}{4} + 2\pi k, \frac{7\pi}{4} + 2\pi k$ **11. a.** $\frac{\pi}{6}$ **b.** $30°$ **13. a.** $\frac{\pi}{6}, \frac{5\pi}{6}$ **b.** no solution **15.** $0°, 180°$
17. $\frac{\pi}{6}, \frac{5\pi}{6}, \frac{7\pi}{6}, \frac{11\pi}{6}$ **19.** $\frac{\pi}{6}, \frac{5\pi}{6}, \frac{\pi}{3}, \frac{5\pi}{3}$ **21.** $0, \pi, \frac{\pi}{3}, \frac{4\pi}{3}$ **23.** $\frac{\pi}{4}, \frac{3\pi}{4}, \frac{5\pi}{4}, \frac{7\pi}{4}$ **25.** $0, \pi, 1.2310, 5.0522$ **27.** $\frac{3\pi}{2}$ **29.** $.4636, .9828, 3.6052, 4.1244$
31. $18.8°$ **33.** $2.2370, 4.0461$ **35.** $\frac{2\pi}{3}$ **37.** $0, \pi, \frac{\pi}{3}, \frac{4\pi}{3}$ **39.** $2.2370, 4.0461$ **41.** 65

Problem Set 4.2, pages 161–162

1. $\frac{\pi}{6}, \frac{5\pi}{6}, \frac{7\pi}{6}, \frac{11\pi}{6}$ **3.** $\frac{\pi}{3}, \frac{2\pi}{3}, \frac{4\pi}{3}, \frac{5\pi}{3}$ **5.** $\frac{2\pi}{3}, \frac{5\pi}{6}, \frac{5\pi}{3}, \frac{11\pi}{6}$ **7.** $\frac{\pi}{12}, \frac{5\pi}{12}, \frac{3\pi}{4}, \frac{13\pi}{12}, \frac{17\pi}{12}, \frac{7\pi}{4}$ **9.** $\frac{5\pi}{12}, \frac{7\pi}{12}, \frac{17\pi}{12}, \frac{19\pi}{12}$ **11.** $0, 1.57, 2.09, 3.14, 4.19, 4.71$
13. $.39, 1.96, 3.53, 5.11$ **15.** 1.57 **17.** no solution **19.** $.81, 2.90, 3.95, 6.04$ **21.** $.02, 1.59, 3.17, 4.74$ **23.** $.52, 2.62, 4.71$
25. $0.00, .52, 2.62, 3.14$ **27.** $0.00, 1.05, 1.22, 1.92, 2.09, 3.14, 3.32, 4.01, 4.19, 5.24, 5.41, 6.11$ **29.** $.52, 1.05, 3.67, 4.19$ **31.** $.0002$
33. 32.5 min and 77.5 min

Problem Set 4.3, page 170

1. a. 0 **b.** $\frac{\pi}{6}$ **c.** $\frac{\pi}{2}$ **d.** 0 **3. a.** $-\frac{\pi}{2}$ **b.** $\frac{3\pi}{4}$ **c.** $-\frac{\pi}{4}$ **d.** $\frac{2\pi}{3}$ **5. a.** $\frac{\pi}{3}$ **b.** $\frac{\pi}{4}$ **c.** $\frac{\pi}{4}$ **d.** $\frac{3\pi}{4}$ **7. a.** $15°$ **b.** $67°$ **c.** $20°$ **9. a.** $164°$
b. $-43°$ **c.** $111°$ **11. a.** $21°$ **b.** $15°$ **c.** $135°$ **13. a.** $.21$ **b.** $.58$ **c.** $.33$ **15. a.** 1.94 **b.** 2.50 **c.** $-.40$ **17. a.** $.17$ **b.** $.34$ **c.** 2.82

19. a. $\frac{\pi}{6}$ **b.** $\frac{1}{3}$ **c.** $\frac{2\pi}{15}$ **21.** $-2\sqrt{2}$ **23.** $\frac{4}{5}$ **25.** $\frac{4}{5}$ **27.** 0 **29.** $\dfrac{2\sqrt{2} + \sqrt{15}}{12}$

31. Let $\theta = \text{Arcsec } x$ where $x \geq 1$ or $x \leq -1$. Then,

$\sec \theta = x \qquad 0 \leq \theta \leq \pi \qquad$ *Definition of* Arcsec x, $\theta \neq \frac{\pi}{2}$

$\frac{1}{\cos \theta} = x \qquad 0 \leq \theta \leq \pi, \theta \neq \frac{\pi}{2}$

$\cos \theta = \frac{1}{x} \qquad 0 \leq \theta \leq \pi, \theta \neq \frac{\pi}{2}$

$\theta = \text{Arccos}\frac{1}{x} \qquad 0 \leq \theta \leq \pi \qquad$ *Definition of* Arccos x

33. Let $\theta = \text{Arctan } x$ where $x \geq 0$. Then

$\tan \theta = x \qquad\qquad 0 \leq \theta < \frac{\pi}{2} \qquad$ *Definition of* Arctan x

$\sec^2\theta = \tan^2\theta + 1 \qquad\qquad$ *Fundamental identity*

$\sec \theta = \sqrt{\tan^2\theta + 1} \qquad\qquad$ *Positive value since* $0 \leq \theta < \frac{\pi}{2}$

By substitution, $\sec(\text{Arctan } x) = \sqrt{x^2 + 1}$

35. Let $\theta = \text{Arctan } x$ where $x \geq 0$. Then,

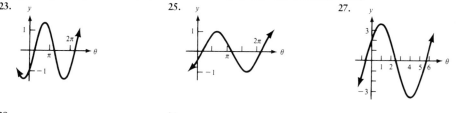

$$\tan \theta = x \qquad 0 \leq \theta < \frac{\pi}{2} \quad \textit{Definition of } \text{Arctan } x$$

$$\sec \theta = \sqrt{x^2 + 1} \qquad \textit{From Problem 33}$$

$$\cos \theta = \frac{1}{\sqrt{x^2 + 1}} \qquad \textit{Since } \sec \theta = \frac{1}{\cos \theta}$$

By substitution $\cos(\text{Arctan } x) = \dfrac{1}{\sqrt{1 + x^2}}$

Problem Set 4.4, page 174

1. $\sqrt{2} \sin(\theta + \frac{\pi}{4})$ **3.** $\sqrt{2} \sin(\frac{\theta}{2} + \frac{3\pi}{4})$ **5.** $\sin(\theta + \frac{\pi}{3})$ **7.** $\sin(\theta + \frac{2\pi}{3})$ **9.** $\sqrt{2} \sin(\theta + \frac{5\pi}{4})$ **11.** $2 \sin(\pi\theta + \frac{2\pi}{3})$ **13.** $\sqrt{13} \sin(\theta + 1.0)$
15. $\sqrt{13} \sin(\theta + 2.6)$ **17.** $5 \sin(\theta + 2.2)$ **19.** $17 \sin(\theta + 4.2)$ **21.** $\sqrt{29} \sin(\theta + 2.0)$

23. 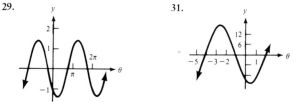 **25.** **27.**

29. **31.**

Chapter 4 Review, pages 175–176

Review of Objectives

1. $1, 1 + \pi$ **2.** $\frac{\pi}{3}, \frac{2\pi}{3}, \frac{4\pi}{3}, \frac{5\pi}{3}$ **3.** $1.25, 2.36, 4.39, 5.50$ **4.** $.31, 2.83$ **5.** $.48, 1.09, 2.05, 2.66, 3.62, 4.23, 5.19, 5.81$
6. $\frac{\pi}{6}, \frac{\pi}{3}, \frac{2\pi}{3}, \frac{5\pi}{6}, \frac{7\pi}{6}, \frac{4\pi}{3}, \frac{5\pi}{3}, \frac{11\pi}{6}$ **7. a.** $30°$ **b.** $45°$ **c.** $60°$ **d.** $45°$ **e.** $\frac{3\pi}{4}$ **f.** $\frac{5\pi}{6}$ **g.** $\frac{\pi}{6}$ **h.** $\frac{\pi}{6}$ **i.** 1.16 **j.** 1.28 **8. a.** I
b. θ between -1.57 and 0 **c.** I **d.** θ between 0 and 1.57 **e.** II **f.** θ between 0 and 1.57 **g.** IV **h.** θ between -1.57 and 0 **i.** II
j. θ between 1.57 and 3.14 **9.** $2\sqrt{2}$ **10. a.** I **b.** III **c.** IV **d.** II **e.** $13 \sin(\theta + 5.9)$
11.

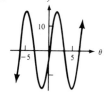

Miscellaneous Problems, page 177

1. $-1 \leq x \leq 1; 0 \leq y \leq \pi$ **3.** all reals; $-\frac{\pi}{2} < y < \frac{\pi}{2}$ **5.** $x \geq 1$ or $x \leq -1$; $-\frac{\pi}{2} \leq y \leq \frac{\pi}{2}, y \neq 0$ **7.** $\frac{4\pi}{3} + 2\pi k, \frac{5\pi}{3} + 2\pi k$
9. $.95 + 6.28k, 2.19 + 6.28k$ **11.** $1.94 + 6.28k, 4.34 + 6.28k$ **13.** $\frac{3\pi}{4}$ **15.** 2.83 **17.** 1.26 **19.** $-\frac{\pi}{4}$ **21.** 1.80 **23.** 2.11 **25.** $-.86$
27. $-.28$ **29.** $.28$ **31.** $\frac{\pi}{6}, \frac{5\pi}{6}$ **33.** 0 **35.** $1.32, 4.97$ **37.** $1.23, 5.05, \frac{2\pi}{3}, \frac{4\pi}{3}$ **39.** $\frac{\pi}{2}, \frac{3\pi}{2}$ **41.** $\frac{\pi}{3}, \frac{5\pi}{3}, .84, 5.44$ **43.** no solution **45.** $.73, 2.41$
47. $.98, 4.12, .46, 3.61$ **49. a.** $\frac{\pi}{3}, \frac{5\pi}{3}$ **b.** $\frac{\pi}{2} + 1$ (or 2.57), $\frac{3\pi}{2} + 1$ (or 5.71)

Application for Further Study: Waves, Harmonic Motion, and Resonance, pages 184–185

1. $y = 3 \sin .03(x - 30.54t)$

3. 24 ft from trough to crest; wavelength $\lambda \approx \frac{2\pi}{1.32} \approx 4.76$ ft; phase velocity is about 4.75 ft/sec ≈ 3.24 mph.

5. $t = .21x - .16 \sin^{-1}(\frac{y}{12})$

7. Since $\omega = 10\pi$, $n = 5$, and $T = \frac{1}{5}$, we know that $y = -\sqrt{3}$ when $t = 0$. Thus, the period is $\frac{1}{5}$, the frequency is 5 cycles per unit of time, and the maximum distance from the origin is $\sqrt{3}$.

9. **a.**   **b.**   **c.**

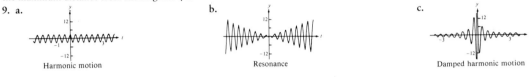

Harmonic motion   Resonance   Damped harmonic motion

11. Maximum value is 60; minimum is -60.

Cumulative Review for Chapters 3–4, pages 186–187

Objective Test

1. See endpaper facing inside front cover.   2.

$$x^2 + y^2 = r^2 \qquad \textit{Pythagorean theorem}$$

$$\frac{x^2}{y^2} + \frac{y^2}{y^2} = \frac{r^2}{y^2} \qquad \textit{Divide both sides by } y^2, y \neq 0$$

$$\left(\frac{x}{y}\right)^2 + 1 = \left(\frac{r}{y}\right)^2 \qquad \textit{Properties of exponents}$$

$$\cot^2\theta + 1 = \csc^2\theta \qquad \textit{Definition of trig functions}$$

3. **a.** $A(1,0), P_\theta(\cos\theta, \sin\theta)$ **b.** $\sqrt{2 - 2\cos\theta}$

4.
$$(\cos\beta - \cos\alpha)^2 + (\sin\beta - \sin\alpha)^2 = \cos^2\beta - 2\cos\alpha\cos\beta + \cos^2\alpha + \sin^2\beta - 2\sin\alpha\sin\beta + \sin^2\alpha$$
$$= (\cos^2\beta + \sin^2\beta) + (\cos^2\alpha + \sin^2\alpha) - 2(\cos\alpha\cos\beta + \sin\alpha\sin\beta)$$
$$= 1 + 1 - 2(\cos\alpha\cos\beta + \sin\alpha\sin\beta)$$
$$= 2 - 2(\cos\alpha\cos\beta + \sin\alpha\sin\beta)$$

5.
$$\frac{\sin 3\theta + \cos 3\theta + 1}{\cos 3\theta} = \frac{\sin 3\theta}{\cos 3\theta} + \frac{\cos 3\theta}{\cos 3\theta} + \frac{1}{\cos 3\theta}$$
$$= \tan 3\theta + 1 + \sec 3\theta$$

6.
$$\sec A - \tan^2 A = \frac{1}{\cos A} - \frac{\sin^2 A}{\cos^2 A}$$
$$= \frac{\cos A - \sin^2 A}{\cos^2 A}$$
$$= \frac{\cos A - (1 - \cos^2 A)}{\cos^2 A}$$
$$= \frac{1}{\cos^2 A}(\cos^2 A + \cos A - 1)$$
$$= \sec^2 A(\cos^2 A + \cos A - 1)$$

7.
$$\tan 2\theta \cos 2\theta = \frac{\sin 2\theta}{\cos 2\theta} \cdot \cos 2\theta$$
$$= \sin 2\theta$$
$$= 2 \sin\theta \cos\theta$$

8. .308, 2.834

9.

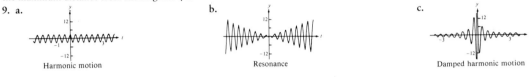

10. $\cos\theta = \frac{4}{5}$; $\sin\theta = -\frac{3}{5}$; $\tan\theta = -\frac{3}{4}$; $\sec\theta = \frac{5}{4}$; $\csc\theta = -\frac{5}{3}$; $\cot\theta = -\frac{4}{3}$

Multiple-Choice Test

1. $\cos\theta\sec\theta = 1$; $\sin\theta\csc\theta = 1$; $\tan\theta\cot\theta = 1$ **2.** $\tan\theta = \dfrac{\sin\theta}{\cos\theta}$; $\cot\theta = \dfrac{\cos\theta}{\sin\theta}$ **3.** $\cos^2\theta + \sin^2\theta = 1$; $1 + \tan^2\theta = \sec^2\theta$; $\cot^2\theta + 1 = \csc^2\theta$

4. $x^2 + y^2 = r^2$ by the Pythagorean theorem   **5.** Multiply by the conjugate   **6.** B   **7.** E   **8.** D   **9.** D   **10.** A   **11.** A   **12.** A   **13.** B
14. D   **15.** E   **16.** C   **17.** B   **18.** D   **19.** B   **20.** B

Problem Set 5.1, pages 195–199

1. $\alpha = 30°$; $\beta = 60°$; $\gamma = 90°$; $a = 80$; $b = 140$; $c = 160$ **3.** $\alpha = 76°$; $\beta = 14°$; $\gamma = 90°$; $a = 29$; $b = 7.2$; $c = 30$
5. $\alpha = 45°$; $\beta = 45°$; $\gamma = 90°$; $a = 49$; $b = 49$; $c = 69$ **7.** $\alpha = 69.2°$; $\beta = 20.8°$; $\gamma = 90.0°$; $a = 26.5$; $b = 10.0$; $c = 28.3$
9. $\alpha = 56.00°$; $\beta = 34.00°$; $\gamma = 90.00°$; $a = 3,484$; $b = 2,350$; $c = 4,202$ **11.** $\alpha = 42°$; $\beta = 48°$; $\gamma = 90°$; $a = 320$; $b = 350$; $c = 470$
13. $\alpha = 67.8°$; $\beta = 22.2°$; $\gamma = 90.0°$; $a = 26.6$; $b = 10.8$; $c = 28.7$ **15.** $\alpha = 28.95°$; $\beta = 61.05°$; $\gamma = 90.00°$; $a = 202.7$; $b = 366.4$; $c = 418.7$
17. $\alpha = 48.3°$; $\beta = 41.7°$; $\gamma = 90.0°$; $a = .355$; $b = .316$; $c = .475$ **19.** 45 m   **21.** 240 m   **23.** 23 m   **25.** 350 ft   **27.** 780 ft
29. The height of the tower is 222.0 ft.   **31.** 995 m   **33.** 6.79×10^7 mi   **35.** 921.1 ft   **37. a.** $m \approx .7$
b. Using exact values, $m = \sqrt{3}$ so the equation is $\sqrt{3}\,x - y - 2\sqrt{3} - 3 = 0$; using approximate values, $m = \tan 60° \approx 1.7$; $1.7x - y - 6.4 = 0$
or $17x - 10y - 64 = 0$
c. Using exact values, $m = \frac{\sqrt{3}}{3}$ so the equation is $\sqrt{3}\,x - 3y - 4\sqrt{3} - 3 = 0$; using approximate values, $m = \tan 30° \approx .6$; $.6x - y - 3.4 = 0$
or $3x - 5y - 17 = 0$
39. Area $= \frac{1}{2}(\text{base})(\text{height})$

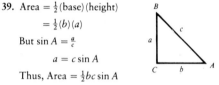

$= \frac{1}{2}(b)(a)$

But $\sin A = \frac{a}{c}$

$a = c\sin A$

Thus, Area $= \frac{1}{2}bc\sin A$

Problem Set 5.2, pages 202–204

1. $54°$ **3.** $49°$ **5.** $80°$ **7.** 13   **9.** 10   **11.** 17   **13.** $78°$ **15.** $46.6°$ **17.** 604   **19.** $a = 4.3$; $b = 5.2$; $c = 3.4$; $\alpha = 55°$; $\beta = 85°$; $\gamma = 41°$
21. $a = 214$; $b = 320$; $c = 126$; $\alpha = 25.8°$; $\beta = 139.4°$; $\gamma = 14.8°$ **23.** $a = 140$; $b = 85.0$; $c = 105$; $\alpha = 94.3°$; $\beta = 37.3°$; $\gamma = 48.4°$ **25.** 249 mi
27. 42 ft (right-triangle solution)   **29.** 308 ft   **31.** Answers vary.   **33.** $d = 2.25$, $q = 1.88$, $p = 2.13$, $\angle D = 68.1°$, $\angle P = 61.2°$, $\angle Q = 50.7°$
35. 50 in.^{2} **37.** Answers vary.   **39.** Answers vary.   **41.** Answers vary.

Problem Set 5.3, pages 207–210

1. $\alpha = 48°$; $\beta = 62°$; $\gamma = 70°$; $a = 10$; $b = 12$; $c = 13$ **3.** $\alpha = 30°$; $\beta = 50°$; $\gamma = 100°$; $a = 30$; $b = 46$; $c = 59$
5. $\alpha = 50°$; $\beta = 70°$; $\gamma = 60°$; $a = 33$; $b = 40$; $c = 37$ **7.** $\alpha = 82°$; $\beta = 19°$; $\gamma = 79°$; $a = 53$; $b = 18$; $c = 53$
9. $\alpha = 35.0°$; $\beta = 81.0°$; $\gamma = 64.0°$; $a = 73.4$; $b = 126$; $c = 115$ **11.** $\alpha = 71°$; $\beta = 85°$; $\gamma = 24°$; $a = 210$; $b = 220$; $c = 91$
13. $\alpha = 45.2°$; $\beta = 21.5°$; $\gamma = 113.3°$; $a = 41.0$; $b = 21.2$; $c = 53.1$ **15.** Cannot be solved since the sum of α and γ is more than $180°$.
17. $\alpha = 14°$; $\beta = 42°$; $\gamma = 123°$; $a = 26$; $b = 71$; $c = 88$ **19.** $\alpha = 74.0°$; $\beta = 81.0°$; $\gamma = 25.0°$; $a = 25.0$; $b = 25.7$; $c = 11.0$
21. $\alpha = 74.7°$; $\beta = 16.1°$; $\gamma = 89.2°$; $a = 80.6$; $b = 23.2$; $c = 83.6$ **23.** $\alpha = 35.60°$; $\beta = 42.10°$; $\gamma = 102.30°$; $a = 16.90$; $b = 19.46$; $c = 28.36$
25. The height of the building is about 260 ft.   **27.** Luke's craft is 970.4 m from the first observation point, and 1,532 m from the second.
29. It is about 60 ft to the top of the building, and the height of the building is about 31 ft.
31. Drop a perpendicular from C and let h be the length. From the definition of the trigonometric functions,

$\sin\alpha = h/b$ and $\sin\beta = h/a$

Solve for h and equate:

$b\sin\alpha = a\sin\beta$

Divide by ab:

$\dfrac{\sin\alpha}{a} = \dfrac{\sin\beta}{b}$

33. From Problem 31 $\dfrac{\sin\alpha}{a} = \dfrac{\sin\beta}{b}$

Multiply by a: $\sin\alpha = \dfrac{a\sin\beta}{b}$

Divide by $\sin\beta$: $\dfrac{\sin\alpha}{\sin\beta} = \dfrac{a}{b}$

35.–39. Proofs vary.   **41.** About 39 mph

Problem Set 5.4, pages 216–218

1. $\alpha > 90°$ and $a \le b$, no triangle formed.   **3.** $\alpha < 90°$ and $a < h < b$, no triangle formed.
5. $\alpha = 125°$; $\beta = 41°$; $\gamma = 14°$; $a = 5.0$; $b = 4.0$; $c = 1.5$ **7.** $\alpha = 75°$; $\beta = 49°$; $\gamma = 56°$; $a = 9.0$; $b = 7.0$; $c = 7.8$
9. $\alpha < 90°$, $a < h < b$, no triangle formed.   **11.** $\alpha = 47.0°$; $\beta = 90.0°$; $\gamma = 43.0°$; $a = 8.63$; $b = 11.8$; $c = 8.05$
13. $\alpha = 52.1°$; $\beta = 115.0°$; $\gamma = 12.9°$; $a = 14.2$; $b = 16.3$; $c = 4.00$ **15.** $\alpha = 147.0°$; $\beta = 15.0°$; $\gamma = 18.0°$; $a = 49.5$; $b = 23.5$; $c = 28.1$
17. $h \approx 43.8 < c < b$; ambiguous case. $\alpha = 90.8°$; $\alpha' = 25.0°$; $\beta = 57.1°$; $\beta' = 122.9°$; $\gamma = 32.1°$; $a = 98.2$; $a' = 41.6$; $b = 82.5$; $c = 52.2$
19. No solution because the sum of the given angles is greater than 180°.
21. No solution because the sum of the two smaller sides must be larger than the third side.
23. $\alpha = 54.5°$; $\beta = 62.5°$; $\gamma = 63.0°$; $a = 179$; $b = 195$; $c = 196$ **25.** He is 5.39 miles from the target.
27. The distance from the first city is 1.926 mi (10,170 ft) and from the second city is .5363 mi (2,832 ft). The altitude is .3503 mi (1,850 ft).
29. The angle of elevation is 19.0°.   **31.** The height of the tower is 985 ft.   **33.** Proof varies.

Problem Set 5.5, pages 222–223

Answers to 1–23 are in square units and are given to the correct number of significant digits.
1. 37   **3.** 83   **5.** 680   **7.** 560   **9.** 6.4   **11.** 35   **13.** 54.0   **15.** 58.2 and 11.3   **17.** 2,150 and 911   **19.** 180   **21.** no triangle formed
23. 20,100   **25.** 7.82 cm   **27.** 18,000 m^3 **29.** 21,000 ft^3 or 760 yd^3 **31.** 400 ft^3 **33.** Proof varies.

Problem Set 5.6, pages 227–230

1. A vector with a direction N37°W and a magnitude of 5.0.   **3.** A vector with a direction N38°E and a magnitude of 23.
5. The horizontal component has a magnitude of 22 and the vertical component a magnitude of 7.1.
7. The boat is traveling at 13 mph with a course of S22°W.   **9.** The horizontal component is 2,090 fps and the vertical component is 368 fps.
11. In one hour it has traveled 590 miles south.   **13.** The direction is S32.8°W with a speed of 22.6 knots.
15. The magnitude is 360 lb, and it makes an angle of 23° with the 220-lb force, and a 29° angle with the 180-lb force.
17. 56° with the 30-kg force, and 30° with the 50-kg force   **19.** The planes are 370 miles apart.
21. At 10 kph it would be there at about 12:25 P.M. and 2:00 P.M.   **23.** The boats are 67.2 miles apart.
25. A force of 4.71 tons is needed to keep it from sliding down the hill. The force against the hill is 51.8 tons.
27. The angle of inclination is 39.2°.   **29.** The heaviest piece of cargo is 12,200 lb.
31. The weight of the astronaut is resolved into two components: one parallel to the inclined plane with length y
and the other perpendicular to it with length x. The weight of the astronaut is $|\mathbf{x}|$.

Problem Set 5.7, pages 236–237

Answers to Problems 1–5 should also include a vector diagram.
1. $\mathbf{v} = 6\mathbf{i} + 6\sqrt{3}\mathbf{j}$ **3.** $\mathbf{v} = 4\cos 112°\mathbf{i} + 4\sin 112°\mathbf{j} \approx -1.4984\mathbf{i} + 3.7087\mathbf{j}$ **5.** $\mathbf{v} = -2\mathbf{i} + 2\mathbf{j}$ **7.** 5   **9.** $\sqrt{85} \approx 9.2195$ **11.** $2\sqrt{2} \approx 2.8284$
13. orthogonal   **15.** not orthogonal
Answers to Problems 17–25 are given in the form $\mathbf{v} \cdot \mathbf{w}$; $|\mathbf{v}|$; $|\mathbf{w}|$; $\cos\theta$
17. 63; 5; 13; $\frac{63}{65}$ **19.** $3\sqrt{5}$; 3; $3\sqrt{6}$; $\frac{\sqrt{30}}{18}$ **21.** 3; $\sqrt{13}$; $\sqrt{61}$; $\frac{3\sqrt{793}}{793}$ **23.** 1; 1; 1; 1   **25.** 0; 1; 1; 0   **27.** 75°   **29.** 45°   **31.** 90°   **33.** $-\frac{8}{5}$
35. Answer varies.

Chapter 5 Review pages 237–241

Review of Objectives

1. a. $\frac{a}{c}$ **b.** $\frac{b}{c}$ **c.** $\frac{b}{a}$ **2.** $a = 5.30$, $b = 12.2$, $c = 13.3$, $\alpha = 23.5°$, $\beta = 66.5°$, $\gamma = 90.0°$ **3. a.** The tower is 2,063 ft.
b. It is 278 ft across the bridge.   **4. a.** $b^2 + c^2 - 2bc\cos\alpha$ **b.** $\dfrac{a^2 + c^2 - b^2}{2ac}$ **5.** $a = 14$, $b = 27$, $c = 19$, $\alpha = 29°$, $\beta = 109°$, $\gamma = 42°$
6. $a = 19$, $b = 7.2$, $c = 15$, $\alpha = 113°$, $\beta = 20°$, $\gamma = 47°$ **7.** It is about 430 ft.   **8.** $\dfrac{\sin\alpha}{a} = \dfrac{\sin\beta}{b} = \dfrac{\sin\gamma}{c}$
9. $a = 92.6$, $b = 273$, $c = 224$, $\alpha = 18.3°$, $\beta = 112.4°$, $\gamma = 49.3°$ **10.** About 113 m of fence is needed.

11. (Ambiguous case) At 12:42 P.M. and at 4:40 P.M.   **12. a.** $\cos\alpha = \dfrac{b^2 + c^2 - a^2}{2bc}$ **b.** $\alpha = \cos^{-1}\left(\dfrac{b^2 + c^2 - a^2}{2bc}\right)$ **c.** $\dfrac{\sin\alpha}{a} = \dfrac{\sin\beta}{b}$

d. $\alpha = \sin^{-1}\left(\dfrac{a\sin\beta}{b}\right)$ **e.** $\alpha + \beta + \gamma = 180°$ **f.** $\alpha = 180° - \beta - \gamma$ **g.** $\dfrac{\sin\alpha}{a} = \dfrac{\sin\beta}{b}$ **h.** $b = \dfrac{a\sin\beta}{\sin\alpha}$ **i.** $b^2 = a^2 + c^2 - 2ac\cos\beta$

j. $b = \sqrt{a^2 + c^2 - 2ac\cos\beta}$ **13. a.** 320   **b.** 51.7   **c.** 1,900   **14.** 130,900 ft^2 **15.** 1,800,000 ft$^3 \approx 66,800$ yd^3 **16.** 7.1; N47°W
17. 44 fps $\approx$ 30 mph; S60°E   **18.** Horizontal component is 2.8 and the vertical component is 3.5.   **19.** .2 ton $\approx$ 420 lb   **20. a.** $\mathbf{v} = \mathbf{i} + \mathbf{j}$
b. $-4.4\mathbf{i} + 8.2\mathbf{j}$ **c.** $-6\mathbf{i} - 5\mathbf{j}$ **21.** $|\mathbf{v}| = \sqrt{13}$; $|\mathbf{w}| = \sqrt{2}$ **22.** 5   **23.** 11°   **24.** $a = -\frac{3}{2}$

Miscellaneous Problems

1. $a = 24; b = 52; c = 61; \alpha = 23°; \beta = 58°; \gamma = 99°$ **3.** $a = 30; b = 46; c = 59; \alpha = 30°; \beta = 50°; \gamma = 100°$
5. $a = 33; b = 40; c = 37; \alpha = 50°; \beta = 70°; \gamma = 60°$ **7.** $a = 15; b = 8.0; c = 10; \alpha = 112°; \beta = 30°; \gamma = 38°$
9. $a = 17; b = 14; c = 12; \alpha = 82°; \beta = 54°; \gamma = 44°$ **11.** $a = 7.0; b = 8.0; c = 2.0; \alpha = 54°; \beta = 113°; \gamma = 13°$
13. no solution   **15.** ambiguous case: $a = 14.5; b = 17.2; c = 24.5$ and $c' = 3.47; \alpha = 35.5°; \beta = 43.5°$ and $\beta' = 136.5°; \gamma = 101.0°$ and $\gamma' = 8.0°$
17. $a = 121; b = 315; c = 255; \alpha = 21.3°; \beta = 108.7°; \gamma = 50.0°$ **19.** $a = 11; b = 9.0; c = 8.0; \alpha = 80°; \beta = 54°; \gamma = 46°$
21. $a = 12.4; b = 32.5; c = 35.1; \alpha = 20.7°; \beta = 67.8°; \gamma = 91.6°$
23. no solution   **25.** $a = 49.5; b = 23.5; c = 28.1; \alpha = 147.0°; \beta = 15.0°; \gamma = 18.0°$
27. ambiguous case: $a = 117$ and $a' = 34.9; b = 82.5; c = 52.2; \alpha = 118.6°$ and $\alpha' = 15.2°; \beta = 38.3°$ and $\beta' = 141.7°; \gamma = 23.1°$
29. ambiguous case: $a = 10.2; b = 11.8; c = 13.5$ and $c' = 2.61; \alpha = 47.0°; \beta = 57.8°$ and $\beta' = 122.2°; \gamma = 75.2°$ and $\gamma' = 10.8°$
31. The tunnel is 979 ft.   **33.** The log should show 12:45 P.M. and 1:11 P.M.
35. It is 7,200 ft from the first tower and 6,220 ft from the second tower.

Application for Further Study: Parking-Lot Design, page 245

1.

θ	c	w
10°	57.6	13.5
20°	29.2	16.6
30°	20.0	19.2
40°	15.6	21.2
50°	13.1	22.5
60°	11.5	23.2
70°	10.6	23.2
80°	10.2	22.4
90°	10.0	21.0

3.

θ	Usable Length
30°	3.7
40°	9.1
50°	12.6
60°	15.2
70°	17.4
80°	19.2
90°	21.0

5. To minimize c, θ should be 90°.

Problem Set 6.1, pages 250–251

1. a. $8 + 7i$ **b.** $11 + i$ **c.** $1 - 6i$ **3. a.** -4 **b.** $8 - 10i$ **c.** $11 - i$ **5. a.** $7 + i$ **b.** $9 + 2i$ **c.** 29   **7. a.** -1 **b.** 1   **c.** -1 **9. a.** 1
b. -1 **c.** 1   **11. a.** -1 **b.** 1   **c.** $-i$ **13. a.** 1   **b.** i **c.** 1   **15.** $32 - 24i$ **17.** $-9 + 40i$ **19.** $-\frac{3}{2} + \frac{3}{2}i$ **21.** $-2i$ **23.** $\frac{-6}{29} + \frac{15}{29}i$

25. $1 + i$ **27.** $\frac{-35}{37} - \frac{12}{37}i$ **29.** $\frac{1}{2} - \frac{1}{2}i$ **31.** $.3131 + 2.2281i$ **33.** $(-1 - \sqrt{3}) + 2i$ **35.** $\left(\dfrac{5 + 2\sqrt{3}}{13}\right) + \left(\dfrac{1 + 3\sqrt{3}}{13}\right)i$

37. $(13 + 8\sqrt{3}) + (12 + 4\sqrt{3})i$ **39.** Answers vary.

Problem Set 6.2, pages 255–256

1. $\sqrt{10}; 5\sqrt{2}$ **3.** $\sqrt{13}; \sqrt{29}$ **5.** $2\sqrt{2}; \sqrt{2}$ **7.** 2; 4   **9.** 2; 5   **11.** $\sqrt{2}$ cis 315°   **13.** 2 cis 30°   **15.** cis 0°   **17.** cis 270°   **19.** 3 cis 55°
21. 4 cis 100°   **23.** $\frac{3}{2} + \frac{3\sqrt{3}}{2}i$ **25.** $-\frac{\sqrt{3}}{2} + \frac{1}{2}i$ **27.** $2\sqrt{3} + 2i$ **29.** $-2.3444 - 5.5230i$ **31.** $9.8163 + 1.9081i$ **33.** Answers vary.

Problem Set 6.3, pages 262–263

1. $6\operatorname{cis}210°$ **3.** $\frac{5}{2}\operatorname{cis}267°$ **5.** $3\operatorname{cis}230°$ **7.** $81\operatorname{cis}240°$ **9.** $-64\,(64\operatorname{cis}180°)$ **11.** $-128+128i\sqrt{3}\,(256\operatorname{cis}120°)$
13. $2\operatorname{cis}80°,\,2\operatorname{cis}200°,\,2\operatorname{cis}320°$ **15.** $2\operatorname{cis}40°,\,2\operatorname{cis}112°,\,2\operatorname{cis}184°,\,2\operatorname{cis}256°,\,2\operatorname{cis}328°$ **17.** $2\operatorname{cis}32°,\,2\operatorname{cis}104°,\,2\operatorname{cis}176°,\,2\operatorname{cis}248°,\,2\operatorname{cis}320°$
19. $3\operatorname{cis}0°,\,3\operatorname{cis}120°,\,3\operatorname{cis}240°$ **21.** $\sqrt[8]{2}\operatorname{cis}56.25°,\,\sqrt[8]{2}\operatorname{cis}146.25°,\,\sqrt[8]{2}\operatorname{cis}236.25°,\,\sqrt[8]{2}\operatorname{cis}326.25°$
23. $2\operatorname{cis}15°,\,2\operatorname{cis}75°,\,2\operatorname{cis}135°,\,2\operatorname{cis}195°,\,2\operatorname{cis}255°,\,2\operatorname{cis}315°$
25. $2^{1/18}\operatorname{cis}15°,\,2^{1/18}\operatorname{cis}55°,\,2^{1/18}\operatorname{cis}95°,\,2^{1/18}\operatorname{cis}135°,\,2^{1/18}\operatorname{cis}175°,\,2^{1/18}\operatorname{cis}215°,\,2^{1/18}\operatorname{cis}255°,\,2^{1/18}\operatorname{cis}295°,\,2^{1/18}\operatorname{cis}335°$
27. $\operatorname{cis}0°,\,\operatorname{cis}36°,\,\operatorname{cis}72°,\,\operatorname{cis}108°,\,\operatorname{cis}144°,\,\operatorname{cis}180°,\,\operatorname{cis}216°,\,\operatorname{cis}252°,\,\operatorname{cis}288°,\,\operatorname{cis}324°$
Problems 29–33 should also be shown graphically.
29. $2,\,2i,\,-2,\,-2i$ **31.** $-.6840+1.8794i,\,-1.2856-1.5321i,\,1.9696-.3473i$
33. $1.9696+.3473i,\,-.3473+1.9696i,\,-1.9696-.3473i,\,.3473-1.9696i$ **35.** $1,\,.3090\pm.9511i,\,-.8090\pm.5878i$
37. $\pm1,\,\frac{1}{2}\pm\frac{\sqrt{3}}{2}i,\,-\frac{1}{2}\pm\frac{\sqrt{3}}{2}i$ **39.** $\pm2,\,\pm2i$ **41.** $-1,\,\frac{1}{2}\pm\frac{\sqrt{3}}{2}i,\,-\frac{1}{2}\pm\frac{\sqrt{3}}{2}i$ **43.** $8\operatorname{cis}0°,\,8\operatorname{cis}72°,\,8\operatorname{cis}144°,\,8\operatorname{cis}216°,\,8\operatorname{cis}288°$
45. Answers vary.

Chapter 6 Review, pages 264–265

Review of Objectives

1. a. $2+5i$ **b.** $-1-3i$ **2. a.** 8 **b.** $-21-20i$ **3. a.** $-\frac{1}{2}+\frac{5}{2}i$ **b.** $4i$
4.

a. $2\sqrt{5}$ **b.** $\sqrt{26}$ **c.** $\sqrt{17}$ **d.** $2\sqrt{5}$ **5. a.** $7\sqrt{2}\operatorname{cis}315°$ **b.** $3\operatorname{cis}270°$
6. a. $-\sqrt{3}+i$ **b.** $2\sqrt{2}-2\sqrt{2}\,i$
7. $\operatorname{cis}154°$ **8.** $-128-128\sqrt{3}\,i$
9. a. $\sqrt{7}\operatorname{cis}165°\approx-2.5556+.6848i,\,\sqrt{7}\operatorname{cis}345°\approx2.5556-.6848i$
b. $\operatorname{cis}0°=1;\,\operatorname{cis}90°=i;\,\operatorname{cis}180°=-1;\,\operatorname{cis}270°=-i.$
Also show these roots graphically.

Miscellaneous Problems

1. $9\sqrt{2}\operatorname{cis}315°$ **3.** $7\operatorname{cis}150°$ **5.** $9\operatorname{cis}0°$ **7.** $16\operatorname{cis}90°$ **9.** $-\frac{9}{2}\sqrt{2}+\frac{9}{2}\sqrt{2}\,i$ **11.** $-\frac{3}{2}-\frac{3\sqrt{3}}{2}i$ **13.** $4.5315+2.1131i$ **15.** $-128-128\sqrt{3}\,i$
17. $\operatorname{cis}228°$ **19.** $\operatorname{cis}30°;\,\operatorname{cis}150°;\,\operatorname{cis}270°$ **21.** $2\operatorname{cis}15°;\,2\operatorname{cis}75°;\,2\operatorname{cis}135°;\,2\operatorname{cis}195°;\,2\operatorname{cis}255°;\,2\operatorname{cis}315°$
23. $2^{1/18}\operatorname{cis}15°;\,2^{1/18}\operatorname{cis}55°;\,2^{1/18}\operatorname{cis}95°;\,2^{1/18}\operatorname{cis}135°;\,2^{1/18}\operatorname{cis}175°;\,2^{1/18}\operatorname{cis}215°;\,2^{1/18}\operatorname{cis}255°;\,2^{1/18}\operatorname{cis}295°;\,2^{1/18}\operatorname{cis}335°$
25. $\operatorname{cis}0°;\,\operatorname{cis}36°;\,\operatorname{cis}72°;\,\operatorname{cis}108°;\,\operatorname{cis}144°;\,\operatorname{cis}180°;\,\operatorname{cis}216°;\,\operatorname{cis}252°;\,\operatorname{cis}288°;\,\operatorname{cis}324°$ **27.** $2\sqrt[3]{4}\operatorname{cis}40°;\,2\sqrt[3]{4}\operatorname{cis}160°;\,2\sqrt[3]{4}\operatorname{cis}280°$
29. $.6848+2.5556i;\,-.6848-2.5556i$

Application for Further Study: Electric Generator, page 268

1. $I=15\sin120\pi t;\,a=15;\,p=\frac{2\pi}{120\pi}=\frac{1}{60}$
3. $\cos\theta=.96;\,\theta\approx16.3°$. Since $\cos\theta$ is in Quadrant I, if $\cos\theta\geq.96$, then $\theta\leq16.3°$. Thus, the maximum current lag (to the nearest degree) is $16°$.

Problem Set 7.1, pages 274–275

Points in Problems 1–12 should also be plotted.
1. Polar: $(4,\frac{\pi}{4}),\,(-4,\frac{5\pi}{4})$; rectangular: $(2\sqrt{2},2\sqrt{2})$ **3.** Polar: $(5,\frac{2\pi}{3}),\,(-5,\frac{5\pi}{3})$; rectangular: $(-\frac{5}{2},\frac{5\sqrt{3}}{2})$
5. Polar: $(\frac{3}{2},\frac{7\pi}{6}),\,(-\frac{3}{2},\frac{\pi}{6})$; rectangular: $(\frac{-3\sqrt{3}}{4},-\frac{3}{4})$ **7.** $(5\sqrt{2},\frac{\pi}{4})$ and $(-5\sqrt{2},\frac{5\pi}{4})$ **9.** $(4,\frac{5\pi}{3})$ and $(-4,\frac{2\pi}{3})$ **11.** $(3\sqrt{2},\frac{7\pi}{4})$ and $(-3\sqrt{2},\frac{3\pi}{4})$
13. a. United States   **b.** India   **c.** Greenland   **d.** Canada   **15.** yes   **17.** no   **19.** yes   **21.** yes   **23.** yes   **25.–39.** Answers vary.

Problem Set 7.2, pages 279–280

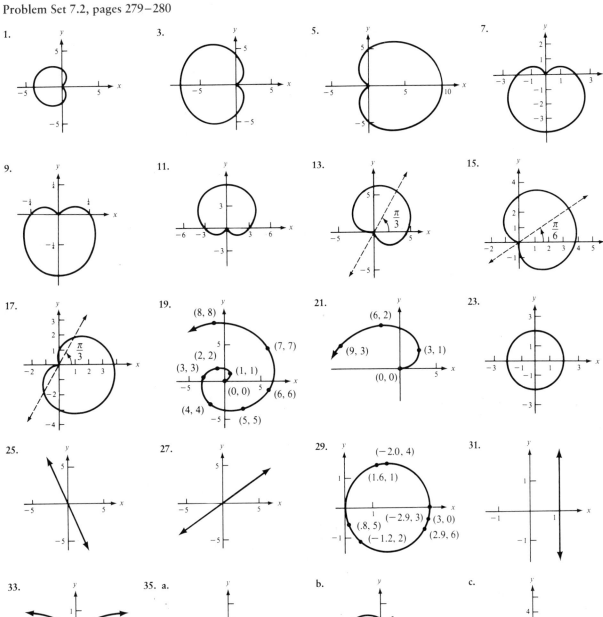

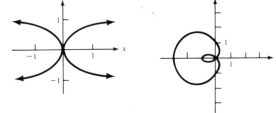

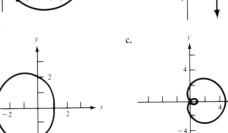

d.

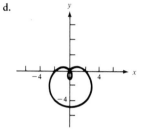

e.

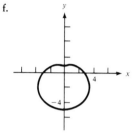

f.

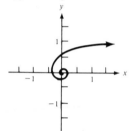

37. a. spiral of Archimedes

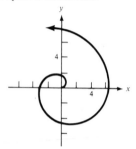

b. spiral of Archimedes

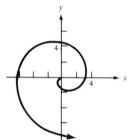

c. hyperbolic spiral

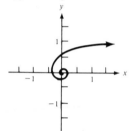

d. hyperbolic spiral

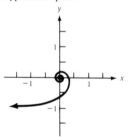

e. logarithmic spiral*

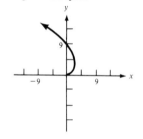

f. logarithmic spiral

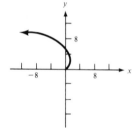

Problem Set 7.3, page 285

1. lemniscate   **3.** three-leaved rose   **5.** cardioid   **7.** none   **9.** none (circle)   **11.** none   **13.** none (it is a line)   **15.** cardioid

17.

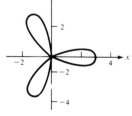

19.

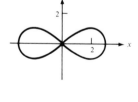

21.

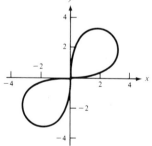

23.

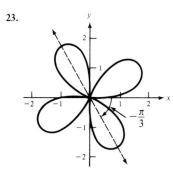

25.

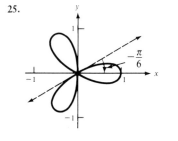

27.

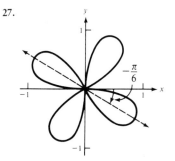

29.

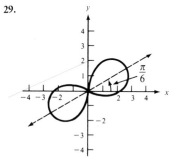

31.

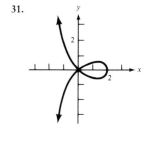

33.

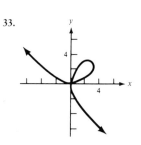

Problem Set 7.4, pages 290–291

1.

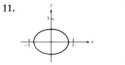

3.

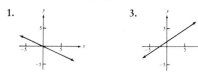

5.

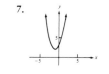

7.

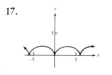

9.

11.

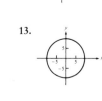

13.

15.

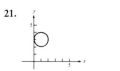

17.

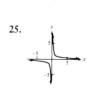

19.

21.

23.

25.

27.

29.

31. $x = a\cos\theta + a\theta\sin\theta, \ y = a\sin\theta - a\theta\cos\theta$
33.–35. Answers vary.

Problem Set 7.5, pages 297–298

Graphs should also be shown for Problems 1–30.

1. $(0,0)$, $(2\sqrt{2}, \frac{\pi}{4})$ **3.** $(1, \frac{\pi}{3})$, $(1, \frac{5\pi}{3})$ **5.** $(3,0)$, $(3, \pi)$ **7.** $(0,0)$, $(2, \frac{\pi}{2})$, $(2, \frac{3\pi}{2})$ **9.** $(2,0)$, $(2, \pi)$ **11.** $(0,0)$, $(2, \frac{\pi}{4})$
13. $(0,0)$ **15.** $(0,0)$, $(a,0)$, (a, π) **17.** $(0,0)$, $(\pi, \frac{\pi}{3})$, $(4\pi, \frac{4\pi}{3})$
19. $(0,0)$, $(\frac{16}{5}, \text{Arccos}(-\frac{3}{5}))$ [*Note:* Arccos$(-\frac{3}{5}) \approx 2.214$ or $126.87°$]
21. $(0,0)$, $(3.2, 2\pi - \text{Arccos} .8)$ [*Note:* $2\pi - \text{Arccos} .8 \approx 5.640$ or $323.13°$]   **23.** $(0,0)$, $(1,0)$, $(.6, \text{Arcsin} .8)$ [*Note:* Arcsin $.8 \approx .927$ or $53.1°$]
25. $(2, \frac{\pi}{2})$, $(2, \frac{3\pi}{2})$ **27.** $(2+\sqrt{2}, \frac{5\pi}{4})$, $(2+\sqrt{2}, \frac{3\pi}{4})$, $(2-\sqrt{2}, \frac{\pi}{4})$, $(2-\sqrt{2}, \frac{7\pi}{4})$ **29.** $(2, \frac{\pi}{2})$

Chapter 7 Review, pages 298–300

Review of Objectives

1.

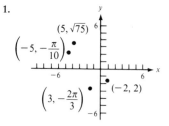

2. a. $(-\frac{3}{2}, \frac{-3\sqrt{3}}{2})$ **b.** $(-4.7553, 1.5451)$ **3. a.** $(3, \frac{4\pi}{3})$, $(-3, \frac{\pi}{3})$
b. $(5, \frac{9\pi}{10})$, $(-5, \frac{19\pi}{10})$ **c.** $(6, \frac{2\pi}{3})$, $(-6, \frac{5\pi}{3})$
d. $(5.3852, 1.1903)$, $(-5.3852, 4.3319)$ **4. a.** no   **b.** yes   **c.** yes   **d.** no

5. a.

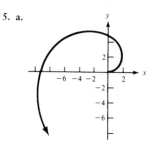

b.

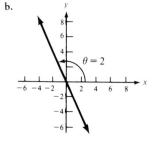

c.

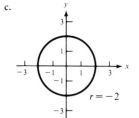

6.

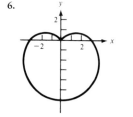

7.

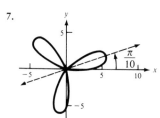

8.

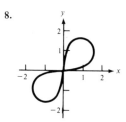

9. a. lemniscate   **b.** cardioid   **c.** five-leaved rose   **d.** circle
10.

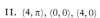

11. $(4, \pi)$, $(0,0)$, $(4,0)$

Miscellaneous Problems

1. rose curve   **3.** circle   **5.** cardioid   **7.** rose curve   **9.** lemniscate   **11.** rose curve   **13.** rose curve   **15.** cardioid

17.

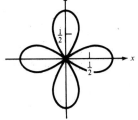

19.

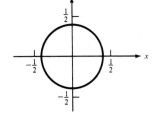

21.

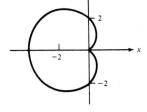

23.

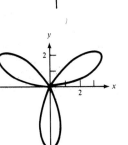

25.

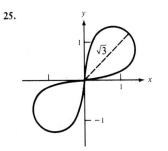

27.

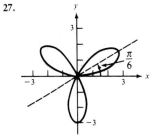

29.

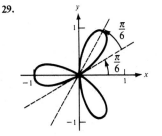

31.

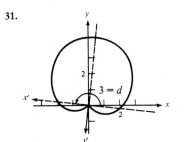

33.

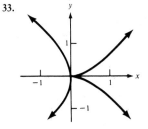

35.

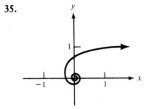

37.

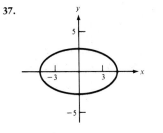

39.

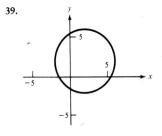

Application for Further Study: Communication of Bees, page 302

1.

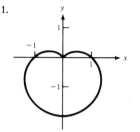

The food source is 90° away from the sun.

3.

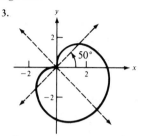

It is 50° away from the direction given in Problem 1. It is also a distance requiring about twice as much energy to get there.

Cumulative Review for Chapters 5–7, pages 303–304

Objective Test

1. **a.** Law of Cosines **b.** Law of Cosines **c.** Law of Sines (ambiguous case) **d.** Law of Sines **e.** no solution **f.** Law of Cosines
g. right-triangle solution **h.** no solution **i.** Law of Sines (ambiguous case) **j.** Law of Sines 2. **a.** rose curve **b.** lemniscate
c. rose curve **d.** cardioid **e.** circle **f.** cardioid **g.** lemniscate **h.** line **i.** cardioid **j.** cardioid
3. $\alpha = 121°$; $\beta = 32°$; $\gamma = 27°$; $a = 83$; $b = 52$; $c = 43$ 4. $\alpha = 57°$; $\beta = 68°$; $\beta' = 112°$; $\gamma = 55°$; $\gamma' = 11°$; $a = 55$; $b = 61$; $c = 53$; $c' = 13$
5. $46°, 102°, 32°$ 6. **a.** polar, $(3, \frac{7\pi}{4})$, $(-3, \frac{3\pi}{4})$; rectangular, $(\frac{3\sqrt{2}}{2}, \frac{-3\sqrt{2}}{2})$ **b.** polar, $(5, \frac{5\pi}{6})$, $(-5, \frac{11\pi}{6})$; rectangular, $(\frac{-5\sqrt{3}}{2}, \frac{5}{2})$
c. polar, $(3, 0)$, $(-3, 3.1416)$; rectangular, $(3, 0)$ 7. $\sqrt[3]{5}\,\text{cis}\,50°$; $\sqrt[3]{5}\,\text{cis}\,170°$; $\sqrt[3]{5}\,\text{cis}\,290°$ 8. $2\sqrt{2}\,\text{cis}\,135°$; $2\sqrt{2}\,\text{cis}\,315°$
9.   10.

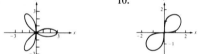

Multiple-Choice Test

1. $a^2 + b^2 - 2ab\cos\gamma$; $\dfrac{a^2 + c^2 - b^2}{2ac}$ 2. $<$; $b <$ opp $<$ adj   3. $r\cos\theta$; $r\sin\theta$ 4. $r_1 r_2 \text{cis}(\theta_1 + \theta_2)$ 5. $r = f(\theta)$; through an angle α
6. C   7. B   8. A   9. C   10. A   11. C   12. A   13. E   14. B   15. D   16. E   17. B   18. C   19. D   20. C

Problem Set 8.1, pages 311–312

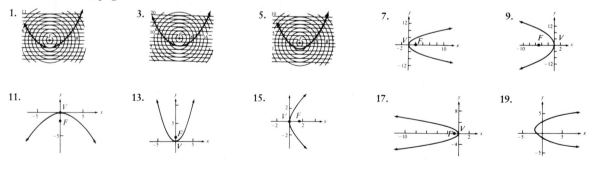

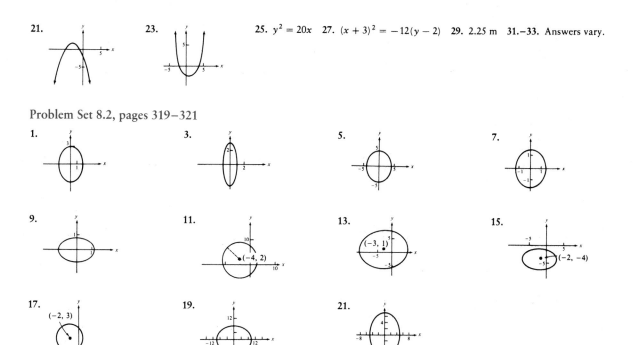

21. **23.** **25.** $y^2 = 20x$ **27.** $(x + 3)^2 = -12(y - 2)$ **29.** 2.25 m **31.–33.** Answers vary.

Problem Set 8.2, pages 319–321

1. **3.** **5.** **7.**

9. **11.** **13.** **15.**

17. **19.** **21.**

23. circle; center, $(-2, 3)$; radius 3; $(x + 2)^2 + (y - 3)^2 = 9$ **25.** horizontal ellipse; center, $(0, 0)$; $a = 10$, $c = 6$, $b^2 = 64$; $\dfrac{x^2}{100} + \dfrac{y^2}{64} = 1$

27. vertical ellipse; center, $(0, 0)$; $a = 7$, $c = 5$, $b^2 = 24$; $\dfrac{x^2}{24} + \dfrac{y^2}{49} = 1$ **29.** horizontal ellipse; center, $(0, 0)$; $a = 10$, $c = 2$, $b^2 = 96$; $\dfrac{x^2}{100} + \dfrac{y^2}{96} = 1$

31. Aphelion is 94,500,000 and perihelion is 91,500,000. **33.** .053 **35.** Answers vary.

Problem Set 8.3, pages 326–327

1. **3.** **5.** **7.**

9. **11.** **13.** **15.**

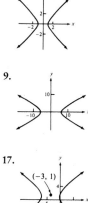

17. **19.** **21.**

23. $\dfrac{y^2}{25} - \dfrac{x^2}{24} = 1$ 25. $\dfrac{x^2}{1} - \dfrac{y^2}{24} = 1$ 27. $\dfrac{x^2}{9} - \dfrac{(y+3)^2}{7} = 1$ 29.–31. Answers vary.

Problem Set 8.4, pages 332–333

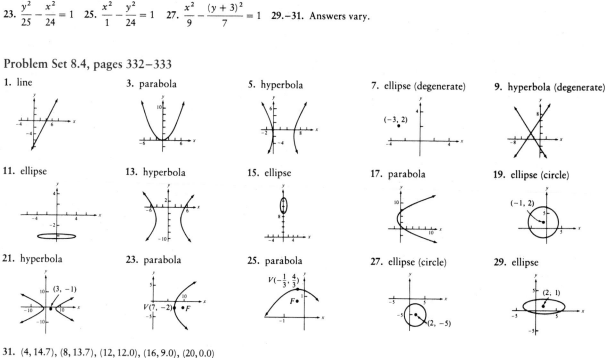

1. line

3. parabola

5. hyperbola

7. ellipse (degenerate)

9. hyperbola (degenerate)

11. ellipse

13. hyperbola

15. ellipse

17. parabola

19. ellipse (circle)

21. hyperbola

23. parabola

25. parabola

27. ellipse (circle)

29. ellipse

31. $(4, 14.7)$, $(8, 13.7)$, $(12, 12.0)$, $(16, 9.0)$, $(20, 0.0)$

Problem Set 8.5, pages 340–341

1.–3. Answers vary.   5. hyperbola   7. ellipse   9. parabola   11. hyperbola   13. parabola   15. ellipse   17. ellipse

19. $\theta = 45°$; $x = \frac{1}{\sqrt{2}}(x' - y')$; $y = \frac{1}{\sqrt{2}}(x' + y')$; $\dfrac{y'^2}{2} - \dfrac{x'^2}{2} = 1$

21. $\theta = 45°$; $x = \frac{1}{\sqrt{2}}(x' - y')$; $y = \frac{1}{\sqrt{2}}(x' + y')$; $\dfrac{x'^2}{16} - \dfrac{y'^2}{16} = 1$

23. $\theta = 45°$; $x = \frac{1}{\sqrt{2}}(x' - y')$; $y = \frac{1}{\sqrt{2}}(x' + y')$; $\dfrac{x'^2}{9} + \dfrac{y'^2}{4} = 1$

25. $\tan \theta = 2$; $\theta \approx 63.4°$; $x = \frac{1}{\sqrt{5}}(x' - 2y')$; $y = \frac{1}{\sqrt{5}}(2x' + y')$; $(x' + 1)^2 = 4(y' + \frac{7}{10})$

27. $\theta = 30°$; $x = \frac{1}{2}(\sqrt{3}\,x' - y')$; $y = \frac{1}{2}(x' + \sqrt{3}\,y')$; $\dfrac{x'^2}{4} - \dfrac{y'^2}{9} = 1$

29. $\theta = 30°$; $x = \frac{1}{2}(\sqrt{3}\,x' - y')$; $y = \frac{1}{2}(x' + \sqrt{3}\,y')$; $x'^2 = -\frac{1}{16}(y' - 4)$

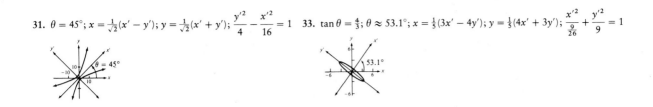

31. $\theta = 45°$; $x = \frac{1}{\sqrt{2}}(x' - y')$; $y = \frac{1}{\sqrt{2}}(x' + y')$; $\dfrac{y'^2}{4} - \dfrac{x'^2}{16} = 1$ **33.** $\tan\theta = \frac{4}{3}$; $\theta \approx 53.1°$; $x = \frac{1}{5}(3x' - 4y')$; $y = \frac{1}{5}(4x' + 3y')$; $\dfrac{x'^2}{\frac{9}{26}} + \dfrac{y'^2}{9} = 1$

Problem Set 8.6, pages 346–347

1. $\varepsilon = 1$; parabola   **3.** $r = \dfrac{\frac{5}{3}}{1 + \frac{5}{3}\cos\theta}$; $\varepsilon = \frac{5}{3}$; hyperbola   **5.** $r = \dfrac{1}{1 - \frac{4}{5}\cos\theta}$; $\varepsilon = \frac{4}{5}$; ellipse   **7.** $r = \dfrac{4}{1 - 2\cos\theta}$; $\varepsilon = 2$; hyperbola

9. $r = \dfrac{4}{1 - \cos\theta}$; $\varepsilon = 1$; parabola   **11.** $\varepsilon = 1$; parabola   **13.** $r = \dfrac{\frac{5}{4}}{1 - \frac{1}{2}\cos(\theta - \frac{\pi}{6})}$; $\varepsilon = \frac{1}{2}$; ellipse   **15.** $\varepsilon = 1$; parabola

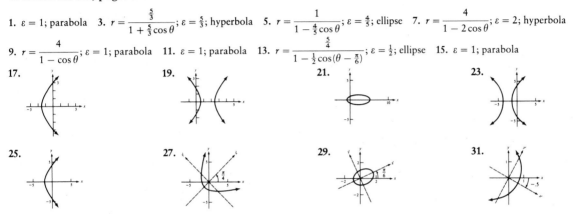

17.   **19.**   **21.**   **23.**

25.   **27.**   **29.**   **31.**

33. Answers vary.

Chapter 8 Review, pages 347–350
Review of Objectives

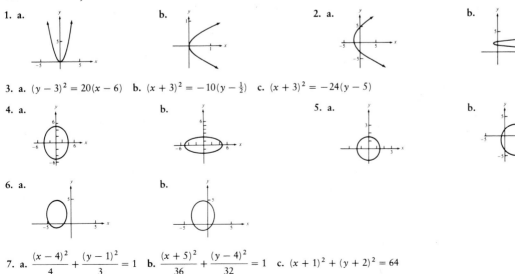

1. a.   **b.**   **2. a.**   **b.**

3. a. $(y - 3)^2 = 20(x - 6)$ **b.** $(x + 3)^2 = -10(y - \frac{1}{2})$ **c.** $(x + 3)^2 = -24(y - 5)$

4. a.   **b.**   **5. a.**   **b.**

6. a.   **b.**

7. a. $\dfrac{(x - 4)^2}{4} + \dfrac{(y - 1)^2}{3} = 1$ **b.** $\dfrac{(x + 5)^2}{36} + \dfrac{(y - 4)^2}{32} = 1$ **c.** $(x + 1)^2 + (y + 2)^2 = 64$

8. a. 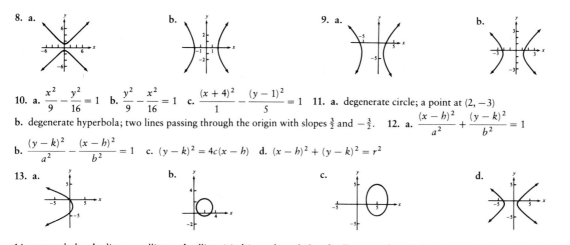 **b.**   **9. a.**   **b.**

10. a. $\dfrac{x^2}{9} - \dfrac{y^2}{16} = 1$ **b.** $\dfrac{y^2}{9} - \dfrac{x^2}{16} = 1$ **c.** $\dfrac{(x+4)^2}{1} - \dfrac{(y-1)^2}{5} = 1$ **11. a.** degenerate circle; a point at $(2, -3)$

b. degenerate hyperbola; two lines passing through the origin with slopes $\frac{3}{2}$ and $-\frac{3}{2}$. **12. a.** $\dfrac{(x-h)^2}{a^2} + \dfrac{(y-k)^2}{b^2} = 1$

b. $\dfrac{(y-k)^2}{a^2} - \dfrac{(x-h)^2}{b^2} = 1$ **c.** $(y-k)^2 = 4c(x-h)$ **d.** $(x-h)^2 + (y-k)^2 = r^2$

13. a. 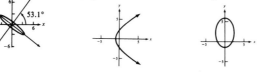 **b.**   **c.**   **d.**

14. a. parabola **b.** line **c.** ellipse **d.** ellipse (circle) **e.** hyperbola **f.** ellipse **g.** hyperbola **h.** parabola **i.** parabola **j.** hyperbola
15. a. $45°$ **b.** $26.6°$ **16.**   **17.** parabola   **18.** ellipse

Miscellaneous Problems

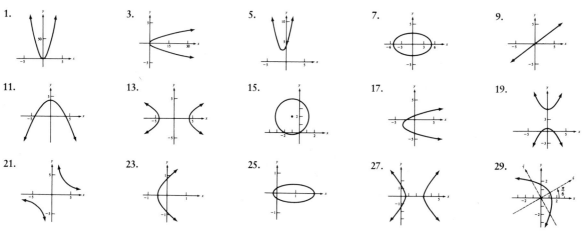

1.   **3.**   **5.**   **7.**   **9.**

11.   **13.**   **15.**   **17.**   **19.**

21.   **23.**   **25.**   **27.**   **29.**

Application for Further Study: Planetary Orbits, Pages 353–354

1. $1.5 \cdot 10^8$; $1.3 \cdot 10^8$ **3.** $3.7 \cdot 10^9$; $3.6 \cdot 10^9$; almost a circular orbit **5.** $5{,}500$ mi **7.** $1.2x^2 + 1.3y^2 = 1.6 \cdot 10^{15}$
9. $8.802x^2 + 8.805y^2 = 7.751 \cdot 10^{16}$

Problem Set B, pages 362–363

1. $4914 \approx 4{,}900$ **3.** $5517840 \approx 5{,}520{,}000$ **5.** $5937 \approx 5{,}940$ **7.** $378193771 \approx 380{,}000{,}000$ **9.** $40 \approx 40.0$ **11.** $7714.00791 \approx 7{,}714.01$
13. 12.2 **15.** $.264$ **17.** $.008353$ **19.** 2.62 **21.** $.310$ **23.** 92.3 **25.** $.24$ **27.** 1.0 **29.** $-.379$

INDEX